Have you been to our website?

For code downloads, print and e-book bundles, extensive samples from all books, special deals, and our blog, please visit us at:

www.rheinwerk-computing.com

Rheinwerk Computing

The Rheinwerk Computing series offers new and established professionals comprehensive guidance to enrich their skillsets and enhance their career prospects. Our publications are written by the leading experts in their fields. Each book is detailed and hands-on to help readers develop essential, practical skills that they can apply to their daily work.

Explore more of the Rheinwerk Computing library!

Metin Keras
Developing AI Applications: An Introduction
2024, 402 pages, paperback and e-book
www.rheinwerk-computing.com/5899

Bert Gollnick
Generative AI with Python: The Developer's Guide to Pretrained LLMs, Vector Databases, Retrieval Augmented Generation, and Agentic Systems
2025, 392 pages, paperback and e-book
www.rheinwerk-computing.com/6057

Thomas Theis
Getting Started with Python
2024, 437 pages, paperback and e-book
www.rheinwerk-computing.com/5876

Johannes Ernesti, Peter Kaiser
Python 3: The Comprehensive Guide
2022, 1036 pages, paperback and e-book
www.rheinwerk-computing.com/5566

Kofler, Öggl, Springer
AI-Assisted Coding: The Practical Guide for Software Development
2025, 395 pages, paperback and e-book
www.rheinwerk-computing.com/6058

www.rheinwerk-computing.com

Joachim Steinwendner and Roland Schwaiger

Programming Neural Networks with Python

Editor Meagan White
Acquisitions Editor Hareem Shafi
German Edition Editors Almut Poll, Lisa Helmus
Translation Winema Language Services, Inc.
Copyeditor Julie McNamee
Cover Design Graham Geary
Photo Credits iStockphoto: 174905315/© inhauscreative; Shutterstock: 2203660561/© Panuwatccn
Layout Design Vera Brauner
Production Graham Geary
Typesetting SatzPro, Germany
Printed and bound in Canada, on paper from sustainable sources

ISBN 978-1-4932-2696-2
1st edition 2025
3rd German edition published 2025 by Rheinwerk Verlag

© 2025 by:
Rheinwerk Publishing, Inc.
2 Heritage Drive, Suite 305
Quincy, MA 02171
USA
info@rheinwerk-publishing.com
+1.781.228.5070

Represented in the E.U. by:
Rheinwerk Verlag GmbH
Rheinwerkallee 4
53227 Bonn
Germany
service@rheinwerk-verlag.de
+49 (0) 228 42150-0

Library of Congress Cataloging-in-Publication Control Number: 2025009124

All rights reserved. Neither this publication nor any part of it may be copied or reproduced in any form or by any means or translated into another language, without the prior consent of Rheinwerk Publishing.

Rheinwerk Publishing makes no warranties or representations with respect to the content hereof and specifically disclaims any implied warranties of merchantability or fitness for any particular purpose. Rheinwerk Publishing assumes no responsibility for any errors that may appear in this publication.

"Rheinwerk Publishing", "Rheinwerk Computing", and the Rheinwerk Publishing and Rheinwerk Computing logos are registered trademarks of Rheinwerk Verlag GmbH, Bonn, Germany.

All products mentioned in this book are registered or unregistered trademarks of their respective companies.

No part of this book may be used or reproduced in any manner for the purpose of training artificial intelligence technologies or systems. In accordance with Article 4(3) of the Digital Single Market Directive 2019/790, Rheinwerk Publishing, Inc. expressly reserves this work from text and data mining.

Contents at a Glance

1	Introduction	17

PART I Up and Running

2	Starter Kit for Developing Neural Networks with Python	43
3	A Simple Neural Network	65
4	Learning in a Simple Network	93
5	Multilayer Neural Networks	127
6	Learning in a Multilayer Network	147
7	Examples of Deep Neural Networks	179
8	Programming Deep Neural Networks Using TensorFlow 2	209

PART II Deep Dive

9	From Brain to Network	241
10	The Evolution of Artificial Neural Networks	253
11	The Machine Learning Process	277
12	Learning Methods	319
13	Areas of Application and Real-Life Examples	351

Contents

Preface 15

1 Introduction 17

1.1	Why Neural Networks?	17
1.2	About This Book	18
1.3	The Contents in Brief	19
1.4	Is This Bee a Queen Bee?	22
1.5	An Artificial Neural Network for the Bee Colony	23
1.6	From Biology to the Artificial Neuron	28
	1.6.1 The Biological Neuron and Its Technical Copy	28
	1.6.2 The Artificial Neuron and Its Elements	29
1.7	Classification and the Rest	32
	1.7.1 Big Picture	32
	1.7.2 Artificial Intelligence	32
	1.7.3 History	34
	1.7.4 Machine Learning	35
	1.7.5 Deep Neural Networks	37
	1.7.6 Transformer Neural Networks	37
1.8	Summary	39
1.9	Further Reading	39

PART I   Up and Running

2 Starter Kit for Developing Neural Networks with Python 43

2.1	The Technical Development Environment	43
	2.1.1 The Anaconda Distribution	43
	2.1.2 Our Cockpit: Jupyter Notebook	47
	2.1.3 Major Python Modules	57

	2.1.4	The Google Colab Platform for Jupyter Notebooks	59
	2.1.5	Additional Jupyter Notebook Cloud Resources	62
2.2	Summary		63

3 A Simple Neural Network 65

3.1	Background	65
3.2	Bring on the Neural Network!	65
3.3	Neuron Zoom-In	68
3.4	Step Function	73
3.5	Perceptron	75
3.6	Points in Space: Vector Representation	76
	3.6.1 Task: Completing Values	77
	3.6.2 Task: Outputting the Iris Dataset as a Scatterplot	79
3.7	Horizontal and Vertical: Column and Line Notation	82
	3.7.1 Task: Determining the Scalar Product Using NumPy	83
3.8	The Weighted Sum	84
3.9	Step-by-Step: Step Functions	85
3.10	The Weighted Sum Reloaded	85
3.11	All Together	86
3.12	Task: Robot Protection	89
3.13	Summary	91
3.14	Further Reading	91

4 Learning in a Simple Network 93

4.1	Background: Plans Are Being Made	93
4.2	Learning in Python Code	94
4.3	Perceptron Learning	94
4.4	Separating Line for a Learning Step	98
4.5	Perceptron Learning Algorithm	99
4.6	The Separating Lines or Hyperplanes for the Example	103

4.7	scikit-learn Compatible Estimator	106
4.8	scikit-learn Perceptron Estimator	113
4.9	Adaline	115
4.10	Summary	125
4.11	Further Reading	126

5 Multilayer Neural Networks   127

5.1	A Real Problem	127
5.2	Solving XOR	129
5.3	Preparations for the Launch	134
5.4	The Plan for Implementation	135
5.5	The Setup ("class")	136
5.6	The Initialization ("__init__")	138
5.7	Something for In-Between ("print")	141
5.8	The Analysis ("predict")	141
5.9	The Usage	143
5.10	Summary	145

6 Learning in a Multilayer Network   147

6.1	How Do You Measure an Error?		147
6.2	Gradient Descent: An Example		149
	6.2.1	Gradient Descent: The Concept	149
	6.2.2	Algorithm for the Gradient Descent	150
6.3	A Network of Sigmoid Neurons		157
6.4	The Cool Algorithm with Forward Delta and Backpropagation		158
	6.4.1	The __init__ Method	158
	6.4.2	The "predict" Method	161
	6.4.3	The "fit" Method	165
	6.4.4	The "plot" Method	167
	6.4.5	The Complete Picture	168

6.5	A "fit" Run	170
	6.5.1 Initialization	172
	6.5.2 Forward	173
	6.5.3 Output	174
	6.5.4 Hidden	175
	6.5.5 Delta W_kj	176
	6.5.6 Delta W_ji	177
	6.5.7 W_ji	177
	6.5.8 W_kj	178
6.6	Summary	178
6.7	Further Reading	178

7 Examples of Deep Neural Networks 179

7.1	Convolutional Neural Networks	179
	7.1.1 The Architecture of Convolutional Networks	181
	7.1.2 The Coding Block	182
	7.1.3 The Prediction Block	188
	7.1.4 Training Convolutional Neural Networks	190
7.2	Transformer Neural Networks	194
	7.2.1 The Network Structure	195
	7.2.2 Embeddings	197
	7.2.3 Positional Encoding	197
	7.2.4 Encoder	200
	7.2.5 Decoder	202
	7.2.6 Training Transformer Neural Networks	203
7.3	The Optimization Method	204
	7.3.1 Momentum Optimization	204
	7.3.2 ADAM Optimization	205
7.4	Preventing Overfitting	205
	7.4.1 Early Stopping	205
	7.4.2 Dropout	206
7.5	Summary	207
7.6	Further Reading	207

8 Programming Deep Neural Networks Using TensorFlow 2 — 209

- 8.1 Convolutional Networks for Handwriting Recognition — 209
  - 8.1.1 The MNIST Dataset — 209
  - 8.1.2 A Simple Convolutional Neural Network — 213
  - 8.1.3 The Results — 217
- 8.2 Transfer Learning with Convolutional Neural Networks — 223
  - 8.2.1 The Pretrained Network — 224
  - 8.2.2 Data Preparation — 226
  - 8.2.3 The Pretrained Network — 227
  - 8.2.4 The Results — 229
- 8.3 Transfer Learning with Transformer Neural Networks — 231
  - 8.3.1 The Transformer Library — 232
  - 8.3.2 Tokenizers and Models — 234
  - 8.3.3 The Model Hub from Hugging Face — 235
- 8.4 Summary — 236
- 8.5 Further Reading — 236

PART II   Deep Dive

9 From Brain to Network — 241

- 9.1 Your Brain in Action — 241
- 9.2 The Nervous System — 242
- 9.3 The Brain — 243
  - 9.3.1 The Parts — 243
  - 9.3.2 A Section — 244
- 9.4 Neurons and Glial Cells — 245
- 9.5 A Transfer in Detail — 247
- 9.6 Representation of Cells and Networks — 249
- 9.7 Summary — 251
- 9.8 Further Reading — 251

10 The Evolution of Artificial Neural Networks — 253

- 10.1 The 1940s — 254
  - 10.1.1 1943 McCulloch-Pitts Neurons — 254
  - 10.1.2 1949: Donald Hebb — 255
- 10.2 The 1950s — 255
  - 10.2.1 1951: Marvin Minsky and Dean Edmonds – SNARC — 255
  - 10.2.2 1955/1956: Artificial Intelligence — 256
  - 10.2.3 1958: Rosenblatt's Perceptron — 256
  - 10.2.4 1960: Bernard Widrow and Marcian Hoff – Adaline and Madaline — 256
- 10.3 The 1960s — 257
  - 10.3.1 1969: Marvin Minsky and Seymour Papert — 257
- 10.4 The 1970s — 257
  - 10.4.1 1972: Kohonen – Associative Memory — 258
  - 10.4.2 1973: Lighthill Report — 258
  - 10.4.3 1974: Backpropagation — 258
- 10.5 The 1980s — 258
  - 10.5.1 1980: Fukushima's Neocognitron — 258
  - 10.5.2 1982: John Hopfield — 260
  - 10.5.3 1982: Kohonen's SOM — 269
  - 10.5.4 1986: Backpropagation — 269
  - 10.5.5 1987: NN Conference — 270
  - 10.5.6 1989: Yann LeCun: Convolutional Neural Networks — 270
- 10.6 The 1990s — 270
  - 10.6.1 1997: Sepp Hochreiter and Jürgen Schmidhuber – Long Short-Term Memory — 271
- 10.7 The 2000s — 271
  - 10.7.1 2006: Geoffrey Hinton et al. — 271
- 10.8 The 2010s — 272
  - 10.8.1 2014: Ian J. Goodfellow et al. – Generative Adversarial Networks — 272
  - 10.8.2 2017: Ashish Vaswani et al. – Attention Is All You Need — 274
- 10.9 Summary — 274
- 10.10 Further Reading — 274

11 The Machine Learning Process   277

11.1 The CRISP-DM Model   277
11.1.1 Business Understanding   278
11.1.2 Data Understanding   279
11.1.3 Data Preparation   279
11.1.4 Modeling   280
11.1.5 Evaluation   280
11.1.6 Deployment   280

11.2 Ethical and Legal Aspects   281
11.2.1 Algorithmic Fairness and Bias   282
11.2.2 Explainability and Interpretability   284
11.2.3 Ecological Aspects   288
11.2.4 Legal Aspects   289

11.3 Feature Engineering   290
11.3.1 Feature Coding   292
11.3.2 Feature Extraction   302
11.3.3 The Curse of Dimensionality   311
11.3.4 Feature Transformation   312
11.3.5 Feature Selection   316

11.4 Summary   317

11.5 Further Reading   318

12 Learning Methods   319

12.1 Learning Strategies   319
12.1.1 Supervised Learning   320
12.1.2 Unsupervised Learning   324
12.1.3 Reinforcement Learning   335
12.1.4 Semi-Supervised Learning   344

12.2 Tools   345
12.2.1 Confusion Matrix   345
12.2.2 Receiver Operating Characteristic Curves   347

12.3 Summary   350

12.4 Further Reading   350

13   Areas of Application and Real-Life Examples — 351

13.1   Warm-Up — 351
13.2   Image Classification — 354
   13.2.1   Definitions — 354
   13.2.2   On Bees and Bumblebees — 356
   13.2.3   Pretrained Networks — 366
13.3   Dreamed Images — 373
   13.3.1   The Algorithm — 373
   13.3.2   Implementation — 376
13.4   Deployment with Pretrained Networks — 382
   13.4.1   A Web Application for a Neural Network to Generate Image Descriptions — 383
   13.4.2   A Web Application for Image Generation — 384
13.5   Summary — 386
13.6   Further Reading — 386

Appendices — 387

A   Python in Brief — 389
B   Mathematics in Brief — 417
C   TensorFlow 2 and Keras — 435

The Authors — 445
Index — 447

Preface

This preface was written in cooperation with a transformer neural network, demonstrating the power of neural networks.

The roots of knowledge, much like the great trees of the Austrian and North American forests, grow deep and strong when nurtured over time. During our computer science studies at University of Salzburg, Austria and Bowling Green State University, Ohio, USA, a long time ago, we were fortunate to immerse ourselves in an academic environment that encouraged curiosity, exploration, and innovation. The spirit of learning we experienced there has stayed with us ever since, shaping the way we approach technology and education. With this book, we hope to share that same sense of discovery with you.

Neural networks and artificial intelligence have undergone a remarkable transformation over the past decades. When we first encountered the field, training even a small neural network required considerable effort, and computational power was a limiting factor. Today, with the rise of deep learning frameworks, cloud computing, and powerful hardware, what was once confined to research labs is now accessible to anyone with a laptop and an internet connection. The impact of AI is everywhere—from medical diagnostics and autonomous vehicles to art, music, and natural language processing.

As authors, we believe that understanding AI should not be limited to specialists but should be approachable for anyone with a passion for learning. That is why we have taken great care to structure this book in a way that is both theoretically sound and practically useful. You will find detailed explanations of key concepts, step-by-step implementations in Python, and hands-on exercises that encourage experimentation. Our goal is not only to teach how neural networks work but also to inspire creative applications of AI for real-world problems.

Bringing this book to life has been a journey filled with challenges and excitement. Along the way, we were privileged to receive invaluable feedback and encouragement from colleagues, students, and fellow researchers. We extend our deepest gratitude to our families and friends, who have supported us with patience and understanding throughout this process. Writing a book is never a solitary endeavor—it is the result of countless discussions, revisions, and shared moments of insight.

As you embark on this journey into the world of neural networks, we hope that this book will be both a guide and an inspiration. AI is not just a technical discipline—it is a field that constantly evolves, challenges assumptions, and opens new possibilities. We invite you to explore, experiment, and most importantly, enjoy the process of learning.

Joachim Steinwendner and Roland Schwaiger
Sünikon, Switzerland, and Bad Dürrnberg, Austria, March 2025

Chapter 1
Introduction

Neural networks are at the heart of modern artificial intelligence (AI) and can be found in areas such as machine learning (ML), deep learning, and large language models. They enable technologies such as image recognition, speech processing, and self-driving cars. In this introductory chapter, we'll take a look at their structure and functionality, and then we'll explain why neural networks are the basis of many AI breakthroughs.

Our motivation is to look behind the curtain and explain neural networks to you from the ground up, in terms of both their use in real life and their historical development, as well as in a simple theoretical approach. Of course, we set certain priorities for you to facilitate access to the topic. This means that while we can't introduce you to all types of neural networks—they are numerous—we can discuss those that are currently the most widely used. As we don't want to beat about the bush for long, we'll provide a quick introduction to the topic of *artificial neural networks* (ANNs). A simple example serves as motivation, while a concrete ANN that can solve the problem becomes our vehicle.

1.1 Why Neural Networks?

Neural networks have proven to be a useful tool in computer science for tasks that require flexibility. They have found their place in subject areas that can't be solved by programs with simple rules due to their complexity. Think, for example, of image recognition, which is almost an ideal discipline for ANNs, as a huge amount of data has to be processed, and these calculations can't be implemented according to simple if-then rules.

Due to the simple basic elements of ANNs, we can intuitively imagine the way in which they process data. In general, most people use the analogy to the brain, which we also like to use. Of course, the more complex structure of the ANN is a bit of a trick for us. The high degree of networking in an ANN creates precisely the technical capabilities that are required for complex tasks.

Neural networks are capable of learning and can adapt to new situations—if the environment changes, the ANNs and their answers change accordingly. For an ANN, the world consists not only of *true* and *false* or—as computer scientists put it—zero and one, but it also knows gradations between 0 and 1, and thus fine gradations of answers are possible. Thanks to the high performance of modern computer modules and sophisticated software tools, ANNs can provide their results very quickly. This means that the calculations for the evaluations in an ANN are programmed efficiently and can now be carried out using smartphones.

Due to the technical solutions we can implement via neural networks, every software developer is expected to have a basic knowledge of neural networks to better adapt an app to its respective users and thus improve the user experience. Who wouldn't be happy about an autocorrect that efficiently eliminates typos? In this book, you can find out what else neural networks are capable of.

1.2 About This Book

This is a hands-on book for getting started with ANNs. Supported by graphics, sample applications, and Python code, we'll demonstrate how ANNs work and how you can program them yourself.

This book consists of two parts. The first part gives you a compact overview from the concept to the implementation of various neural networks. We explain to you in a practice-oriented and clear way what you need to do to build a neural network. You'll learn about different neural approaches as well. By the end of the first part, you will have programmed and understood multiple functioning networks. You'll also learn you how to find solutions that work in real-life scenarios.

The second part provides background and context for you because you now want to take a broader approach to the topic and add some theory to your first steps. We discuss the historical development, biological comparisons, different types of networks, and finally different learning algorithms for networks. You'll get to know more technical terms, learn about more advanced concepts, and be able to classify the individual steps to the finished network into a larger system. Specifically, the *Neocognitron network* by Kunihiko Fukushima, a pioneer in the field of ANN research, is mentioned here, which in our view was the pioneer for modern *convolutional neural networks* (CNNs) and ought to receive the appropriate recognition.

To make it as easy as possible for you to get started, we've reduced the necessary prior knowledge to a minimum. In two crash courses in the appendixes of the book, we explain important basic principles of Python (Appendix A) and mathematics (Appendix B), and you get a short introduction to TensorFlow version 2 and Keras (Appendix C). We explain all other basics as simply and comprehensibly as possible directly in the practice-oriented chapters. If you have little experience with programming, you should definitely work through Chapter 2 and Appendix B and, if necessary, refer to Python books or tutorials. Rheinwerk Publishing offers fantastic books on this topic. If you already do have programming experience, you can skim through Appendix A to get to know the Python syntax better. The same applies to math skills and Appendix B.

Using summaries and task sections, we want to help you deepen your overall knowledge. However, the tasks also serve to illustrate the concepts described. We also explain where the fields of application for neural networks can be found and provide a brief overview of the most important terms in *artificial intelligence* (AI). References to the original articles or further sources at the end of the individual chapters provide additional opportunities to deepen the knowledge you've already acquired and to delve into the various areas of knowledge.

In this book, we use Python with and without libraries to program ANNs, whereby we rely on TensorFlow 2 with integrated Keras for the libraries. TensorFlow is an open-source Python library for neural networks from Google that is designed to make the best possible use of graphics processing unit (GPU) resources.

We've stored the code listings in this book in Jupyter Notebooks (see Chapter 2 for details), which are available via the download materials. These notebooks consist of cells in which the listings are entered. Each cell is preceded by the listing number in the book. There are also instructions in the notebooks that deactivate Python warnings. This may seem strange at first glance, but development is so fast that the Python community has developed a strategy that warns against obsolete functions or function parts. This is sometimes annoying and can be switched off by deactivating the warnings. Throughout the book, the output of the code listings is introduced using `# Output:`. Despite deactivating the warnings, a warning appears here immediately: some code listings build on each other, so they should be processed in Jupyter Notebook in the specified order, as error messages are of course not switched off.

1.3 The Contents in Brief

In this introduction, we want to whet your appetite for neural networks, Python, and experimenting with our book. In this section, we provide a thematic classification of neural networks in the broad field of AI. We explain concepts, contexts, and terms such as *artificial intelligence*, *machine learning* (ML), and *data science*.

This section provides a bird's-eye view of our topic and shows you the most important ways to train networks. You'll also find out how networks learn. We present the most important *learning methods*, such as *supervised*, *unsupervised*, and *reinforcement* learning, with their areas of application, by using specific examples.

This is followed by the practical **Part I**, which is intended to be hands-on, and we'll also take the time to link the topics with theory in an easily digestible way. It's important that you work out the results yourselves and thus gain intimate, intuitive access to knowledge about the internal processes that take place inside networks. For this purpose, we'll install the necessary tools for programming, program the setup and analyzing calculations in the network and implement the visualization of the results. Step-by-step and in the form of easy-to-understand tutorials, we'll even guide you through the programming of the *learning algorithms* of neural networks—sometimes without and sometimes with the help of the TensorFlow library.

Chapter 2 describes everything you need to know about installing and setting up the required development tools. You'll get to know the *Anaconda* data science platform, the *Jupyter* programming environment and the first lines of *Python*. This will be an important basic chapter, as the skills you acquire in using the web-based, interactive *Jupyter Notebook* programming environment will make the subsequent programming steps easier for you. To ensure a uniform computing platform for all readers, we recommend that you use Google's web platform for Jupyter Notebooks: *Google Colab*.

In **Chapter 3**, we describe simple neural networks and demonstrate step-by-step how you can perform calculations with them. You'll of course implement the code for these networks yourself using a Jupyter Notebook.

Chapter 4 is dedicated to the *perceptron*, a neuron model that still forms the basis of the ANN today. For this purpose, we set ourselves tasks that—put simply—consist of categorizing things into groups, such as determining whether a bee is a queen bee or whether an email is spam. During the description of the perceptron, you'll discover certain weaknesses of this neuron model, such as the fact that a possible solution to a task can't be learned in a *stable* manner. In other words, although the perceptron has already learned the correct answer to a question once, it will "forget" this answer again if it's given new examples. In many cases, this is unsatisfactory, which is why we describe the implementation of another network, the *Adaline*, as a possible countermeasure to the memory loss of the perceptron. Not only does Adaline remain stable, but it also memorizes the learned responses. Being able to memorize something is a prerequisite for learning, and so Adaline is a simple example of how learning also works with large neural networks. You'll learn a lot of new things, for example, what a *square error* is and how it can be useful for learning.

Chapter 5 focuses on solving a historical problem for perceptrons, the *XOR problem*. XOR, the *exclusive OR*, stands for a specific logical operation, such as AND or OR. This much can already be revealed at this point: the XOR problem is an unsolvable problem for the perceptron. As a result, research in the field of neural networks was put on hold for a while. However, a solution has emerged: take multiple perceptrons, arrange them in layers, and stack them on top of each other. This is exactly how you'll get to know and implement the solution to the XOR problem.

With this adaptation, the ANNs learn to process even more complex tasks where no clear assignment to a group is possible. Consider, for example, the distinction between three types of bees: In the hive you have the queen bee, the drones, and the worker bees. If you measure the length of the bee bodies, groups will emerge based on length; however, there will always be bees that can't be clearly assigned to a group.

Chapter 6 has it all, as we discuss learning in multilayer networks. This will give you a deep understanding of the internal processes taking place in ANNs. This requires some math, which, as promised, is explained in more detail in Appendix B and discussed in parts in this chapter. In the course of this, you'll learn the meaning of such beautiful terms as *gradient descent*, *feed-forward calculation*, *backpropagation* up to the output layer, backpropagation up to the hidden layer, correction of weightings, and backpropagation in matrix form. All of these are exciting things that you'll implement using NumPy, a Python module for mathematical calculations.

Chapter 7 takes you to the *deep neural networks* or *deep learning*. From the thousands of network architectures, we pick out and explain two architectures that we consider to be essential: (1) *convolutional neural networks* (CNNs), which are mainly used in the current machine processing of image and audio files, and (2) *transformer neural networks*, which are mainly used in text and speech processing and form the basis for *large language models* (LLM).

While Chapter 7 focuses on theoretical explanations, **Chapter 8** is dedicated to the application of this knowledge. A simple example of how the layers are laid down is used to deepen your understanding. Among other things, we use the classic image recognition task: the recognition of handwritten numbers.

Chapter 8 thus implements the knowledge you gained in Chapter 7. Here, we use Python libraries to guide you to your own modern neural networks. Because the development of a completely self-implemented Python solution for neural networks would be very complex, we rely on the aforementioned TensorFlow open-source library.

In **Part II**, you'll learn even more about neural networks. This includes the biological background, the research history of neural networks and an overview of the various areas of application of neural networks.

In **Chapter 9**, we'll explain the components of ANNs based on biology: *soma*, *dendrites*, *axon*, *cell body*, *action potential*, and so on. Our plan is to enable you to carry out your own experiments that change elements of neural networks. For example, perhaps you want to model neurons more biologically or delve deeper into biology—let us pave the way for you.

Chapter 10 describes a wide variety of network architectures. Deep learning and CNNs are currently the talk of the town, but there are other networks that should stimulate your imagination and encourage independent experimentation. For this purpose, we'll take an excursion into the history of networks, placing what we've discussed in its historical context. We didn't miss the opportunity to implement another classic, the *Hopfield network*. This type of network belongs to the class of *networks with feedback* (or *feedback networks*). It has a wide range of applications; for example, it helps to recognize objects in noisy images.

Chapter 11 explains the aspects of the machine learning (ML) process in connection with neural networks. This chapter addresses an essential aspect—handling data in an ethical and technical sense. The algorithms of AI have reached such a level of maturity that they have a significant impact on people's lives, which is why we can't avoid discussing ethical aspects. In the life of a data scientist, however, preparing data for ML algorithms or neural networks is also part of the daily, often tedious, grind. This preparation is also known as *feature engineering*. Reading this chapter will give you a small but well-sorted toolbox for dealing with different data types.

In **Chapter 12**, we take a very practical look at the different *learning methods*. We describe what learning strategies there are and what they're used for. We'll also implement two beautiful variants. You'll learn details about supervised, unsupervised, and reinforcement learning, that is, the main representatives of the learning methods for ANNs. Currently, supervised learning is mainly used where the network is also informed of an expected result for an input, such as the body length of a bee: for example, whether it's a queen bee, a drone, or a worker bee. If the network determines an output for an input that doesn't match the desired result, it must learn. Specifically for unsupervised and reinforcement learning, which aren't often the focus of attention, we've worked hard on our programming and, among other things, trained software agents. A *software agent* is, for example, a simulated *robot* that can move around in an environment. The agent should develop a strategy on its own initiative so that it can move safely in unfamiliar terrain. The chapter is rounded off with a description of tools that serve to ensure the quality of ANN training. We also describe how models can be compared with each other.

Chapter 13 uses interesting examples to show the different areas of application of neural networks in real life.

We know how difficult it can be to find the right information on the web for specific questions. Most of the time you're overwhelmed by the amount of information, and it's even more difficult to assess the relevance of the information when you're starting out in a new field of learning. That is why, as already mentioned, we've compiled what we consider to be the key information for a quick start in Python, math, and TensorFlow in **Appendix A**, **Appendix B**, and **Appendix C**.

Here's another piece of good news: You can download every code, every Jupyter Notebook, and the data records used in the book from the book website, which we strongly recommend. To do this, go to *www.rheinwerk-computing.com/6059*, scroll down to the **Product supplements** section, click on **Supplements list >** link, and then click on the **Download** button.

1.4 Is This Bee a Queen Bee?

Let's suppose you want to automatically distinguish the queen bee from the worker bees and the drones in a beehive. You've probably received an order from a beekeeper. According to your client, there are several criteria for this, but we only use the following here: The queen bee is the longest bee, followed by the drones, while the worker bees are the shortest, as shown with artistic license in Figure 1.1. We apologize to all beekeepers among our readers for this didactic simplification of reality.

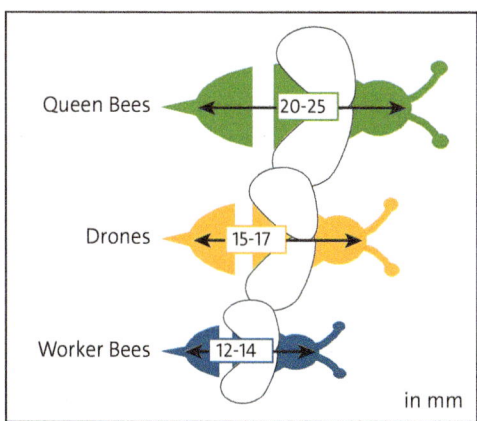

Figure 1.1 Classification of Bees

In our little experiment, a technical system provides the length of a buzzing bee. This means that we have a value, that is, the length, to determine which group each bee belongs to. Admittedly, this classification task can be solved by an if-then rule, for example, by the following pseudocode:

```
if( length(bee) greater than or equal to 20mm ) then
    is_queen = true
else
    is_queen = false
```

On the other hand, this example can also be used wonderfully to explain how an ANN works, which we'll do in the next section.

1.5 An Artificial Neural Network for the Bee Colony

We now present a simple network that solves this task. Please take a close look at the schematic visualization in Figure 1.2 as it will guide you through the book. Note that further elements will be added as you get to know more complex neural networks.

You can already see it now: We've deliberately created *layers* in such a way that related elements are marked with the same color. We'll go through and explain the individual elements of the drawing step-by-step in the rest of this section.

The task of this network is to receive a length and carry out a classification,, that is, a division into two groups: queen and not-queen. Thus, the length x_1 is the *input value* of the network—sometimes just referred to as *input*. The *output value* or *output* for short, is y_1, and we specify that the value 1 should indicate that a bee is the queen. Otherwise, the network should output the value 0.

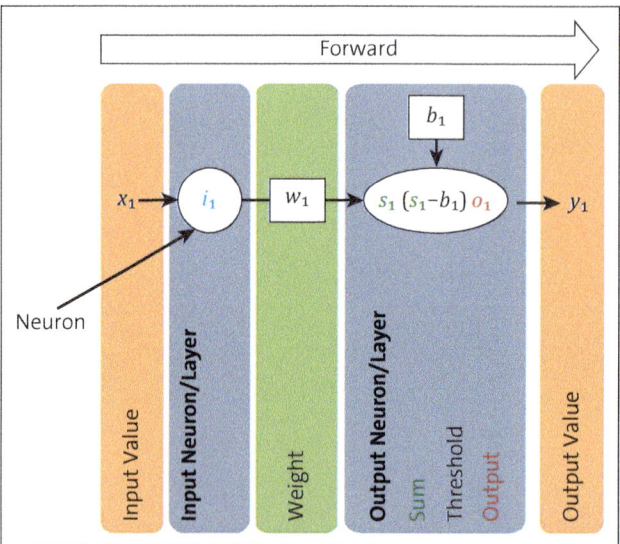

Figure 1.2 A Simple ANN

The *input layer* consists of as many *input neurons* as are needed for the specific task. In our case, this is an input for the length. However, the weight of the bee could also be added, for example, in which case, we would already have two input neurons. The same applies to the *output layer*. For us, this means that only one input neuron is required, as we only have the length of the bee available. The output also contains exactly one neuron that provides information about whether or not the bee is a queen bee.

Neurons are small *computational units* that perform computational steps depending on the type of neuron. In our example, these are the following steps:

1. An *input neuron* passes the input value on to the next layer, for example, 21 mm. In the bee classification, that would be a queen bee. This means that the input neuron has the value 21, as you can see in Figure 1.3. We've written 21 in the input neuron instead of its name, i_1. The 1 in the designation of the input neuron is an *index*. If you had multiple inputs, you could label them with the lowercase letter *i* and simply increment the counter. Then, for example, the input for the weight of the bee would have the index 2.

2. Each individual *output neuron* adds up the values of all its inputs in such a way that each input value is multiplied by the weight of the input connection. This sum is then obviously referred to as a *weighted sum*. In this calculation, a *weight* regulates the proportion that the input value has in the input sum. If the weight of a connection is 0, for example, then the input value has no influence, as 0 multiplied by a

value always results in 0. In our network, we only have one input to the output neuron with the weight w_1, so this results in the weighted sum:

$s_1 = i_1 \cdot w_1$

For example, if we set the weight w_1 to the value 1, as shown in Figure 1.4, the result is

$s_1 = i_1 \cdot w_1 = 21 \cdot 1 = 21$

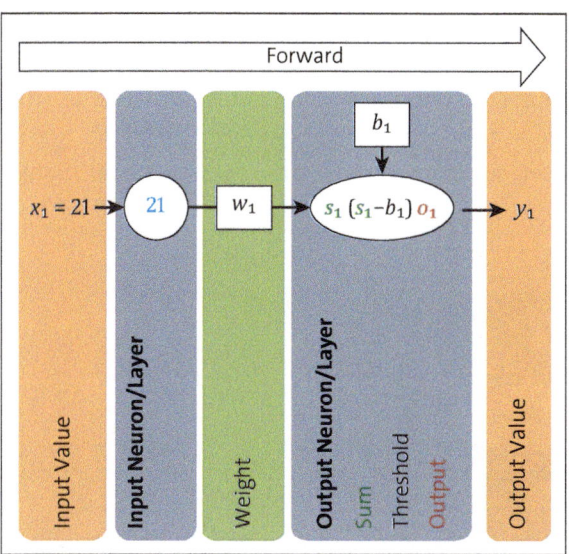

Figure 1.3 Input Value 21 Transferred to the Network

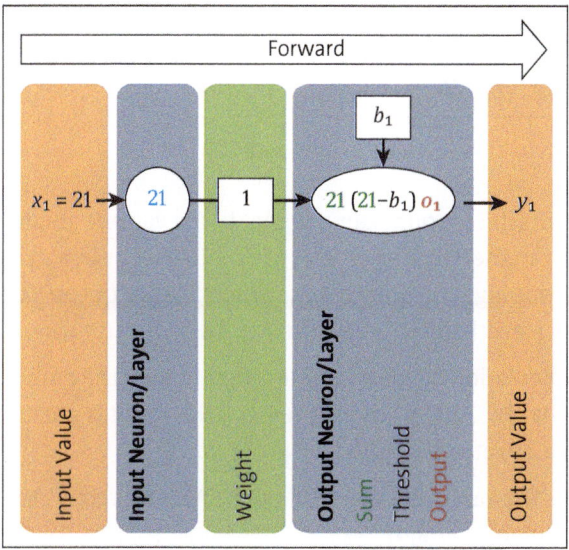

Figure 1.4 Calculating the Weighted Input for the Output Neuron

In the bee classification network, we only have one input to the output neuron, but in more complex networks, the output neurons have many more inputs, which are also referred to as *predecessors*.

3. Each output neuron takes the calculated sum and compares it with a *threshold value*, also known as *bias* and referred to as b_1 in Figure 1.4. This threshold value comparison is a central step in the calculation algorithm, which we'll motivate biologically in Chapter 9. The comparison can be written as $s_1 >= b_1$ or can be converted to $s_1 - b_1 >= 0$. For our example, we've set the threshold value b_1 to 20 and replaced b_1 with 20 in Figure 1.5. Why 20? That's an important and very legitimate question. You'll see later how this value can be calculated, but for our example, it's chosen so that the differentiation of the queen bee from the rest of the bees can be calculated.

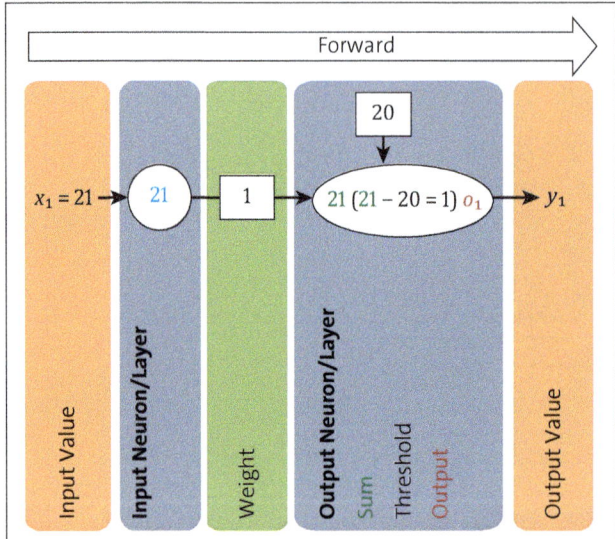

Figure 1.5 Threshold Calculation

The threshold value calculation results in 1 for our example; that is, the value is greater than 0. The result of the calculation can therefore be greater than, equal to, or less than 0. However, we want to achieve a grouping into two groups, namely 0 ("not queen") or 1 ("queen"). This leads us to the final step, that is, the calculation of the output.

If the result of the threshold value calculation has any value less than 0, then the output neuron should return the value 0, that is, "not queen". However, if the result is greater than or equal to 0, the value 1 should be returned, as shown in Figure 1.6.

At the start, we defined the value 1 as the value for the queen, so the ANN correctly classified the 21 mm bee as a queen based on its length.

1.5 An Artificial Neural Network for the Bee Colony

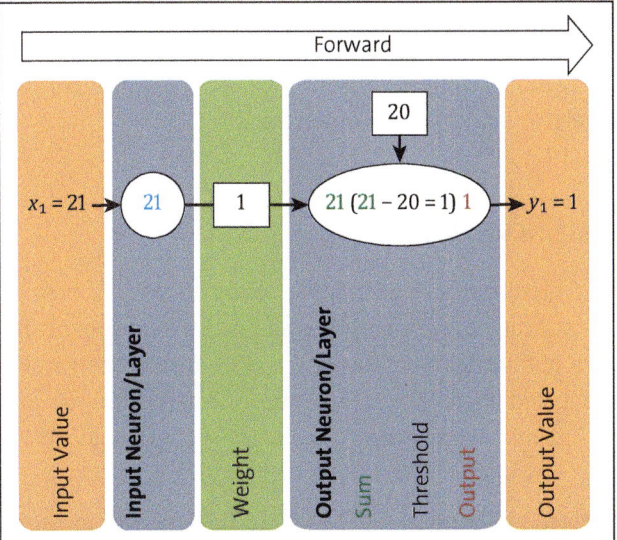

Figure 1.6 Output Calculation

To summarize, we've defined the following components for our ANN:

- We've assigned the value 1 to the weight from the input layer to the output layer.
- For the output function, we've used a step function, which is also known as the *Heaviside function* (you'll see this later in the chapter in Figure 1.10).
- We've set the threshold value to 20.

The more complicated the task, the more neurons are required, which is why the *hidden neurons* are placed between the input and output. When you're dealing with neural networks, you pass data to the input neurons and receive results from the output neurons; you don't come into contact with the neurons in between, so they're called "hidden."

On the way from input i_1 to output o_1, the network has passed on the value from layer to layer—and always in a forward direction. This type of ANN is referred to as a *feed-forward network*. Later in the book, we'll also introduce you to *recurrent networks*, in which there is feedback between the same or previous layers.

Task
Now it's your turn. Why not try it with an input of 14. Is the bee a queen bee? What is the ANN's answer?

Solution:

Correct, the answer is 0, so it's not a queen bee.

Although this ANN is so simple, we've already used it to show you the basic components of networks that are currently at the top of international competitions for image recognition, natural language processing (NLP), or games. However, this involves thousands of neurons, numerous layers, and millions of connections.

1.6 From Biology to the Artificial Neuron

The term *neuron* is actually motivated by a biological model and conceived according to this model: *bionics* is the name given to the scientific discipline that is responsible for the technical implementation of findings from biological systems and their analyses. The basic idea is a tried and tested approach: take a system from nature that has been optimized and developed over thousands of years and try to replicate it as closely as possible.

Just as the flight of birds encouraged mankind to fly, the Velcro fastener was invented based on the example of burdocks, or suction cups, which were copied from octopuses or beetles, neural networks are a good example from bionics. Take a (not necessarily human) brain, analyze it, and convert it into a mathematical algorithm that can solve complex tasks.

The idea of neural networks is already several years old and dates back to the 1940s. At that time, it was introduced by the neurophysiologist Warren McCulloch and the mathematician Walter Pitts. The development of ANNs has had an eventful history, which you'll learn more about in the second part of this book, and it has experienced an immense upswing in recent years under the concept of *deep learning*.

ANNs are powerful, scalable, and used for many complex ML tasks such as the automatic classification of a vast number of images in the Google network, for speech recognition on the iPhone, for video recommendation systems on YouTube, or to learn to beat the world champion in Go, the strategic board game from China.

1.6.1 The Biological Neuron and Its Technical Copy

The human brain is a truly unique object. It weighs an average of 1400 grams and essentially consists of around 86 billion *nerve cells* (i.e., neurons). These enable us not only to see, hear, and control complex movements of our body but also to be creative: writing books, making films, painting, discussing, and so on.

Neurons (aka nerve cells) communicate by transmitting signals to each other by chemical means, while incoming signals are transmitted electrically within the neurons (see Figure 1.7). That doesn't sound too complicated, but it becomes fascinating when the 86 billion neurons in the brain communicate with each other within a few seconds to respond to visual (seeing), auditory (hearing), olfactory (smelling), gustatory (tasting), and tactile (touching) stimuli, for example, by quickly running away when a roaring tiger attacks. This works incessantly with a thoroughly impressive success rate; otherwise, humanity wouldn't exist.

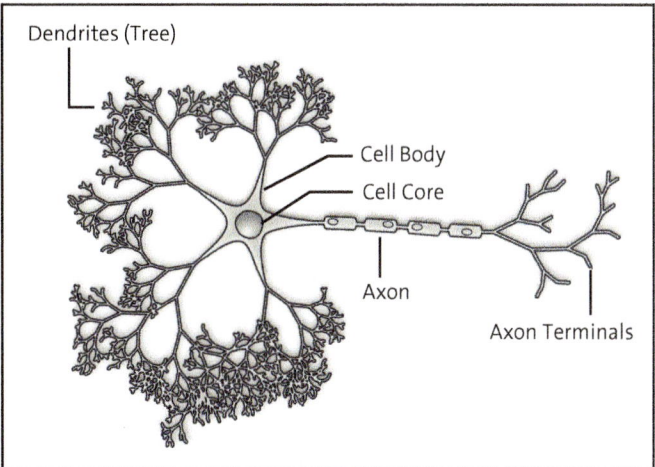

Figure 1.7 Schematic Visualization of a Neuron (By Nicolas.Rougier/CC-BY-SA-3.0, https://commons.wikimedia.org/w/index.php?curid=2192116)

The incoming chemical signals from other neurons reach the *cell body* of the neuron via mostly highly branched *dendrites*. Information is transmitted via *axons*, which can be up to 1 meter (39 inches) long. Most neurons have exactly one axon, at the tree-like end branch of which the information is passed on to multiple neurons. The dendrites and axons dock on to the neurons or their cell bodies via *synapses*. These synapses are a kind of interface: they convert the chemical signal (between the neurons) into an electrical signal (within the neuron) and have an inhibitory or excitatory effect on the signals.

A special feature of signal transmission within the neuron is the *all-or-none law*, which means that the signals transmitted to the cell body via the dendrites will be passed on to the axon only if a certain total signal strength is reached. In the bee example, we've replicated this by using a threshold value.

> **All-or-None Law**
> This law describes a peculiarity of nerve cells, namely, that a response to a stimulus either occurs in its entirety or not at all. A certain threshold value must be exceeded to trigger a response.

1.6.2 The Artificial Neuron and Its Elements

Now that we've taken a look at the biological neuron and know its elements and tasks, let's try to build an artificial neuron and represent it graphically. Because we'll also use this artificial neuron for computational tasks later on, we'll use the popular language of mathematics to describe the individual components.

We orientate ourselves at Figure 1.8, proceed from left to right and start with the dendrites, via which the input signals are sent to the cell body. We assume a *dendrite tree* with n branches. The first input signal is therefore referred to as x_1, the second as x_2, and so on, and the last one (the nth one) as x_n. We summarize this series of numbers in an *input vector* $\vec{x}$.

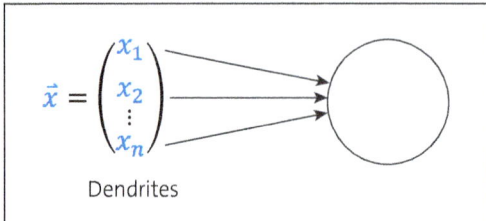

Figure 1.8 Artificial Dendrites

The dendrites transmit the signals from other neurons to the cell body of our neuron via an *inhibitory* or *amplifying* interface, the synapse. We form these synapses as numerical values $w_1, w_2, \ldots, w_n$ for each input signal. These values are also referred to as *weights*, and the total of the values is the *weight vector* $\vec{w}$. The weights are finally multiplied by the input signals and added together so that the entire *stimulus*, which arrives in the cell body, can be represented as the sum total of the input signals multiplied by the *synapse weights* (see Figure 1.9):

$$\sum_{i=1}^{n} x_i \cdot w_i$$

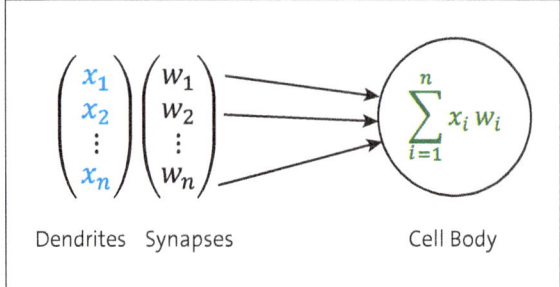

Figure 1.9 Artificial Dendrites with Synapses (Weights)

Note on Multiplication

Sometimes, the multiplication between the weight and the input is represented with a dot, and sometimes the dot is omitted.

Now we could simply pass this sum on to the next neurons as an output signal via the artificial axon, if it weren't for the all-or-none law. This means that we can only pass on the signal if the sum exceeds a certain threshold.

Here, we help ourselves mathematically by passing this (*weighted*) *sum* to a function that fulfills the all-or-none law. This function is also known as the *activation function* f_{act} and ultimately provides us with the output

$$y = f_{act}\left(\sum_{i=1}^{n} x_i w_i\right)$$

of the cell body, which is transferred via the axon. The activation function can now have various forms, which turn the all-or-none law into a kind of catch-all clause, an almost-all-or-nearly-none law. We've outlined a few examples in Figure 1.10.

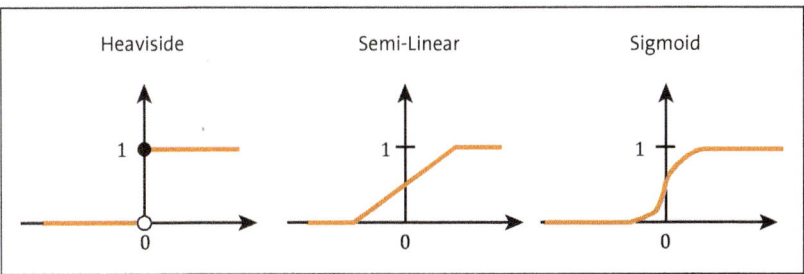

Figure 1.10 Examples of the fact Activation Function

The *Heaviside* function is the classic step function with the values 0 or 1, whereby the function returns 1 as the result if the input to the function is greater than or equal to 0. The *semilinear* function increases from 0 to 1 using a line and can therefore assume intermediate values, just like the *sigmoid function*, which we'll use frequently later in the book. Due to these different activation functions, there's no hard threshold value, but it moves in a value interval. In this way, we can see once again the ability of the ANN to take into account fine gradations in the answers. We've thus reproduced all the correspondences and show our finished artificial neuron in Figure 1.11.

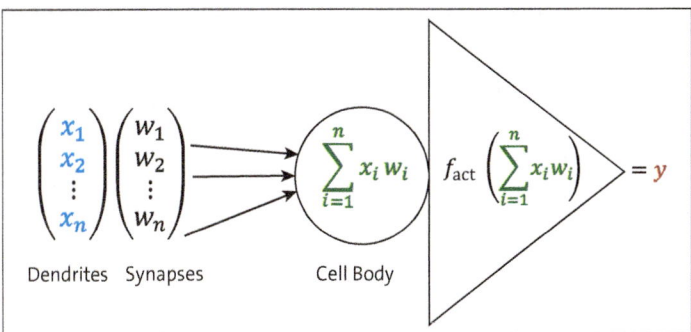

Figure 1.11 The Artificial Neuron

1.7 Classification and the Rest

In the previous section, you've seen the basic mode of operation and the structure of neural networks. Are you disillusioned by the simplicity of the components? Nevertheless, you'll find superlatives in the news that range from the end of the world due to AI to the cooperative relief of mankind from mindless, noncreative work through AI.

To enable you to make a qualified assessment of the current situation, we've set ourselves the task of showing you the basic mechanisms and techniques behind "smart" approaches, thereby restoring some normality to the assessment of the possibilities. So, please take a deep breath and stay focused!

We'll focus our energy on putting ANNs into the right context. If you're the type of learner who likes to start with an overview and work your way up to the details, you've chosen the right book.

1.7.1 Big Picture

It's not easy to show the development of the concepts and their dependencies. Let's take a look at the highly focused and context-optimized *AI* (see Figure 1.12).

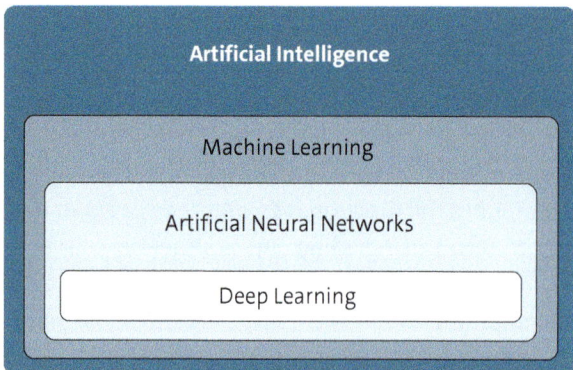

Figure 1.12 Diagram of Terms for "Artificial Intelligence"

Here, you can see the thematic dependencies of the different terms, which we want to explain in more detail in the following subsections. A wide variety of disciplines, such as psychology, neurology, and mathematics, contribute to the foundations and content of the topics presented, with computer science playing a major and central role.

1.7.2 Artificial Intelligence

Artificial intelligence (*AI* for short) refers to a branch of computer science that deals with the simulation of behavior that is interpreted as "smart" by humans. We can already hear the first people shouting: "Wait a minute, what does 'smart' mean?" Unfortunately, we can't provide a definition for this either, but rather symptoms on which we would hang the term, such as independent problem-solving or interactive behavior like that of a human being with the help of a combination of hardware, software, and networks.

The *Turing test*, named after and defined by the mathematician Alan Turing, provides an operational definition of intelligence (refer to Section 1.9 at the end of the chapter). Roughly summarized, this means that an algorithm passes the intelligence test if a human who has asked written questions can't determine whether the answer provided originates from a human or from the algorithm. From a programming perspective, this requires a number of things:

- **Natural language processing (NLP)**
  NLP refers to the processing, interpretation, generation, and output of natural language because the questions are asked by a person in natural language and the person expects the answer in this form.

- **Knowledge representation**
  The content of the question is understood and the answers are drawn from the knowledge context of the question.

- **Logical reasoning**
  Using "If . . . then . . ." is an example of logical reasoning: If you've bought a movie ticket and are on your way to the movie theater, then you're probably going to see a movie.

- **Machine learning (ML)**
  ML refers to adapting to new contexts, generalizing, and recognizing basic structures, for example, the context-dependent interpretation of terms.

This list already provides the main areas and main tasks of AI, which can be derived from the original Turing test. However, humans themselves still serve as a measure of intelligence, with the desire for AI to be able to solve tasks autonomously and offer support functions for humans, especially for super boring routine tasks such as sorting letters by ZIP code.

The Turing Test

The Turing test, which was originally proposed by Alan Turing in 1950, was long regarded as the benchmark for determining whether a machine can *imitate* human intelligence. Today, the Turing test is increasingly considered insufficient to assess true intelligence or consciousness due to the following:

- **Imitation ability vs. understanding**
  The test only measures a machine's ability to *imitate* human behavior, not whether it actually *understands* what it's "saying." AI can have human-like conversations, but that doesn't mean it has consciousness or real understanding.

- **Progress in AI**
  Today's language models and AI can pass the Turing test under certain conditions

without possessing "real" intelligence or an understanding of the world. They process huge amounts of data to generate human-like responses without "understanding" the content.

- **New assessment approaches**
  Researchers are developing more sophisticated tests that go beyond pure imitation. They focus on aspects such as creativity, problem-solving, moral judgment, and dealing with complex real-life scenarios.
- **Ethics and consciousness**
  The Turing test ignores important questions about consciousness and ethical behavior. A machine can pass the test but still act without an understanding of morality or ethical responsibility.

Overall, the Turing test is often considered outdated today, as it only assesses the surface performance of AI and doesn't capture deeper concepts such as understanding, consciousness, or emotional intelligence.

There are two main groups within AI: strong AI and weak AI. Representatives of *weak AI* want to master concrete application problems of human thinking. You could also say that it's about *simulating* intelligent behavior. Significant progress has been made in the field of weak AI in recent years, for example, in image and speech recognition and speech generation. Representatives of *strong AI*, on the other hand, aim to create an intelligence that mechanizes human thinking, that is, at least achieves the same intellectual abilities as humans.

1.7.3 History

The official beginning of AI is considered to be in 1956; we'll look at what happened before that in the second part of this book. In 1956, John McCarthy, then working at Dartmouth College, organized a two-month workshop with Marvin Minsky, Claude Shannon and Nathaniel Rochester, which dealt with seven topics of AI. Ten people were asked to work on a study of these topics for a period of two months. What is interesting here is the working hypothesis that every aspect of learning or intelligence in general can be described so precisely that it can be simulated by a machine. This meeting gave rise to a field of research that has offered a broad field of activity for many researchers ever since and continues to do so today, but is also characterized by sometimes exaggerated announcements and hopes.

Fun Fact

Incidentally, the term *artificial intelligence* was used for the first time in McCarthy's call to the workshop. The authorship of the term can therefore be attributed to him (see *www-formal.stanford.edu/jmc/history/dartmouth.pdf*).

In the 1950s and 1960s, the approaches of symbolic AI were strongly promoted, alongside the earlier subsymbolic or *connectionist* approaches. In this list, the term *connectionism* is most appropriate for the ANN approach, as it deals with the behavior of networked information processing units.

But what do "symbolic" and "subsymbolic" actually mean, and what is the real difference? The distinction stems from the way in which knowledge is represented. *Symbolic AI* explicitly represents knowledge, that is, examples and derived rules. You could also say that knowledge exists explicitly and is processed using mathematical logic. *Explicit* here means that it provides explanatory possibilities and that we as humans understand how decisions were made. As examples for the representation of knowledge, simply think of if-then rules, such as, "If it rains, then I will take an umbrella with me" or "If the length is >= 20 mm, then it's a queen bee."

In *subsymbolic AI*, such as ANNs, knowledge is represented *implicitly*, which means there are no explicit rules about the representation of knowledge and therefore the problem for humans is to make knowledge comprehensible to humans. On the other hand, the daily sensational news about subsymbolic systems shows that there is considerable power in this approach. You'll find plenty of examples of this approach in the course of reading this book.

As we've seen, history has (currently) developed in favor of subsymbolic systems. The problem of the actual complexity of reality is being addressed with new algorithms and increased computing power. Some of the models suited for this, namely *convolutional neural networks* (CNNs) and *transformer networks*, are dealt with prominently in this book.

As mentioned earlier, one of the subareas of AI is ML, which we'll look at in more detail in the next section.

1.7.4 Machine Learning

A vast amount of data—terabytes of data—is produced every day, for example, through phone calls, purchases, electricity readings, and sensors. How should this data be analyzed? What "knowledge" can be gained from it?

> **Terabyte**
>
> If one page of text is recorded for every billion people, which in turn corresponds to approximately 1 KB of data, then a total of 1 TB of data is recorded.

These are precisely the questions addressed by *machine learning* (ML), a term coined by Arthur Samuel in 1959. Everyone who uses online retail is aware of the effects of this technological discipline. We're constantly being suggested things and being followed by them across the web. Rules are applied when selecting the items that follow us, for example: "Customer X is similar (at least in terms of purchasing behavior) to customer Y and therefore will probably be interested in similar things to those that Y has already bought."

ML is understood as a branch of AI that aims to recognize patterns in data, to generally elicit new insights from data that aren't immediately apparent to humans, and to generalize previously unknown cases. Because the findings aren't clear from the outset, it's not possible to develop explicit programs with hard-coded logic, as the internal structure of the program to be created wouldn't be known.

Thus, ML is busy generating *knowledge from experience*. A system learns patterns from examples and can generalize as yet unknown examples based on them. This type of learning is also referred to as *inductive learning* or *learning from observations*.

As trained data scientists, we talk about how models are learned that work sufficiently well for their intended field of application. A *model* is a (simplified) representation of reality, such as a neural network. The distinction between the different methods in ML is based on the way in which learning takes place. A distinction is mainly made between *supervised* learning, *unsupervised* learning, and *reinforcement* learning, as shown in Figure 1.13:

- **Supervised learning**
  This is learning on the basis of examples. The desired result is also made available to the algorithm for an example. If the algorithm is presented with new, unlearned examples, it should find an answer by analogy with the learned examples.

- **Unsupervised learning**
  This is learning by discovering patterns. The algorithm is provided only with input data and no desired output. The objective is for the algorithm to find a distribution of the data and thus provide new insights into the data.

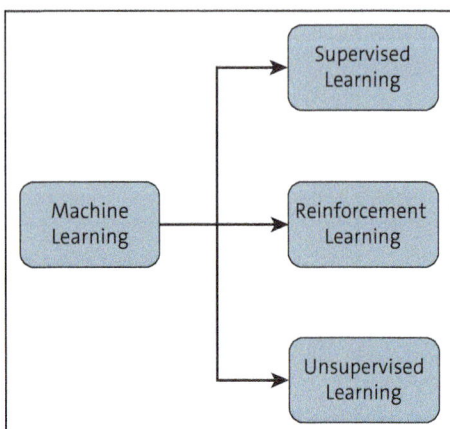

Figure 1.13 Learning Strategies in Machine Learning

- **Reinforcement learning**
  This is learning with rewards. This approach uses methods that enable a software agent (a computer program that exhibits a certain degree of autonomous and self-dynamic behavior) to learn strategies so that the reward achieved gets maximized. A good example of where this approach is used is in video game learning.

1.7.5 Deep Neural Networks

As we've indicated, a wide variety of ANN types are used in ML approaches. In recent years, a special category of ANN has developed called the *deep neural network*. We want to highlight these in particular, as great successes have been achieved with this type of network and there's no end in sight. Figure 1.14 shows a very simplified diagram of a deep neural network.

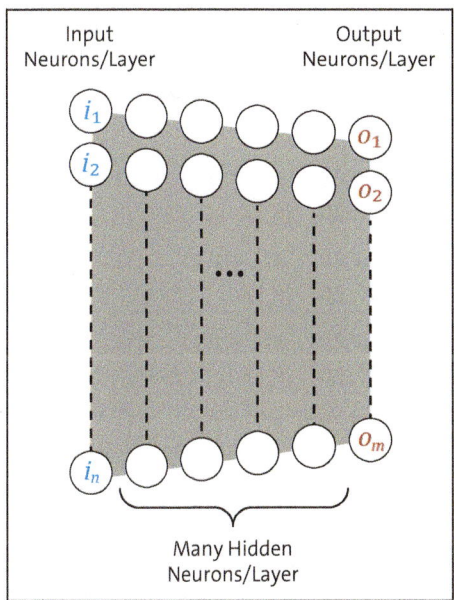

Figure 1.14 Diagram of a Deep Neural Network

1.7.6 Transformer Neural Networks

A *transformer neural network* or *transformer network* is a special model that is primarily used for the processing of language in AI. Unlike older methods that read words one after the other, a transformer can look at all the words in a sentence at the same time and recognize connections between them. This enables the network to understand more quickly and accurately what's important in a sentence. It consists of two parts: an *encoder* that reads the text and a *decoder* that generates a meaningful response. This system is used for many modern AI applications, for example, for translations or LLMs such as *GPT* (*generative pretrained transformers*). We'll talk more about transformer networks later.

> **Note on the Notation**
>
> For the queen bee, we've used i_1 for the input neurons and o_1 for the output neurons. We then used the *subscript*, that is, the lower index, for the input vector $\vec{x}$ to designate the different inputs, such as x_1 for the first input. In Figure 1.14, we use the same notation as for the queen bee, with the addition of a large number of hidden neurons, which aren't described in detail.

Deep neural networks differ from conventional networks, as you've seen with the queen bee, for example, in that they are much more complex in terms of their structure. They have a high number of internal processing units, the *hidden neurons* in the *hidden layers*. Through these many hidden layers, which lead to a deep layering of the network, it's possible for the network to learn very complex mappings from input to output. We'll look at some practical examples of the performance of these networks in Chapter 8.

If you're wondering why more complex networks weren't used in practice earlier, the very simple answer is that we lacked the computing power and the data.

> **CPU, GPU, and TPU**
>
> *Central processing units* (*CPUs*) are optimized to provide a high processing speed (gigahertz, GHz), whereas *graphics processing units* (*GPUs*), that is, the computing units on the graphics cards in computers, are optimized for throughput. CPUs only have a small number of *cores* (main processor cores in a chip)—in contrast to GPUs, which have hundreds to thousands of cores.
>
> A *tensor processing unit* (*TPU*), in turn, is a specialized processor developed by Google to execute ML and AI algorithms more efficiently and faster, especially those based on TensorFlow.

It was the change in hardware in particular that revived the ANN and triggered the current hype. Calculations are preferably performed on graphics cards, and this allows complexities to be achieved that were unimaginable just a few years ago—to be more precise, before 2011, when NVIDIA teamed up with researcher Andrew Ng from Stanford to compete against CPUs with NVIDIA GPUs and was able to hold its own. This drastically reduced the hardware entry hurdle for the calculation of highly complex networks; we're talking about networks with billions of neurons and connections.

> **Fun Fact**
>
> The number of neurons in the brain is estimated at around $8.6 \cdot 10^{10}$. For comparison, a total of around 10^{10} bacteria live on the surface of our skin.

1.8 Summary

This concludes the introduction, in which we hope to have given you a first overview of neural networks and their context. It was important to present a small concrete example and thus a little reality in dealing with neural networks. We've discussed the following key points:

- A simple ANN to determine the queen bee
- The derivation of the artificial neuron from the biological neuron
- The classification of neural networks in the field of AI

Thank you for staying with us. You're cut from the wood needed for this book as you've shown perseverance, stamina, and a thirst for knowledge. We hope we've made it as easy as possible for you to get to grips with the subject matter of this book.

Right, now for the first practical topic in the first part, which is about installing the framework we're using: Python, TensorFlow, and Jupyter Notebook.

1.9 Further Reading

- Alan M. Turing, "Computing Machinery and Intelligence," 1950, *https://phil415.pbworks.com/f/TuringComputing.pdf*
- Stuart Russell and Peter Norvig, *Artificial Intelligence*, Pearson, 2012
- Volodymyr Mnih et al., "Playing Atari with Deep Reinforcement Learning," *DeepMind*, 2013. *https://arxiv.org/pdf/1312.5602v1.pdf*

PART I
Up and Running

This part is for doers and makers. We'll give you a quick introduction and then jump into the nuts and bolts you need to know to see results, such as setting up a development environment, programming your first perceptron, and more. In the form of an easy-to-understand tutorial, we take you by the hand and guide you through the programming of a finished neural network.

Chapter 2
Starter Kit for Developing Neural Networks with Python

In the end, even a fool sees the goal, the more clever man sees it in the middle, and only the wise man sees it at the first step.
—Friedrich Rückert (German writer, 1788–1866)

To make the initial step easier, we'll first create a working environment in this chapter. From a technical point of view, this is a development environment for programming neural networks in the Python programming language.

2.1 The Technical Development Environment

The first step is to install an *integrated development environment* (IDE), which contains tools for editing the source code, testing, debugging, and, of course, compiling (i.e., translating into computer-understandable code).

There are countless ways to install Python on computers with different operating systems. On the Python Software Foundation website (*https://docs.python.org/3/using/index.html*), you'll find installation instructions for the three most important operating systems: macOS, Windows, and Linux.

2.1.1 The Anaconda Distribution

The installation instructions on the Python Software Foundation website are very instructive, but we recommend using the Anaconda distribution (*www.anaconda.com*). Anaconda is one of the most popular data science platforms and offers some very helpful features:

- A large number of important data science modules, including the NumPy, scikit-learn, TensorFlow, and Keras modules, which are all important for this book
- The Virtual Environment Manager for Windows, Linux and macOS, which allows you to work in development environments with very specific versions of modules
- The Conda package for installing and upgrading modules

It's important to mention that we use Python version 3 throughout this book, so you should also download the Anaconda version for Python 3.x (*x* stands for any number greater than or equal to 0). The detailed installation instructions can be found at *https://docs.anaconda.com/anaconda/*.

This website contains additional information and thousands of packages that can be installed very easily if required.

Task

Before you continue reading, you should first install the Anaconda distribution on your computer now. The distribution is designed in such a way that the installation requires little effort and runs automatically. You can also download the *cheat sheet* for Anaconda (*https://docs.anaconda.com/anaconda/user-guide/cheatsheet*), which provides a compact overview of the Anaconda distribution. Then, start the Anaconda Navigator.

Anaconda Navigator

For managing and starting development environments, the Anaconda distribution provides the **Anaconda Navigator** screen (see Figure 2.1). This program enables you to administrate the various Python packages, and you can use it to start our main tool for Python programming, the Jupyter Notebook, among other things.

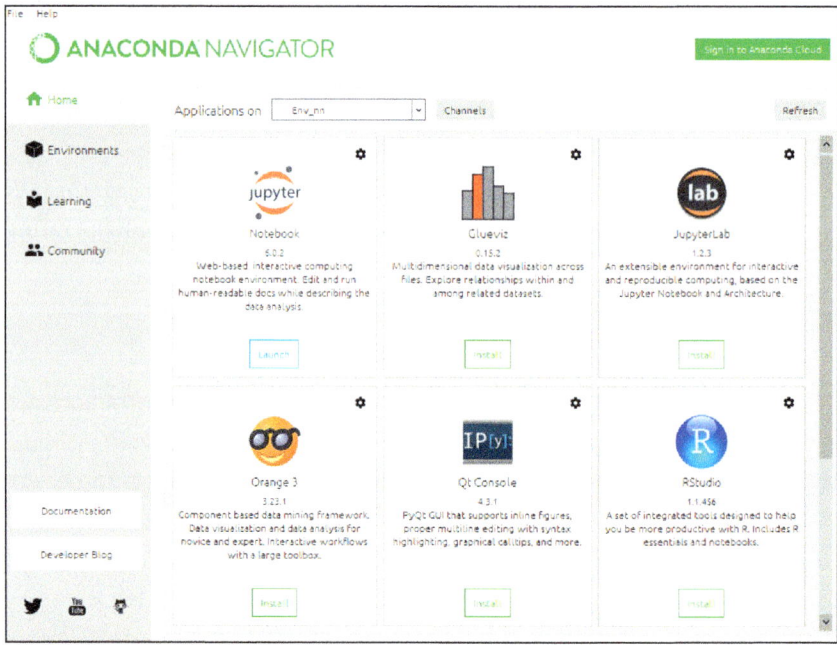

Figure 2.1 Anaconda Navigator

As soon as the Anaconda Navigator is started, it's possible to start different applications in the current environment. Using the example of Figure 2.1, this includes Qt Console, JupyterLab, Glueviz, Orange 3, and RStudio.

The current environment is shown in the top line when the **Home** menu gets selected. In our example in Figure 2.2, this is the environment named **Env_nn**.

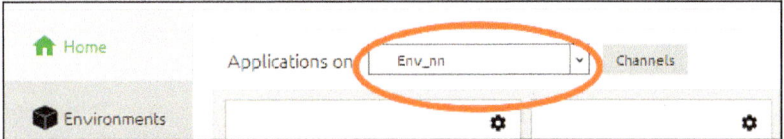

Figure 2.2 The Current Environment

Creating New Environments

The following examples illustrate the importance of using your own environments:

- For some projects, you only need a small number of modules, so it's pointless to install all existing packages in one environment.
- The open-source community is very fast, which means different versions of the same module exist. Sometimes, modules that build on each other need very specific versions.
- Some modules have device-dependent versions (e.g., *TensorFlow* and *TensorFlow-gpu* for PCs with one or more GPUs). You can also set up separate environments for this.

In the Anaconda Navigator, the **Environments** menu item is located in the left-hand area, which shows an overview of the environments in the central area and a complete list of the packages installed per environment in the right-hand area (see Figure 2.3).

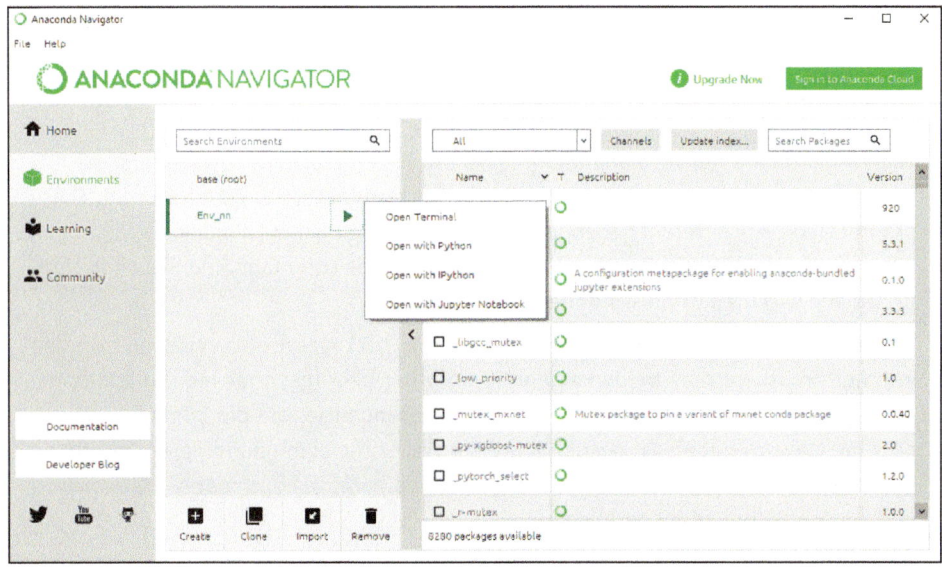

Figure 2.3 The Environment Window

By clicking on the **Create** button (at the bottom of the screen), you can create a new environment and give it a suitable name.

Installing New Packages

There are several methods for installing new packages in an environment; we'll describe two options here. One of them is aimed more at fans of graphical user interfaces (GUIs), while the other is intended for advocates of the command-line interface (CLI):

- **Installation via the GUI**

  First, you must select or create an environment. To do this, select the **Environments** menu item in the **Explorer** section, which is located on the left-hand side of the **Anaconda Navigator** screen (see Figure 2.3). Then, click on an environment (or create a new one) in the central section that now appears. Select or search for a package to be installed in the right-hand area. After selecting a package (or multiple packages), an **Apply** button appears at the bottom right. Clicking this button starts the installation.

- **Installation via the CLI**

  Another installation option is to enter the following command in a *terminal*. You can open the terminal by left-clicking on the triangle icon next to an environment name in the **Environment** view. A submenu will then open, as you can see in Figure 2.3. The first menu item, **Open Terminal**, opens a command-line window. The current environment can also be recognized in the command prompt.

  To install a package (scipy in our example), enter one of the following commands in the terminal:

  - `conda install scipy` (for the latest scipy version)
  - `conda install scipy=0.15.0` (for scipy version 0.15.0)

Terminal = Command-Line Interface (CLI)

The window for entering command lines is often referred to as the *terminal*, as it is here in Anaconda Navigator. The input line in the terminal starts with a *prompt*. These prompts look very different depending on the operating system (Windows: `C:\>`, Unix: `user@machine:~$`). A command must be entered in the command line, which is completed and executed by pressing the `Enter` key.

This installation process can take a few minutes, as packages are downloaded from the internet and there may be dependencies on other packages that require additional packages to be installed. But don't worry, the dependencies are predefined, and each package knows what else it needs. Figure 2.4 shows the command-line version of the installation of scipy, and the additionally required packages or updates are displayed in the lower section of the terminal.

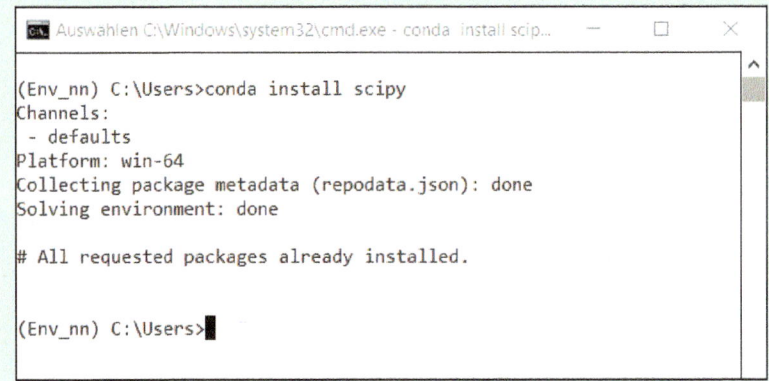

Figure 2.4 Installation via Terminal in the Env_nn Environment

> **Task**
> Create a new environment named "Env_nn". Install the numpy package via the GUI and the pandas package via the CLI.

2.1.2 Our Cockpit: Jupyter Notebook

A *Jupyter Notebook* is a web application that allows you to create and share documents with a community. The application contains Python code, equations, visualizations, and explanatory text. Jupyter Notebooks are therefore a kind of digital notepad. In addition to the advantage of combining documentation and code, these notepads can also be easily exchanged. Notebooks are stored in *JavaScript Object Notation* (JSON) format. So, it's the ideal tool or development environment for experimenting with neural networks and Python code, as well as for developing serious applications.

> **The Jupyter Concept**
> The Jupyter Notebook is a result of the Jupyter project (*http://jupyter.org*). The project name is derived from the three programming languages, Julia, Python, and R.

Starting Jupyter Notebook

You can start the Jupyter Notebook in the Anaconda Navigator either in the **Home** area or in the **Environment** area:

- In the **Home** area (refer to Figure 2.1), first select the desired environment under **Applications on**, and then click on the **Jupyter Notebook** icon (top left in the image).
- In the **Environment** area, click on the triangle next to the corresponding environment, and then select **Open with Jupyter Notebook** (refer to Figure 2.3).

Jupyter Notebook in the Cloud

With Anaconda, you have Jupyter Notebook installed locally on your computer and are therefore more or less independent of a connection to the internet. However, as Jupyter Notebook is a web application, it makes sense to make this tool available in the cloud. This has the advantage that you can use more computing power, which is particularly beneficial when working with deep neural networks with large amounts of data.

Fortunately, a plethora of providers make resources available via Jupyter Notebook, although these are subject to a charge once a certain level of usage has been reached. You also need to register and often provide a credit card. In Section 2.1.5, you'll find a partial list of providers.

The Cells of a Jupyter Notebook

A document created with Jupyter Notebook consists of a list of consecutive cells. These cells can take on various forms (see Figure 2.5).

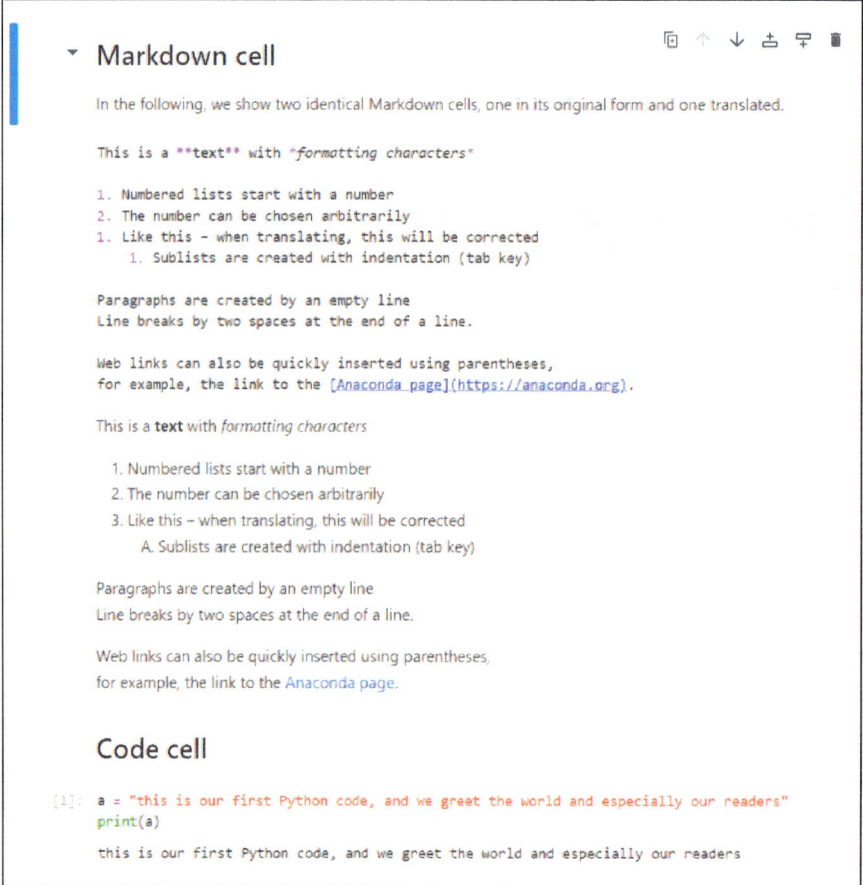

Figure 2.5 Markdown and Code Cells in a Jupyter Notebook Document

We're only interested in the two most important ones here:

- **Markdown cell**
  This cell type allows you to enter an explanatory text. However, it's much more powerful because you can also format the text to change the font or display bold type, formulas, web links, images, and so on. Markdown is a markup language that was developed with the aim of making the source form easy to read, which consists of plain text that includes formatting characters. For example, entering **Bold** in a Markdown cell creates **bold print**.

- **Code cell**
  Cells of this type enable you to program. Here, you can enter Python code. Of course, you can also enter Python comments, but without formatting options.

What both cell types have in common is that a text gets entered (either text with formatting characters or Python code). To output the content as formatted text or to run the code, the cell must first be executed. This means that the process for translating into formatted text or for executing the code statement must first be triggered, either by a menu selection (**Cell • Run Cells**) or by a key combination (Shift + Enter).

The Jupyter Notebook document gets saved in JSON format and has the file extension, *.ipynb*, which means that the document can be easily passed on. Information on the JSON format can be found at *www.json.org*. We won't go into this in detail here, except that it's a lean data exchange format that can easily be read and written by humans and translated and generated by computers.

> **Task**
>
> And off you go! Start a Jupyter notebook and try to recreate the contents of Figure 2.5. Play around with this development environment. If you're new to programming or have no experience with Python, you can now take a look at Appendix A and try out some code examples in the Jupyter Notebook.

Interactive Python

Jupyter is designed in such a way that it can be used with different programming languages, hence the origin of the name (refer to the "The Jupyter Concept" information box earlier in this section). To be able to use other languages, the corresponding kernel must be installed. We use the Python kernel or, more precisely, the Interactive Python kernel (IPython kernel). This IPython kernel provides even more advanced functionalities than we have available in the notebooks. These are usually commands that are called or combined with special characters such as %, ?, !, or a tab. The commands called with the % prefix are also referred to as *magic functions*. The %quickref magic function, which is simply entered into a code cell of a Jupyter Notebook, provides a very good overview of the interactivity functionalities of Jupyter. In the following subsections, we describe a few important examples for accessing the documentation of Python functions and debugging Python code in Jupyter Notebooks.

Access to Help and Documentation

There's an incredible amount of information available in the data science field, which makes it almost impossible even for an experienced data science expert to keep everything in mind. Finding information quickly and effectively, whether on the internet, in operating instructions, or within the development environment, is much more important.

Jupyter Notebooks provides some simple and quick methods to answer questions like these:

- What arguments does this function call need?
- How was a function or class implemented?
- What functions, methods, and so on does the imported module contain?

A quick and useful source of information is the *docstring*, which contains a concise and precise summary of an object. The following code examples show how you can obtain this information quickly:

```python
# The docstring can be output via the help function
help(print)
# Output:
Help on built-in function print in module builtins:

print(...)
    print(value, ..., sep=' ', end='\n', file=sys.stdout, flush=False)

    Prints the values to a stream, or to sys.stdout by default.
    Optional keyword arguments:
    file:  a file-like object (stream); defaults to the current sys.stdout.
    sep:   string inserted between values, default a space.
    end:   string appended after the last value, default a newline.
    flush: whether to forcibly flush the stream.
```

Listing 2.1 Docstring Output

Instead of calling the `help()` function, you can also simply place a question mark at the end of the function name. Depending on the development environment, the description is displayed in an output cell, in a popup window, or in a separate output area in Jupyter Notebook (see Figure 2.6).

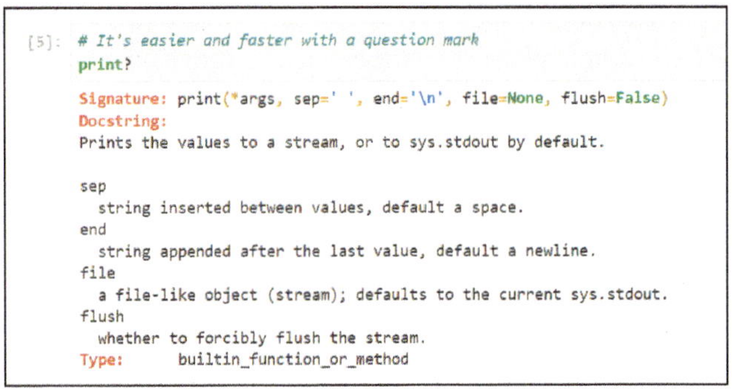

```
[5]: # It's easier and faster with a question mark
     print?

     Signature: print(*args, sep=' ', end='\n', file=None, flush=False)
     Docstring:
     Prints the values to a stream, or to sys.stdout by default.

     sep
       string inserted between values, default a space.
     end
       string appended after the last value, default a newline.
     file
       a file-like object (stream); defaults to the current sys.stdout.
     flush
       whether to forcibly flush the stream.
     Type:      builtin_function_or_method
```

Figure 2.6 Output Area or Pager in the Jupyter Notebook

The question mark can be used in many different ways. It can be attached to functions or objects, as shown in Listing 2.2. In the first case, it outputs information about the myList.append function in docstring format; in the second case, it outputs information about the myList object, such as type, form, or length.

```
# the question mark ? can be used in a variety of ways
myList = ['NumPy', 'scikit', 'Keras', 'TensorFlow']
myList.append?
# Output: (in the pager window - at the bottom of the web browser)
Signature: myList.append(object, /)
Docstring: Append object to end of the list.
Type:      builtin_function_or_method

# or simply to obtain information about an object
myList?
# Output: (in the pager window - at the bottom of the web browser)
Type:         list
String form: ['NumPy', 'scikit', 'Keras', 'TensorFlow']
Length:       4
Docstring:
Built-in mutable sequence.

If no argument is given, the constructor creates a new empty list.
The argument must be an iterable if specified.
```

Listing 2.2 The Question Mark

> **Note**
>
> Because the open-source world changes very quickly, the output shown here may look slightly different on your computer depending on the version of Python you've installed. This listing's outputs are in Python v3.7.

Of course, Jupyter Notebook also provides a way to quickly access information about functions and objects via the GUI. If you place the cursor on the function or object name and press the Shift+Tab key combination, a balloon help appears, as shown in Figure 2.7.

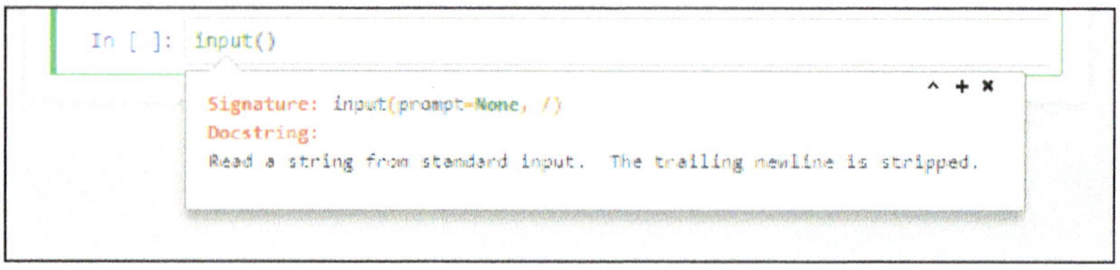

Figure 2.7 Context Help for the Built-In input() Function

You can also use the Tab key to explore the content of modules, classes, or objects. In Figure 2.8, you can see a list of the functions of the math module when entering m. Tab in a code cell of Jupyter.

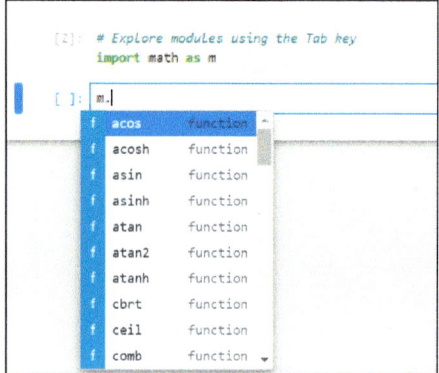

Figure 2.8 Using the Tab Key to List the Attributes and Functions of Modules

Of course, this also applies to our myList list object from Listing 2.2. In Figure 2.9, you can see the list that you get by entering myList. Tab.

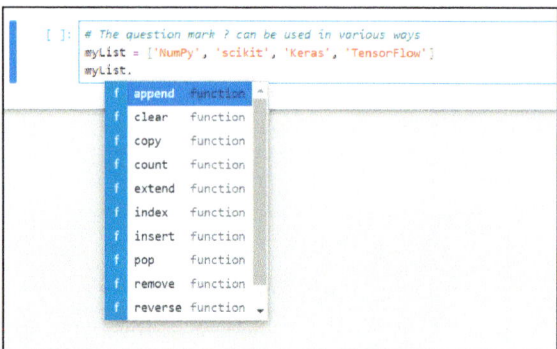

Figure 2.9 Using the Tab Key for Objects

2.1 The Technical Development Environment

Errors and Debugging

For a simple error analysis, there is the %xmode magic function, which allows you to control the amount of error information output. This function requires one of the arguments Plain, Context, or Verbose, which control the level of detail the debug information is provided in if there is an error.

Debugging: Fun Fact

In the world of programming, *debugging* means tracking the application process. This is mainly done when the application doesn't do what it should or generates an error (i.e., a *bug*). The origin of the word is often dated to September 9, 1947, when a *bug*, or more precisely, a moth, was found in a relay of a computer. This moth was responsible for an error that caused the computer to malfunction. Figure 2.10 shows the logbook at the time with the moth inserted. However, the term had already been used in this sense before, but the employee at the time found it funny, as did we, that a real bug led to a computer failure.

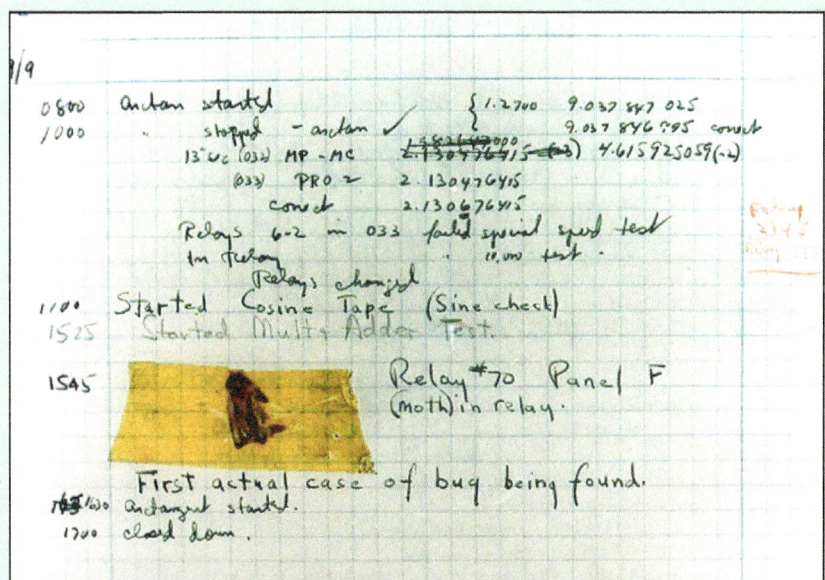

Figure 2.10 The First Bug in Computer History (Source: U.S. Naval Historical Center Online Library)

In the following examples, you can see the effects of the arguments. We're already getting a bit ahead of ourselves here with Python programming. If you still feel unsure about this, please refer to Appendix A.

In Listing 2.3, we first define a simple function named avg_beelength(), which has the task of calculating the mean values of the lengths of our bees. Within this function, we then call the errorfunc() function, knowing full well that the call could generate an error.

```python
def errorfunc(x,y):
    return x/y

def avg_beelength(valuelist):
    total = sum(valuelist)
    count = len(valuelist)

    return errorfunc(total,count)

bees = (3, 3.5, 4.1, 5.2, 5, 2)
nobees = ()
avg_beelength(bees)
# Output:
3.8000000000000003
```

Listing 2.3 The avg_beelength Function without Error

So far, so good: Listing 2.3 defines our two functions. With the list of bee lengths, `bees`, we get a reasonable result and above all, no error.

Let's now take a look at Listing 2.4 to see which error message is displayed (depending on the %xmode set) if we call the same function with `nobees`, that is, a list without values.

```
# This is the default setting
%xmode Context
avg_beelength(nobees)
# Output:
---------------------------------------------------------------------------
ZeroDivisionError                         Traceback (most recent call last)
<ipython-input-11-be3dced12b16> in <module>
      1 # This is the default setting
      2 get_ipython().run_line_magic('xmode', 'Context')
----> 3 avg_beelength(nobees)

<ipython-input-10-185b8ad2cde6> in avg_beelength(valuelist)
      6     count = len(valuelist)
      7
----> 8     return errorfunc(total,count)
      9
     10 bees = (3, 3.5, 4.1, 5.2, 5, 2)

<ipython-input-10-185b8ad2cde6> in errorfunc(x, y)
      1 def errorfunc(x,y):
----> 2     return x/y
      3
      4 def avg_beelength(valuelist):
```

```
    5      total = sum(valuelist)
```

ZeroDivisionError: division by zero
`# slightly more compact error message`
`%xmode Plain`
`avg_beelength(nobees)`
`# Output:`
Exception reporting mode: Plain
Traceback (most recent call last):

  File "<ipython-input-12-fe49876a1c72>", line 3, in <module>
    avg_beelength(nobees)

  File "<ipython-input-10-185b8ad2cde6>", line 8, in avg_beelength
    return errorfunc(total,count)

  File "<ipython-input-10-185b8ad2cde6>", line 2, in errorfunc
    return x/y

ZeroDivisionError: division by zero

`# detailed error message`
`%xmode Verbose`
`avg_beelength(nobees)`
`# Output:`
Exception reporting mode: Verbose

ZeroDivisionError   Traceback (most recent call last)
<ipython-input-13-3d86e356f330> in <module>
      1 # Detailed error message
      2 get_ipython().run_line_magic('xmode', 'Verbose')
----> 3 avg_beelength(nobees)
        global avg_beelength = <function avg_beelength at 0x0000000004FE1B88>
        global nobees = ()

<ipython-input-10-185b8ad2cde6> in avg_beelength(valuelist=())
      6   count = len(valuelist)
      7
----> 8   return errorfunc(total,count)
        global errorfunc = <function errorfunc at 0x000000000500BDC8>
        total = 0
        count = 0
      9
     10 bees = (3, 3.5, 4.1, 5.2, 5, 2)

```
<ipython-input-10-185b8ad2cde6> in errorfunc(x=0, y=0)
      1 def errorfunc(x,y):
----> 2     return x/y
        x = 0
        y = 0
      3
      4 def avg_beelength(valuelist):
      5     total = sum(valuelist)
```

`ZeroDivisionError: division by zero`

Listing 2.4 Error Messages with Varying Degrees of Detail

In all modes, a *stack trace* or *traceback* gets displayed, which shows the error history. It starts with the function at the top level and goes down to the last element that triggered the error. In our case, the `return x/y` is contained in the `errorfunc()` function.

However, the standard `Context` mode provides not only the function sequence as in `Plain` mode but also the immediate environment in the program code. This makes it easier to find the program locations, and the cause of the error is often in the immediate vicinity.

`Verbose` mode shows us even more, namely the value of the local variable. In the last lines, it shows that the values for `x` and `y` are 0. The cause is already clear to us: we've divided by the number 0 (*division by zero*), which is of course not possible. However, if the code gets more complex, it can be very difficult to analyze the error message, as the traceback can be extremely long.

> **Traceback (or Stack Trace)**
>
> The traceback is a tool for tracing a program error. Starting from the occurrence of the error, all the functions involved get displayed, allowing the error to be traced back to its cause. The more complex and nested the program code is, the longer such stack traces can get.

Sometimes, the code doesn't return an error, but still doesn't quite do what you wanted it to do. To obtain more precise information about what happens when the program code is executed, you want to use a *debugger*. In the Jupyter Notebook, the interactive debugger can be started using the `%debug` command.

If we now call this interactive debugger, a command-line window opens with the `ipdb>` prompt. In this command line, we can enter commands that display the contents of our variables, among other things. This will hopefully enable us to identify where the error is located. Listing 2.5 shows how you can use the internal debugger.

In our traceback, we recognize at least two stages. The last one shows where exactly in the `errorfunc(x,y)` function the error occurred. Here, we have access to all variables within the function, that is, x, y, and, of course, to any global variables. If we want to go one level higher, in this case, to the `avg_beelength()` function, we must enter the up command within the debugger. With the output of the `count` and `total` variables at this level, this results in 0 both times and leads us to the actual error, namely the input of an empty list.

%debug

```
# Output:
> <ipython-input-1-78d6430f450b>(2)errorfunc()
      1 def errorfunc(x,y):
----> 2     return x/y

ipdb> print(x, y)
0
ipdb> print(count)
*** NameError: name 'count' is not defined

ipdb> up
> <ipython-input-10-185b8ad2cde6>(8)avg_beelength()
      6     count = len(valuelist)
      7
----> 8     return errorfunc(total,count)
      9
     10 bees = (3, 3.5, 4.1, 5.2, 5, 2)

ipdb> print(total, count)
0 0
ipdb> quit
```

Listing 2.5 Using the Interactive Jupyter Debugger

Of course, as always, entire books could be written on the subject of error analysis and debugging. However, the tools presented here are sufficient for our purposes.

2.1.3 Major Python Modules

In this section, we briefly describe the Python modules that are important for this book. For each of these modules, there are *cheat sheets* available, which show examples of the most important functions and classes for a module on an A4 page. You can find them at *www.datacamp.com/community/data-science-cheatsheets*. However, note that we'll use elements from modules throughout the book that we don't present or describe in detail. This way of working is quite typical in Python projects, as there is a library for almost every task that is easy to use and saves us a lot of work; in other words, we don't have to program ourselves.

NumPy

NumPy (*www.NumPy.org*) stands for *Numerical Python* and is of great importance for scientific computing. This module contains an optimized data structure for multidimensional matrixes and the corresponding linear algebra functions. The core element of this module is the `ndarray` class for multidimensional matrixes.

Matplotlib

The Matplotlib (*www.matplotlib.org*) module enables the creation of scientific diagrams and visualizations. It provides functions for line charts, histograms, and scatter plots. If Matplotlib is used with Jupyter Notebook, the visualizations can be displayed directly in the Notebook.

scikit-learn

scikit-learn (*https://scikit-learn.org*) is a very popular tool and contains a large number of modern machine learning algorithms. It's used both in industry and in the university sector.

pandas

pandas (*https://pandas.pydata.org*) allows you to preprocess data and merge data from different sources. In particular, data can be imported using an SQL-like language to formulate queries and merge tables into the `Series` (one-dimensional), `DataFrames` (two-dimensional), and `Panels` (three-dimensional) structures. The `DataFrames` are particularly important for us.

TensorFlow

TensorFlow (*www.tensorflow.org*) is actually a whole package of libraries, online resources, and so on, which are described in more detail in Appendix C. It's thanks to Google that we now have machine learning libraries developed for deep learning. The calculations are carried out using *data flow graphs*. A graph in turn consists of nodes that are connected by edges. Each algorithm calculated with TensorFlow consists of mathematical operations (represented as *nodes*) and data (represented as *edges*). All calculations from a simple sum to highly complex matrix operations can be performed and displayed.

In the book, we use TensorFlow v2, which includes a Keras library developed by François Chollet for easier handling of deep neural networks.

Keras

Keras (*http://www.keras.io*) is a standalone library that can work with various neural network modules, such as TensorFlow, Apache MXNet, or Theano. These modules are referred to as *backends*. With the release of TensorFlow 2, Google has fully integrated Keras. However, Keras will also continue to operate as an independent library, although François Chollet also suggests that users only work with TensorFlow 2 in the future. You can find the Keras documentation at *www.tensorflow.org/guide/keras*.

2.1.4 The Google Colab Platform for Jupyter Notebooks

You now know that Jupyter Notebooks is a web-based development environment. The web server required for this can run locally on your own computer. Anaconda helps you provide the necessary applications. As a web-based application, you can also run this development environment in the cloud. We recommend Google Colab, not because we're paid by Google, but because it offers the following advantages:

- **Standardized platform**
  All readers have the same platform and installation. Important libraries such as TensorFlow, Keras, pandas, and so on are already installed in their current versions.
- **Graphics processing unit (GPU)/tensor processing unit (TPU) support**
  Google Colab provides free access to powerful hardware accelerators (GPUs/TPUs), which are essential for training neural networks.
- **Release and collaboration**
  Similar to Google Docs, Jupyter Notebooks can be shared and edited by several people at the same time. Can reader groups be found to tackle a project together?
- **Integration in Google Drive**
  Notebooks and data can be saved in Google Drive and opened or used there directly.

Let's first take a look at Figure 2.11 to see what Figure 2.5 looks like in Google Colab.

This illustration shows that Google Colab has integrated an editor for the Markdown cells. This means we can use Markdown elements as usual, but we don't have to memorize them because the Markdown commands will be inserted automatically with the editor.

We advise you to explore all the features and functions of Google Colab on your own. Here, we only point out the most important elements that simplify working with this tool.

Directory and Google Drive Connection

The vertical toolbar on the far left contains a button with a file folder icon (see Figure 2.12 ❷). This button shows or hides the directory structure in which this Jupyter Notebook is running.

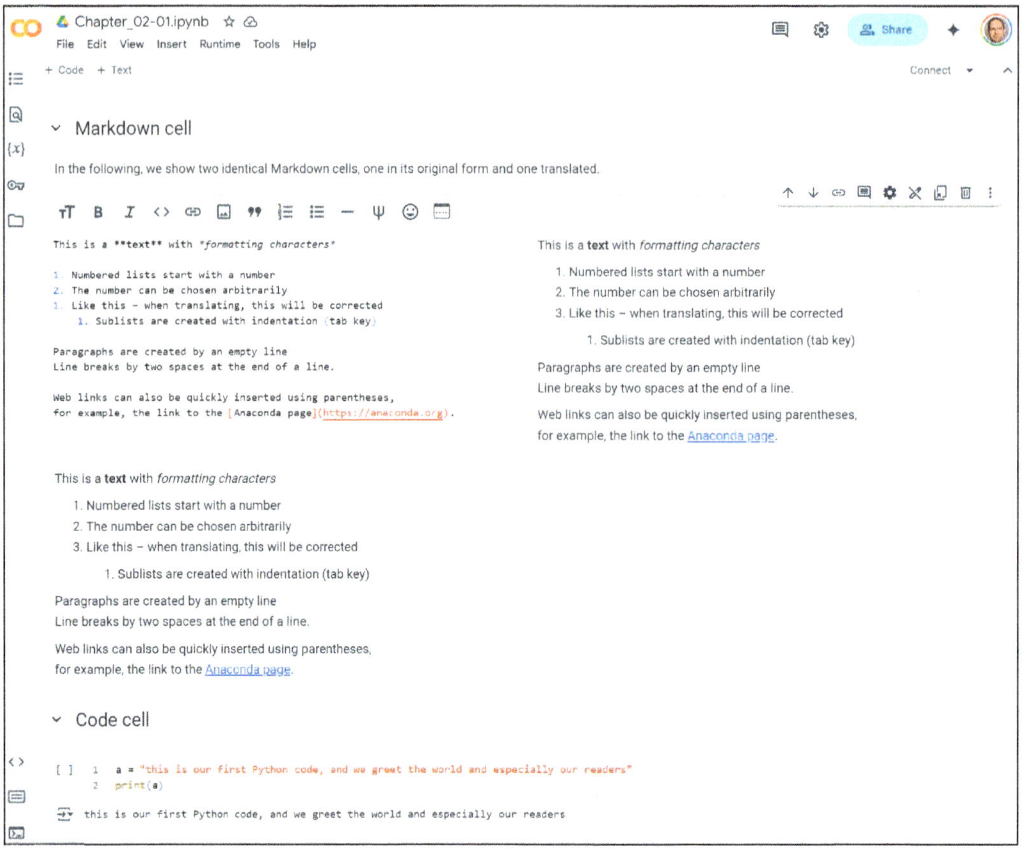

2 Starter Kit for Developing Neural Networks with Python

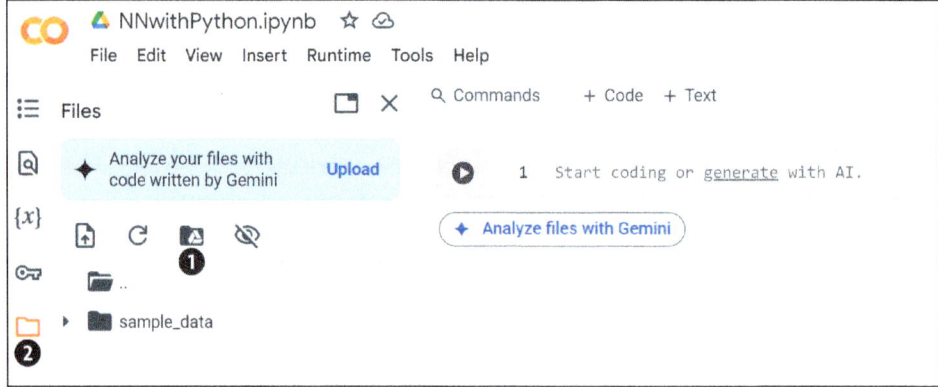

Figure 2.11 Jupyter Notebook in Google Colab

Figure 2.12 Buttons for This Directory Structure and the Google Drive Connection

To insert your own data into this directory structure, simply drag the desired file into the expanded area of the directory structure. You can also achieve the same using the following program lines in a code cell:

```
from google.colab import files
uploaded = files.upload()

for fn in uploaded.keys():
  print('File "{name}" with length {length} bytes'.format(name=fn, length=
len(uploaded[fn])))
```

Listing 2.6 Opening a Selection Window for Loading Local Files

Saving files from the Google Colab environment to the local computer works in a similar way: either manually by right-clicking on the file and selecting the **Download** menu item, or via a code cell in the program:

```
from google.colab import files

with open('Example.txt', 'w') as f:
    f.write('Neural networks are impressive!')

files.download('Example.txt')
```

Listing 2.7 Saving Files from the Google Colab Environment to the Local File System

If you want to connect your notebook to Google Drive, click on the button marked in Figure 2.12 ❶. This connection is indicated by the *drive* folder in the directory structure, which is shown in Figure 2.13.

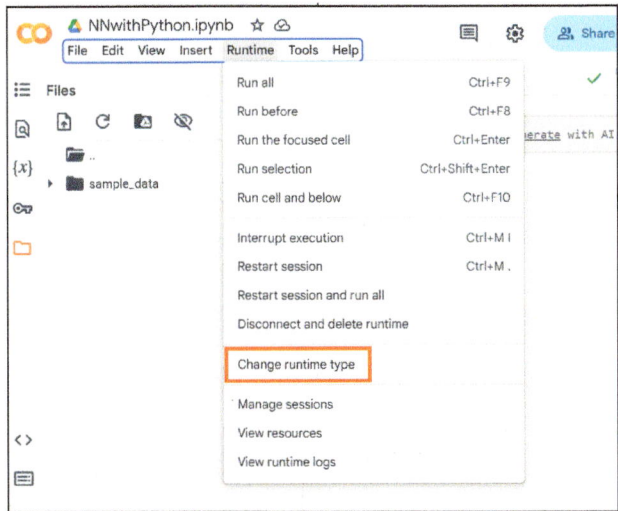

Figure 2.13 Menu Item for Setting GPU or TPU

Now you have plenty of storage space available for the often large amount of training data for neural networks.

Free GPU or TPU

If you want to benefit from a free GPU or TPU, call the **Runtime** menu item in the horizontal menu bar, which allows you to change the runtime type (see Figure 2.13). It's important to make this change before using the notebook. If the runtime type is changed, the notebook restarts and deletes all previous processes and executions.

Note that the GPU is only available to you for limited periods of a few hours. It's therefore not possible to train neural networks unless you choose a paid version of Google Colab. However, the free version is completely sufficient for the examples in this book.

Executing Code Cells

Figure 2.13 shows the submenu of the **Runtime** menu item. In its upper area, you can execute individual cells or all cells by selecting the corresponding menu item or using the corresponding key combination. Each code cell is also equipped with a button with the start symbol, which is located to the left of the code cell—another way to execute a code cell.

2.1.5 Additional Jupyter Notebook Cloud Resources

The following list is a good starting point, but it's incomplete for several reasons. First, there is an almost unmanageable number of providers, and secondly, the websites listed provide much more than just Jupyter Notebooks. We've selected a few that provide resources free of charge to a certain extent.

Jupyter Community

First of all, of course, the *Jupyter Community* website (*https://jupyter.org/try*) provides Jupyter resources, albeit without GPU support and actually only for testing.

Kaggle

Kaggle (*www.kaggle.com/notebooks*) is a data science platform that originally specialized in machine learning competitions. It's now a must if you want to acquire expert knowledge in data science.

Microsoft Azure

Microsoft is also very strongly represented in the machine learning scene with *Azure* (*https://notebooks.azure.com*).

Amazon SageMaker

Like many cloud providers, Amazon also has Jupyter Notebooks in its portfolio with Amazon Web Services (AWS), one of which is SageMaker (*https://aws.amazon.com/sagemaker*).

2.2 Summary

This chapter provided you with an introduction to Anaconda Navigator, Jupyter Notebook, and related Jupyter Notebook cloud resources, which together provide a convenient technical development environment for the Python programming language. For a brief introduction to Python, you can now skip ahead to Appendix A.

Python provides a huge number of libraries. We've briefly described the most important ones for programming neural networks here, with references to very good tutorials. A list of other providers of Jupyter Notebooks in the cloud concluded this chapter.

Chapter 3
A Simple Neural Network

In this chapter, we take our first step by introducing the perceptron, a fundamental type of neural network designed for binary classification. We will explore how it processes inputs, applies weights, and utilizes an activation function to distinguish between two categories based on learned patterns.

This chapter describes the simplest form of neural networks, the *perceptrons*, which are also referred to as *linear classifiers*. We'll work out step-by-step how the perceptrons work using images and codes, as well as simple examples to explain the subject matter. At the end of the chapter, you'll teach a moving robot to recognize holes or avoid driving into the wall.

3.1 Background

Imagine you no longer have to plan because a neural network does it for you. Let's call this type of planning *automatic planning*. The scenario is as follows: Staffing should be planned according to individual scheduling requirements and business needs in such a way that the business purpose can be realized with the help of these same people. Examples of this include the *staffing plan* of a corner store or the timetable in a school.

3.2 Bring on the Neural Network!

Before we get started, consider one question first: How would you separate the two groups in Figure 3.1 if you were allowed to use a pencil and a ruler? And once you've done that, can you find any other options?

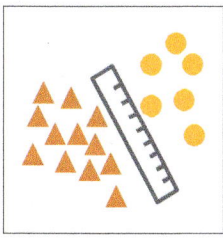

Figure 3.1 Linear Separation Using a Ruler

For humans, it's very easy to solve this task. We can immediately see that there are two groups that we can simply separate by a *straight line*. But how can this process be carried out automatically, and what approaches are available? Depending on the specialist group, the answer is a bit unwieldy:

- Computer scientist: By means of a *classifier*.
- Statistician: By means of a *linear discriminator*.
- Data scientist: By means of a *perceptron*, a special artificial neural network (ANN).

You can add to this list with your own specialist group and approach, of course. We'll continue to pursue the data scientist approach.

But first, what is meant by *perceptron*? This question will keep us busy until the end of this chapter via the following example: Mrs. Apple, the store manager of a small store, has two employees to schedule: Mr. Leek and Mrs. Carrot. Basically, both have time to work from Monday to Friday. The store is closed on Saturdays and Sundays. Mrs. Apple wants to have *at least one* employee at her side every day, for example, on Monday, here's three ways it could play out:

- Only Mrs. Carrot
- Only Mr. Leek
- Both of them, but certainly at least one of the two

When you hear it this way, it sounds a little complicated. The problem is easier to understand if we write it down in the form of Table 3.1.

Mrs. Carrot	Mr. Leek	→	Monday
Not present	Not present	→	Not okay
Present	Not present	→	Okay
Not present	Present	→	Okay
Present	Present	→	Okay

Table 3.1 First Staffing Plan

If Mrs. Carrot and Mr. Leek aren't present, then Mrs. Apple's wish isn't fulfilled. There's no one there to help, and after a day like that, Mrs. Apple is "not okay" because she has to do everything on her own. It's better if, for example, Mrs. Carrot is present, because that is definitely "okay" for Mrs. Apple.

This table makes it much easier to understand the possible combinations, but it's still far from ideal. An ANN can't use the terms "not present, "present," "not okay," and "okay" for its calculations, so we have to think of a different representation with numbers.

> **A Side Note: Data Preparation**
>
> This step of data preparation is one of the most difficult steps in the overall process of using ANNs. How are you supposed to know which format to transform the data into for the ANN? How are you supposed to know what "present"/"not present" should become? We'll look at this topic in more detail in Part II of this book.

Without developing a big theory of data preprocessing for using an ANN, we nevertheless need to know how to appropriately prepare text such as "okay" or "not okay" as ANN input. Do you have any ideas? Think for a moment about what could preprocessing look like from your point of view.

We've come up with the following suggestion: If an employee is *present*, we write "1", and if the employee isn't *present*, "0". This definition may seem arbitrary, and it is. However, the value 1, when something is "on" or "good," is appropriate for the positive event, and the value 0 for "off" or "bad" is also intuitively understandable. We therefore believe that this classification fits well in this context. In addition, we replace "not okay" with "0" and "okay" with "1" to show what the desired result of the planning should look like.

If we display our considerations in table format again, we get the contents shown in Table 3.2.

Mrs. Carrot	Mr. Leek	→	Monday
0	0	→	0
1	0	→	1
0	1	→	1
1	1	→	1

Table 3.2 Second Staffing Plan with ANN-Compatible Code

We can now this for our calculations, which are necessary to teach the ANN that it isn't *okay* if *no employee* is present and that the situation is always *okay* in all other attendance constellations. This means that we have to come up with a calculation to get the desired result. For the time being, we'll do without an ANN, that is, with real thinking and calculations by ourselves. This is super important for understanding the actually very simple calculations that are carried out in an ANN.

The first attempt could be that Mrs. Carrot and Mr. Leek are added together—it isn't really the people who are added up, but the presence of the two. The result is then derived from this sum using a simple rule: "If the sum is greater than or equal to 1, then the result is 1, otherwise the result is 0."

Now let's take a look at the whole thing in Table 3.3.

3 A Simple Neural Network

Mrs. Carrot	Calculation	Mr. Leek	→	Total	Result
0	+	0	=	0	0
1	+	0	=	1	1
0	+	1	=	1	1
1	+	1	=	2	1

Table 3.3 Staffing Plan Calculated from the Attendances

We could of course present these considerations in the form of a network, as shown in Figure 3.2, to obtain a different view than that of the table representation because we want to approach the ANN visualization step-by-step.

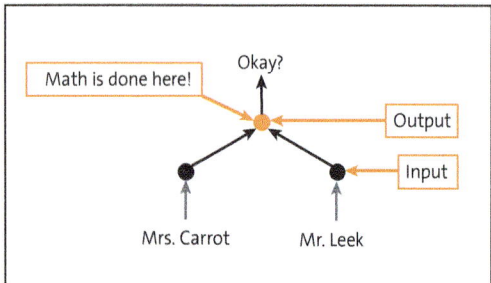

Figure 3.2 Calculating the Decision for Mrs. Carrot and Mr. Leek

This type of representation isn't far removed from the representation of an ANN. We still need to add a few details, but in principle, we're already finished—almost.

3.3 Neuron Zoom-In

One of the details that still needs to be added involves the method of calculation. Let's take a closer look at the neuron. We know from Table 3.3 that we need a calculation and then a decision. We supplement this knowledge in the representation of the calculation node in Figure 3.3.

To put the details of the calculation node in a form that can be handled by a Python program, we need a formula or a calculation rule, which could look like the following:

$Sum = Mrs.\,Carrot + Mr.\,Leek$

and

$Result = Decision(Sum)$

where

$Decision = \begin{cases} 1, & \text{if } Sum \geq 1 \\ 0, & \text{if } Sum < 1 \end{cases}$

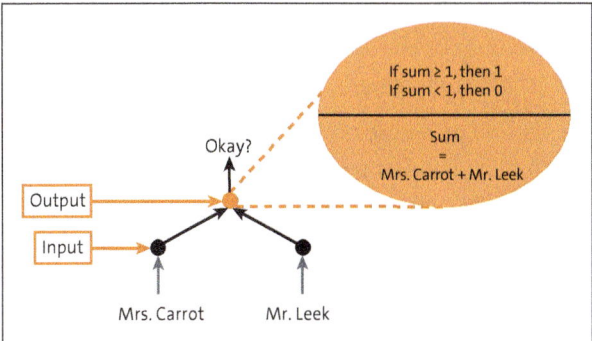

Figure 3.3 Calculation of the Results for Mrs. Carrot and Mr. Leek in Detail

Regarding the line, *Result = Decision (sum)*, we've used the *function notation* to calculate the decision. This only states that the result of the `decision` function depends on the `sum`; that is, the result of the `decision` function is determined as a function of the `sum`. The functions are very often referred to as $f(x)$, for example, $f(x) = x^2$. This only means that the f function is defined as a function of x.

The function notation can be used directly in Python programming. That's what we do now and define a function with the functionality just described. If you haven't yet started your Jupyter Notebook, please do so now. You can also go back to Chapter 2 and read the corresponding details. We'll organize the program examples in different notebooks, with one notebook per chapter. Your next step is therefore to create a new notebook (see Figure 3.4).

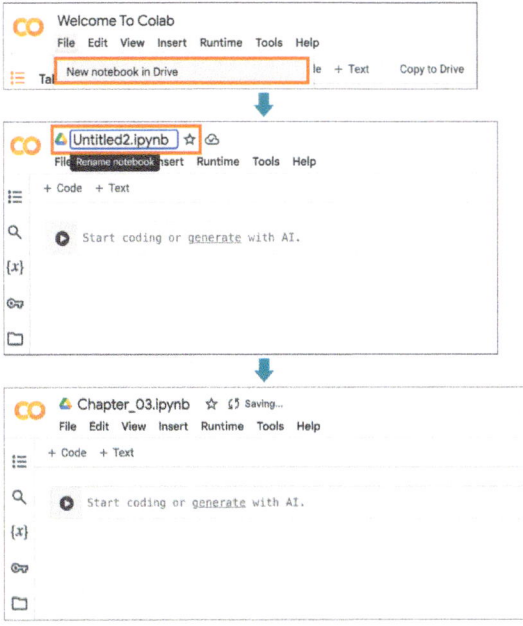

Figure 3.4 Creating the Chapter_03 Notebook

To do this, choose **File** • **New notebook in Drive** in the menu bar. Colab then creates the new notebook, and you can assign a name for the notebook by clicking on the **Untitled.ipynb** text and entering "Chapter_03", for example.

You can directly overwrite the text. This renames the notebook to Chapter_03, and you can start programming. We'll use two cells; the first cell contains the heading, and the second cell contains the code. We'll insert the cell as a **text** cell (see Figure 3.5) to give the code in the subsequent cell a heading. To do this, you need to click on the **+ Text** button.

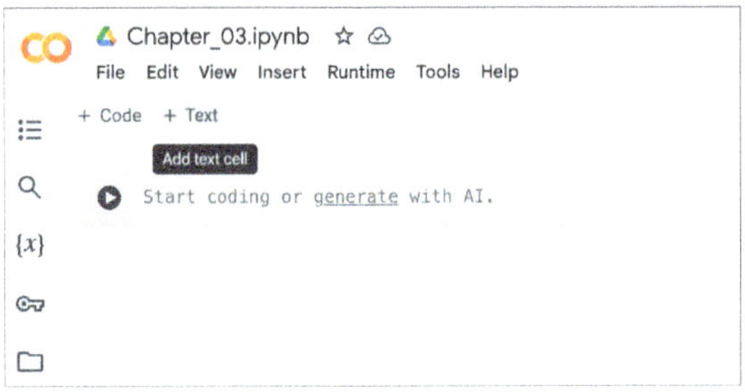

Figure 3.5 Inserting a Cell by Choosing "Text"

Once this has been done, you can enter the markup elements for a heading in the cell, such as a "#" for a level 1 heading and then, separated by a space, enter the text for the heading (see Figure 3.6).

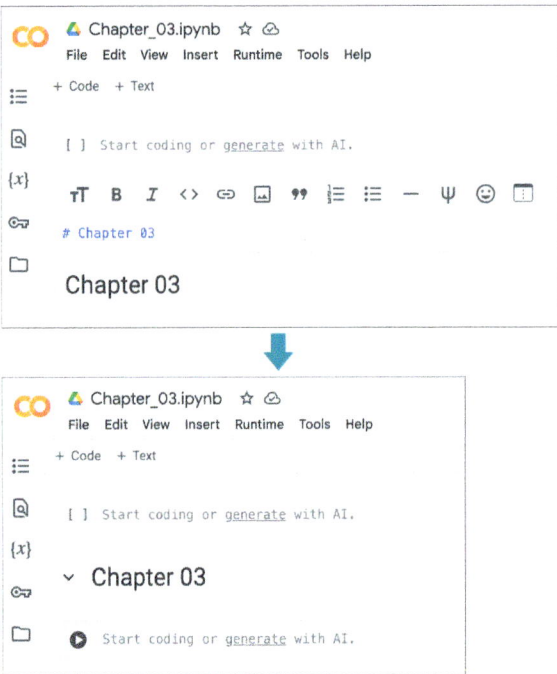

Figure 3.6 Notebook with Heading

Of course, you can also use the Colab text editor toolbar to carry out the formatting. The preview of the text with the formatting effects already appears to the right of the input.

If you confirm your entry with the key combination [Shift] + [Enter], the text is displayed formatted in the Markdown cell.

> **Recurring Header**
>
> We've added two more aspects to the notebooks for download that are found in almost all notebooks:
>
> - A note to call the **Runtime • Run All** menu item when starting for the first time because it ensures that the scripts are executed in all code cells in sequence and that the outputs discussed here will be generated.
> - A **Markdown** cell and a **code** cell are included that suppress any warning regarding future deprecated statements, for example. If you want to see the warning, simply place a comment character in front of the code in the cell.

Now click on a cell and move the mouse to the top or bottom edge of the cell. This will display a toolbar that you can use to insert a code or text cell very easily. Then, place further Markdowns or your code in this cell, for example, the code from Listing 3.1.

```python
def decision( sum ):
    """ Calculation of the decision on the sum value
    Input:  sum
    Output: 1, if sum >= 1,
            0 else
    """
    if sum >= 1:
        return 1
    else:
        return 0
#-----------------------------
# Calculation of the decision
result = decision(1)
# Output in cell as string
print('Input: ', 1)
print('Output:', result)
# Output:
Input:  1
Output: 1
```

Listing 3.1 The Decision as a Step Function

3   A Simple Neural Network

We map the previously discussed calculation of the decision as a `decision` function with the `sum` parameter. This is followed in the code by a function *docstring* (documentation string), which explains the function as described in Chapter 2. Just take a look at Figure 3.7.

```
Chapter 03

WARNING: Please start by running Runtime • Run all from the menu.

To analyze the iris.csv file, you need to upload it from the chapter folder to Colab. Use the "Files" function from the toolbar on the left.

Disabling Warnings

[1]   1  import warnings
      2  warnings.filterwarnings('ignore')

The Decision

Listing 3.1, Figure 3.7

      1  def decision(sum_value):
      2      """ Calculation of the decision based on the value sum_value
      3      Input:  sum_value
      4      Output: 1 if sum_value >= 1,
      5              0 otherwise
      6      """
      7      if sum_value >= 1:
      8          return 1
      9      else:
     10          return 0
     11
     12  # --------------------------------
     13  # Calculation of the decision
     14  result = decision(1)
     15  # Output in cell as string
     16  print('Input: ', 1)
     17  print('Output:', result)
     18

Input:  1
Output: 1

Documentstring

[3]   1  # Output of Docstring via help-function
      2  help(decision)

Help on function decision in module __main__:

decision(sum_value)
    Calculation of the decision based on the value sum_value
    Input:  sum_value
    Output: 1 if sum_value >= 1,
            0 otherwise
```

Figure 3.7 Program and Outputs for the "Decision" Function

The docstring is followed by the code for calculating the decision. If the transferred value is greater than or equal to 1, then 1 is returned as the result of the calculation; otherwise, it's 0.

We've inserted the call of the function directly after the definition and tested it with the value 1. The return value is saved in the `result` variable and output at the end via the print command after the code cell. We'll place the output of the programs after the code and mark them in bold. The result in your Jupyter Notebook should then look similar to the one shown in Figure 3.7.

These detailed steps will prepare you well for the next programming session, in which we'll briefly discuss the use of the matplotlib module. This module can be used to create diagrams, as we already mentioned in Chapter 2. Obviously, we'll visualize the output of the decision function with it.

3.4 Step Function

In the decision function, we've distinguished whether the sum is less than 1 or not (in which case, it's greater than or equal to 1). The value 1 is referred to as the *threshold* because, as with a door threshold, a height jump—or depth jump, if coming from above—is built in. The calculated value of the function suddenly changes from 0 to 1.

Mathematicians have come up with the term *step function* for such a function. The threshold can assume any value. If the threshold is 0, the step function is referred to as the *Heaviside function*, named after the mathematician *Oliver Heaviside*.

We use exactly this step function for the simple ANN we're about to create. We've already programmed the Python function decision for this purpose. Before we turn to the ANN, however, we'll program another Python script to practice creating graphics. The Python module matplotlib is ideal for this, as it provides functions for line charts, histograms, scatterplots, and so on. As Jupyter Notebook users, we also have the advantage that the visualizations of the diagrams are displayed directly in the notebook.

Create another Markdown cell with the heading "Step function" for the new script, just as you've already done before, and also a code cell for the subsequent script in Listing 3.2.

Before we get started, note that we'll reuse the decision function in this script. For this to work, you need to execute the cell using the decision function. Only then can your script in a code cell call a function in another code cell.

```
# Import of the modules for plots
import matplotlib.pyplot as plt
# Very important, otherwise the plot will not be displayed
%matplotlib inline

def decision( sum ):
    """ Calculation of the decision on the sum value
    Input:  sum
    Output: 1, if sum >= 1,
            0 else
    """
```

```
        if sum >= 1:
            return 1
        else:
            return 0
#-----------------------------
# x values of the graph
x = [-1,0,0.999,1,2]
# Calculate y values with the function decision and calculate them using
# a list comprehension to create a new list (see Appendix A)
y = [ decision(i) for i in x ]
# Create the graph with an orange-colored step and the label step
plt.step(x, y, color='Orange', label='step')

# Set the axes
plt.grid(True)
# Draw the horizontal and vertical 0-axis slightly thicker in black
plt.axhline(0, color='black', lw=1)
plt.axvline(0, color='black', lw=1)

# Axis labeling and title
plt.xlabel('Sum')
plt.ylabel('Result')
plt.title('Step function')

# Define legend placement
plt.legend(loc='center right')

# Display the graph
plt.show()
```

Listing 3.2 Use of pyplot

In Figure 3.8, you can see the resulting step function.

First, of course, it's necessary to import `matplotlib.pyplot` and define a suitable name in the local namespace; `plt` has become the standard name for this. The `%matplotlib inline` statement is a called a *magic function*. It's used to specify that the output of the `plot` statement is executed inline directly under the producing cell in the notebook. (You can find out more about the magic functions by referring to Chapter 2.)

The data then gets compiled in lists. We use a list with x values and one with y values. The values of the y list are calculated by the `decision` function by calling the function for all elements of the x list. The `plt.step` function creates the step, whereby, for example, the color can be specified for drawing. In between, the `figure` is edited in detail and finally displayed via `plt.show()`.

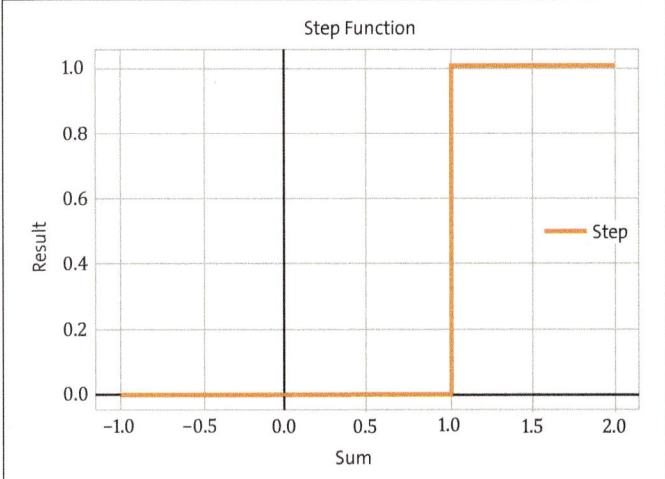

Figure 3.8 Step Function

3.5 Perceptron

Our preliminary considerations now make it very easy to head for the *perceptron*. The perceptron is one of the oldest and simplest ANN models. It was created in 1957 and invented by Frank Rosenblatt (see the Section 3.14). The nodes in the perceptron are referred to as *linear threshold unit* (LTU) because the perceptron uses the step function as the output function and thus performs a linear separation of the input data. For better illustration, we've put together a chart in Figure 3.9 with all the components of the perceptron and the calculation modules. It's like a small experimental box from which we'll now gradually take out and discuss parts.

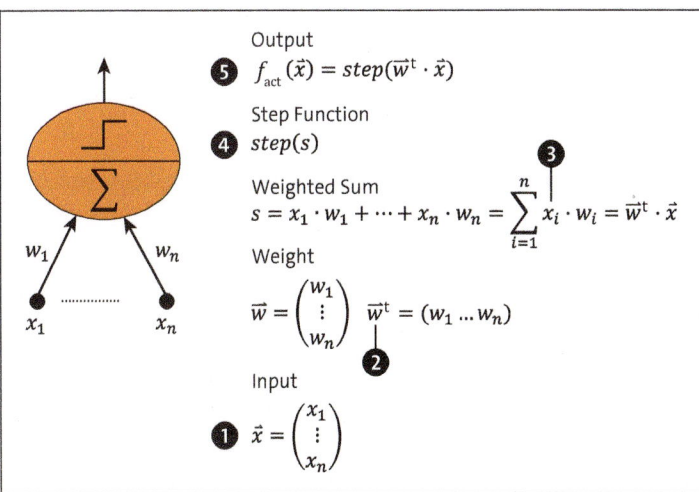

Figure 3.9 Perceptron: Components and Calculation Modules

3 A Simple Neural Network

Are you getting sweaty just looking at all these mathematical symbols? In our opinion, the compact and elegant mathematical notation disguises the simplicity of the model.

We've written down what we've done so far regarding the staffing plan for Mrs. Carrot and Mr. Leek, but in a more general way. To prove this to you, we'll go through the individual components from the diagram shown in Figure 3.9 item by item in the following sections:

- Module ❶, input $\vec{x}$ (Section 3.6)
- Module ❷, $\vec{w}^{"t"}$ (Section 3.7)
- Module ❸, $\sum_{i=1}^{n} x_i \cdot w_i$ (Section 3.8)
- Module ❹, $step(s)$ (Section 3.9)
- Module ❺, the output (Section 3.10)

3.6 Points in Space: Vector Representation

Let's rock and roll! The input x_1 stands for the *presence* of Mrs. Carrot; it can have the values 0 (she isn't present) or 1 (she can help). The input x_2 represents the presence of Mr. Leek and can of course be assigned the same values. For example, if Mrs. Carrot is present, and Mr. Leek is still on vacation, we could write it as follows:

$$\vec{x} = \begin{pmatrix} x_1 \\ x_2 \end{pmatrix} = \begin{pmatrix} 1 \\ 0 \end{pmatrix}$$

Incidentally, the notation for the combined values is called *vector notation*. The number of values is called the *dimension*; in our case, the *dimension* is = 2, which is also referred to as *two-dimensional*. Thus, the point is located in a two-dimensional plane if you look down from above, that is, from the third dimension (Section 3.14 for a nice story about dimensions). Points can also be displayed in diagrams, such as the previous example in Figure 3.10.

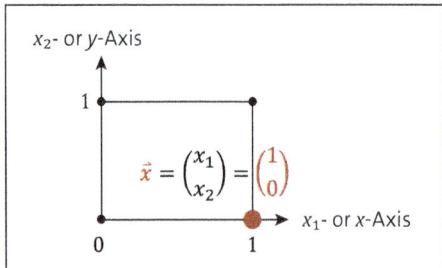

Figure 3.10 Point in the Cartesian Coordinate System

We plot the values in a *coordinate system*. (Graph paper would be great now, as there are already nice little boxes printed on it for inserting the points.) One dimension is the x_1 or simply x dimension, whereby the possible values of this dimension are represented in the coordinate system on the x-axis. And the second axis, the x_2 or y-axis, is drawn on it at a right angle. With these two specifications, x_1 and x_2, the position of a point is clearly defined.

3.6.1 Task: Completing Values

Now it's your turn again: draw the other possible points for our planning example with Mr. Leek and Mrs. Carrot in the coordinate system, and label the points using vector notation, as previously discussed.

Solution:

The result is shown in Figure 3.11.

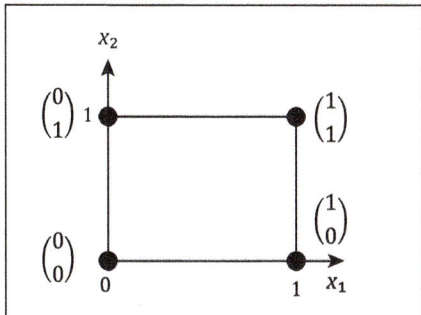

Figure 3.11 All Points for the Planning Example in Vector Notation

As a reward for successfully completing the exercise, you can now write a small Python script that draws the points for you (see Listing 3.3). There, you can see how to declare a vector in Python, and you can also use a scatterplot for visualization. A *scatterplot* represents pairs of values as points in a diagram in the Cartesian coordinate system.

```python
# Mathematics
import numpy as np
# Import the pyplot functions
import matplotlib.pyplot as plt
# Very important, otherwise the plot will not be displayed
%matplotlib inline

# The array function converts a Python list into a numpy array
# The x1 = x coordinates
x1    = np.array([0, 0, 1, 1])
# The x2 = y coordinates
x2    = np.array([0, 1, 0, 1])
# The colors for the points
color = np.array(['black', 'black', 'red', 'black'])
```

3 A Simple Neural Network

```python
# The point size for each point
size = np.array([100, 100, 500, 100])

# Output vector x1 with all x1 coordinates
print('Vector x1: ', x1)

# Set the axes
plt.grid(True)

# Draw the plot for the x1 and x2 coordinates, color, and size
plt.scatter(x1, x2, c=color, s=size)

# Set the axes
plt.xlabel('x1')
plt.ylabel('x2')

# Set the axis divisions
plt.xticks([0.0,1.0])
plt.yticks([0.0,1.0])

# Finally, output the diagram
plt.show()
# Output:
Vector x1:  [0 0 1 1]
```

Listing 3.3 Scatterplot

The result is shown in Figure 3.12.

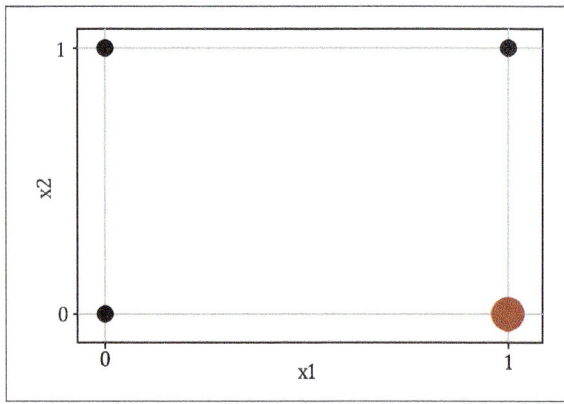

Figure 3.12 Scatterplot for the Planning Points

The script starts with the import of the numpy module for mathematical operations. The module is particularly suitable for calculating with vectors and other mathematical constructs, which we'll discuss later (in Appendix B). We then import the `matplotlib.pyplot` module and rename it to `plt`. This is followed by a section in the program that requires some further explanation. In the previous explanations, we represented points with a vector consisting of the two dimensions, x_1 and x_2. The use of the `scatter` function requires the coordinates of the individual vectors to be split into two `np.array` objects, that is, one array per dimension and one entry in the array per point. To check, we output the array for the x_1 dimension using the `print` function.

We then set the desired properties for the plot, for example, the axis labels using `plt.xlabel('x1')` and `plt.ylabel('x2')`, and the axis divisions using `plt.xticks()` and `plt.yticks()`. At the end of the script, the plot gets output and drawn using `plt.show()`.

3.6.2 Task: Outputting the Iris Dataset as a Scatterplot

The following task is designed to show how you can import data from a file into Python, process it line by line, and use it for visualization. Of course, we have to explain the first steps relating to importing in detail before we get started.

As data material for the plot, we've chosen the famous *Iris dataset*, which is a typical test case for classification techniques. (Visit the book's website and download the resources, including the *iris.csv* dataset [*Chapter_03* directory].) This dataset contains 150 observations of four properties each of irises. The four properties are the width and length of the sepal (*sepalum*) and the width and length of the petal (*petalum*), measured in centimeters, as shown in Figure 3.13. Furthermore, the type of iris is indicated for each existing combination of properties. These are *Iris-setosa*, *Iris-virginica*, or *Iris-versicolor*.

Sepal Length	Sepal Width	Petal Length	Petal Width	Class
5.1	3.5	1.4	0.2	Iris setosa
7.0	3.2	3.5	1.0	Iris versicolor
6.3	3.3	6.0	2.5	Iris virginica
...	...	...	...	...

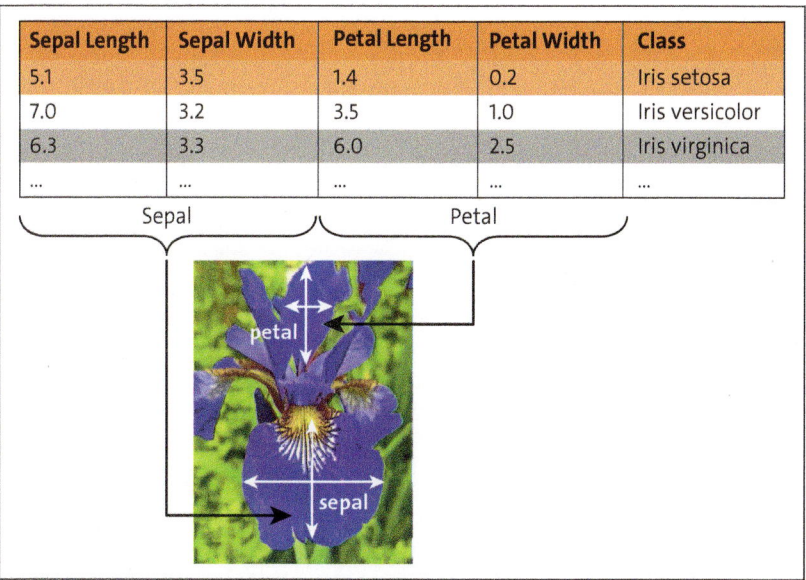

Figure 3.13 Leaf Dimensions for Irises (© Kaggle)

Your task now is to draw a scatterplot using the Python tools we've discussed so far:

1. Before you start coding, upload the *iris.cvs* file from our download to the session memory of your Colab session so that the file can then be imported without specifying a path.

2. Import the *iris.csv* file. To do this, use the `with open("iris.csv", "r") as fobj:` statement, which returns a file handle for reading (`"r"` parameter). You can use the `for line in fobj:` statement to process the dataset line by line; once you've processed all the lines, the `fobj` file handle will get closed automatically.

3. Create three Python lists, `x1`, `x2`, and `colors`, to store the data for sepal length, sepal width, and data point colors. To be able to determine this data, you still need the column structure of the dataset per line, which you can see in Figure 3.13.

    The lengths in the columns are given in centimeters, and the `Class` column can have the following values:
    - Iris setosa
    - Iris versicolor
    - Iris virginica

    The individual columns are separated by a comma.

4. Use `matplotlib.pyplot` to generate the scatterplot.

Listing 3.4 shows the solution.

```
# Mathematics
import numpy as np
# Import the pyplot functions
import matplotlib.pyplot as plt
# Very important, otherwise the plot will not be displayed
%matplotlib inline

# x1 are the coordinates of the x-axis, x2 those of the y-axis
x1    = []
x2    = []
# Colors for the data points
colors = []
# Mapping the irises to colors using
# a Python dictionary
iris_colors = { 'Iris-setosa' : 'red',
                'Iris-versicolor' : 'green',
                'Iris-virginica' : 'blue'
              }
```

```python
# Read file content
# Prerequisite: The iris.csv file has been uploaded to the session memory
# File handle fobj gets closed automatically
with open("iris.csv", "r") as fobj:
    # Process the dataset line by line
    for line in fobj:
        # Split into single words
        words = line.rstrip().split(",")
        # Omit blank lines
        if len(words) != 5:
            continue
        # SepalLength
        x1.append(float(words[0]))
        # SepalWidth
        x2.append(float(words[1]))
        # Color
        colors.append(iris_colors[words[4]])
# Draw grid in scatterplot
plt.style.use('seaborn-v0_8-whitegrid')
# Axis labeling and title
plt.xlabel('Sepal Length')
plt.ylabel('Sepal Width')
plt.title('Scatter Plot')
# Output plot
plt.scatter(np.array(x1), np.array(x2), color=colors )
# Show plot
plt.show()
```

Listing 3.4 Sepal Length and Sepal Width as a Scatterplot

We start by importing the required modules for plots and arrays again. The dataset is then read and processed line by line. The `line.rstrip()` function is used to create a copy of the `line` string from which the spaces at the end have been removed. Using `split(",")`, the string is then split into its individual values—always separated by a comma—and the individual values are combined to form the words array. The first two values in the words array, `words[0]` and `words[1]`, define the coordinates for the plot. The last value in the array at position 4 is the name of the iris that is used to determine the color. The Python dictionary `iris_colors` is used to assign the names to the colors. Then, we set the output of a grid in the plot, transfer the data to the scatterplot, and get the output shown in Figure 3.14.

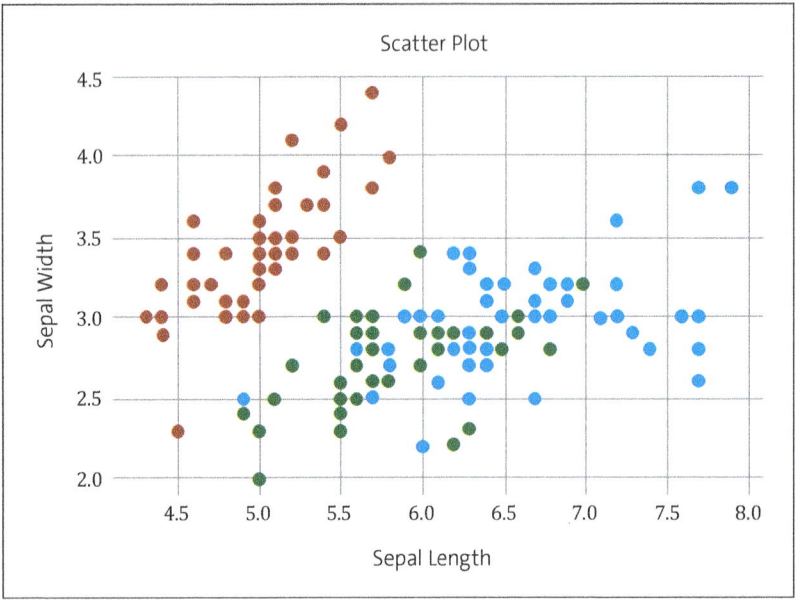

Figure 3.14 The Irises as a Scatterplot with the Sepal Length and Sepal Width Coordinates

3.7 Horizontal and Vertical: Column and Line Notation

$\vec{w}$ is referred to as the *weight vector*. When we talk about *learning* in neural networks, we're referring to the *adaptation of weights*. To understand what a weight is, imagine a connection between two neurons. One neuron, let's call it A, sends a signal to the second neuron, B. How strong the signal arrives in B controls the weight of the connection. If the weight has a value between 0 and 1, then the signal strength is reduced; if the value is greater than 1, the signal is amplified; and if the value is less than 0, the signal becomes negative. The weight thus regulates the signal strength that reaches neuron B. You'll see details of this in Chapter 4 when we discuss perceptron learning.

In addition to the input vector, in which we've written the values on top of each other (the *column notation*), we've also listed the weight vector in *line notation*. The values are written next to each other in a line. To make it easy to recognize the difference from a column vector, the $\vec{w}$ is simply given a t, and so we know that it's represented in line notation. "Given" is a bit unmathematical now, so let's call it *transpose*. The reason we're performing all this mumbo-jumbo is that we can simply use it to mathematically describe the multiplication between the input values and the weights, thus optimally implementing the calculation in Python, namely using $\vec{w}^t \cdot \vec{x}$ instead of $w_1 \cdot x_1 + w_2 \cdot x_2 + \cdots + w_n \cdot x_n$, whereby the number of components in $\vec{w}$ and $\vec{x}$ must be the same. That's much nicer, isn't it? In any case, it's short and sweet.

3.7 Horizontal and Vertical: Column and Line Notation

This is how you perform the multiplication between the vectors: w_1 times x_1 plus w_2 times x_2 plus .... Please complete this by yourself:

$$\vec{w}^t \cdot \vec{x} = (w_1 w_2 \ldots w_n) \cdot \begin{pmatrix} x_1 \\ x_2 \\ \vdots \\ x_n \end{pmatrix} = w_1 \cdot x_1 + w_2 \cdot x_2 + \cdots + w_n \cdot x_n$$

3.7.1 Task: Determining the Scalar Product Using NumPy

Python allows for the multiplication of vectors using functions of the numpy module, more precisely by using the dot function. The multiplication is referred to as *dot product*, *scalar product*, or *inner product*. It turns two sequences of numbers into one number, as you saw before.

Fun Facts

1. The multiplication between the vectors is referred to as the *scalar product* because the result is a scalar, that is, a value, and not a vector.
2. The name *dot product* originates from the fact that there is a dot between the vectors during multiplication.

Now it's time to calculate the scalar product using Python. To do this, you'll again need a code cell in your Jupyter Notebook and, if you wish, a Markdown cell for a heading.

Before starting the implementation, you should read the Python script in Listing 3.5 carefully line by line and analyze the calculations. After your analysis, we'll ask you for two ways to calculate the inner product using NumPy.

```python
# Multiplication of vectors with numpy
#  Dot product, scalar product, inner product
#   turns two sequences of numbers into one number (algebraic)
#  It is the cosine of the angle between two vectors (geometric),
#   multiplied by their lengths
# Mathematics
import numpy as np

# create numpy array
x = np.array([0,1])
# create numpy array
w = np.array([0.5,0.7])

# Output vector x
print("x =", x)
# Output vector w
print("w =", w)
```

```
# numpy arrays are not matrixes, and
# the *, +, -, / operators work element by element
print("w*x =",w*x)

# Inner product. You do not have to transpose the vector,
# that's what numpy does for us
print("np.dot(w,x) =", np.dot(w,x))

# Alternative syntax for the dot product
print("w.dot(x) =", w.dot(x))
# Output:
x = [0 1]
w = [0.5 0.7]
w*x = [0.  0.7]
np.dot(w,x) = 0.7
w.dot(x) = 0.7
```

Listing 3.5 Scalar Product with Python

What two options would you consider for determining the inner product with NumPy? Of course you've found the NumPy dot() variants to be the right candidates!

Let's now continue with our explanation of the other elements. The third element is the representation of the *weighted sum*.

3.8 The Weighted Sum

Mathematicians borrow a symbol from the Greek alphabet for the sum of $w_i \cdot x_i$—the capital sigma: Σ. In addition, some other "tags" are attached to the sigma:

$$\sum_{i=1}^{n}$$

The tags mean that the running variable *i* takes on all values in succession from an initial value to a final value. (It's said to *pass through* the area in whole steps.) In our example, it runs from the start value 1 to the end value *n*. Assuming that $n = 3$, these are steps 1, 2, and 3. With this knowledge, you can now write the sum of all $w_i \cdot x_i$—let's call it *s*—in a pretty compact way:

$$s = \sum_{i=1}^{n} w_i \cdot x_i$$

This results in the third building block:

$$s = \sum_{i=1}^{n} w_i \cdot x_i = \vec{w}^t \cdot \vec{x} = w_1 \cdot x_1 + w_2 \cdot x_2 + \cdots + w_n \cdot x_n$$

Let's move on to the fourth component in the modular system, the step function. Although we've already discussed it several times, we want to present it here in a mathematical representation.

3.9 Step-by-Step: Step Functions

We define the following function for step function $step(s)$:

$$step(s) = \begin{cases} 0, & \text{if } s < \theta \\ 1, & \text{if } s \geq \theta \end{cases}$$

Regardless of which value you use to call the step function, the result is either 0 or 1, but it depends on the threshold value *theta*, the Greek letter θ, when the jump from 0 to 1 takes place. Now that the value of the threshold is no longer fixed, it can also be changed by learning.

Now that we've generalized the step function, we can look at the fifth element, the calculation of function $f_{act}(\ )$.

3.10 The Weighted Sum Reloaded

Now, we'll consider the weighted sum and the step function. This is easier to write than this complicated distinction, for example, with the Heaviside function: "if greater than θ, then this, else that ...":

$$w_1 \cdot x_1 + w_2 \cdot x_2 + \cdots + w_n \cdot x_n \geq \theta$$

If we transform the inequality, that is, simply move the θ to the other side, we obtain the extended weight vector, which is extended by one dimension for the threshold value. This new dimension should be inserted at index 0 so that the original vector with its indexes is retained:

$$w_1 \cdot x_1 + w_2 \cdot x_2 + \cdots + w_n \cdot x_n - \theta \geq 0$$

Here's another little trick from the magic box: to be able to insert the threshold at position 0, we now simply rename $-\theta$ to w_0. Isn't that great? This allows us to write the following:

$$w_1 \cdot x_1 + w_2 \cdot x_2 + \cdots + w_n \cdot x_n + w_0 \geq 0$$

We've thus found a nice standardized form for writing $\vec{w}^t \cdot \vec{x}$ again. Of course, vector $\vec{x}$ must also be extended by one element. But this has a very simple value. Can you imagine which one? Just think about it.

We can answer this using Figure 3.15, where you can see that the neuron for input x_0 always has the value 1. It's referred to as a *bias neuron*.

If you've finished thinking, you'll find the extensions to the vectors and the weighted sum marked in red in Figure 3.15.

3 A Simple Neural Network

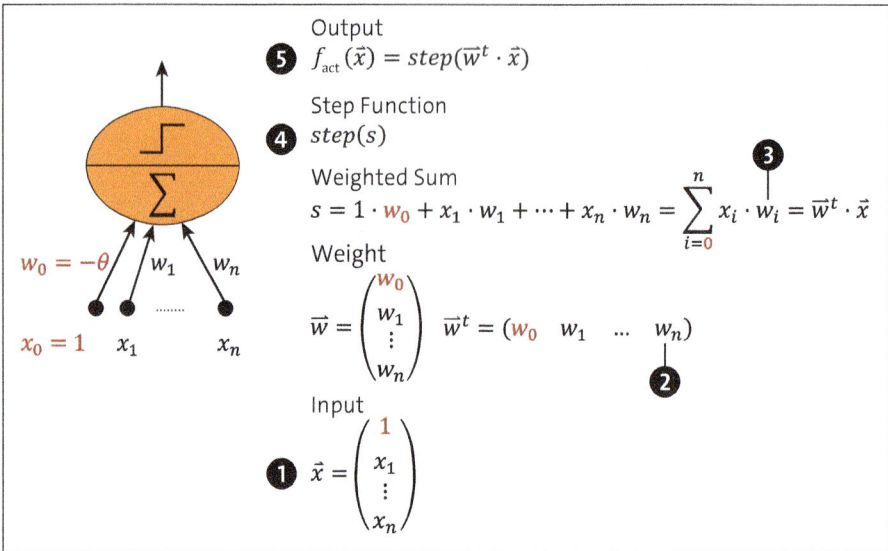

Figure 3.15 Components of the Perceptron

3.11 All Together

It's time to write a small Python program to convert our findings on the perceptron into program code. First, take a look at the following explanations and lines of code and transfer them to a new code cell in your Jupyter notebook to produce the output shown.

We use the attendances and absences of Mr. Leek and Mrs. Carrot as input. Each line in Table 3.4 represents an input vector with three values. Here, 1 stands for the bias neuron, the presence of Mrs. Carrot, and the presence of Mr. Leek.

Bias Neuron	Mrs. Carrot	Mr. Leek	→	Monday	Input Vector
1	0	0	→	0	1
1	0	1	→	1	2
1	1	0	→	1	3
1	1	1	→	1	4

Table 3.4 The Extended Input Vector for the Planning Problem

In the example (1, 0, 1), 1 stands for the bias neuron, 0 for Mrs. Carrot means "not present," and 1 for Mr. Leek means "present," resulting in a total of 1 for an acceptable planning situation. We have a total of four input vectors, which are most easily managed using a two-dimensional array. Because the input vectors are already labeled with x, we use an uppercase X as the name for the array. We summarize the desired outputs, which you can find in the table in the "Monday" column, as vector y. In the following implementation, we've implemented the Heaviside function for the step function, that is, with the threshold value 0, as we've previously integrated the threshold value into the weighted sum.

We've also added the aspect of *error calculation*. For each input, the calculated output is compared with the desired output. Because the perceptron can only output 0 or 1 due to the Heaviside function and the desired output is only 0 or 1, the difference between the desired and the calculated value can only assume the values -1, 0, or 1. To add up the individual errors and thus be able to provide an overall statement about the accuracy of the perceptron's determination of the data, we apply the amount to the individual error in Listing 3.6. Otherwise, an error of -1 would reduce the total error.

```python
# Mathematics
import numpy as np
# Plot
import matplotlib.pyplot as plt

# 3-dimensional input = bias neuron, Mrs. Carrot, Mr. Leek
# 4 input vectors
X = np.array([
    [1,0,0],
    [1,0,1],
    [1,1,0],
    [1,1,1]])
# The 4 desired result values
y = np.array([0,1,1,1])

# Heaviside function
def heaviside( sum ):
    """ Calculation of the decision on the sum value
    Input:  sum
    Output: 1, if sum >= 0,
            0 else
    """
    if sum >= 0:
        return 1
    else:
        return 0

# Perceptron calculation (forward path)
def perceptron_eval(X,y):
    """ Perceptron calculation
    Input:  X, input vector
            y, the desired output
```

```
        Output: The total error, i.e. sum of the amount of the difference
                of calculated and desired output
        """
        # The total error
        total_error = 0;
        # Select the weights so that the OR problem can be solved
        w = np.array([-1,1,1])
        # Index i and element x Determination of array X
        for i, x in enumerate(X):
            # x = Use line by line
            # Inner product between x and w
            sum = np.dot(w,x)
            result = heaviside(sum)
            # Error
            error = np.abs(result - y[i])
            # Total error
            total_error += error
            # Output
            print("Mrs. Carrot = {}, Mr. Leek = {},
                desired result = {}, calculated result = {}, error = {}".
                format(x[1], x[2], y[i], result, error))
        # Total error per epoch over the entire training dataset
        return total_error
#-------------------------------
# Core function for analyzing the input
total_error = perceptron_eval(X,y)
print("Total error = %1d" % (total_error))
# Output:
Mrs. Carrot = 0, Mr. Leek = 0, desired result = 0, calculated result = 0, error
= 0
Mrs. Carrot = 0, Mr. Leek = 1, desired result = 1, calculated result = 1, error
= 0
Mrs. Carrot = 1, Mr. Leek = 0, desired result = 1, calculated result = 1, error
= 0
Mrs. Carrot = 1, Mr. Leek = 1, desired result = 1, calculated result = 1, error
= 0
Total error = 0
```

Listing 3.6 Solution to the Simple Staffing Plan Problem

The error in our example is of course 0, as we've chosen the weight vector optimally. Experiment with the weights in weight vector w, look at the possible errors, and try to reproduce them using the previously discussed and implemented calculations.

3.12 Task: Robot Protection

We've prepared another task for you—protecting a robot from falling. This little robot drives along and is up to no good. But the environment isn't kind to the robot, and large dangerous holes don't mean it well. Fortunately, the robot has two bright/dark sensors, mounted on the front left and right, which can determine the brightness of the ground in front of it (see Figure 3.16).

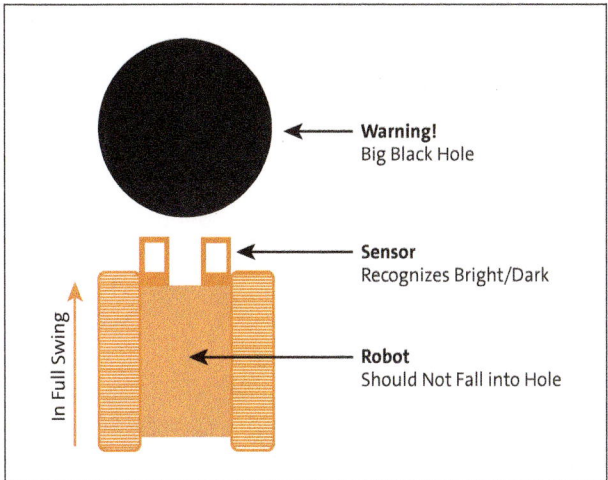

Figure 3.16 Simple Robot Sensors

If a sensor detects *bright*, it returns the value 0, if it detects *dark*, the sensor returns 1. To protect the life of the robot, you need to develop a simple perceptron that recognizes if there is a hole directly in front of the robot or if a safe ride is possible.

To implement the script, please create a new code cell in your Jupyter Notebook, and develop your solution there. Use the example in Listing 3.6 as a guide.

Solution:

If we look at the brightness readings from the sensors, then sometimes the left-hand sensor can respond, the right-hand sensor can respond, both can respond, or none can respond. We illustrate the combinations of brightness statements using Table 3.5.

Left-Hand Sensor	Right-Hand Sensor	Hole
0	0	0
1	0	0
0	1	0
1	1	1

Table 3.5 Hole Detection Using the Sensor Values

We've used the implementation from Listing 3.6 in Listing 3.7, as we recommended, and adapted the desired output y = np.array([0,0,0,1]) and the weight vector to w = np.array([-2,1,1]), so that the desired result from Table 3.5 gets calculated. In addition, we've adapted the text of the output to the task.

```python
# Mathematics
import numpy as np
# Plot
import matplotlib.pyplot as plt
# 3-dimensional input = bias neuron, sensor left, sensor right
# 4 input vectors
X = np.array([
    [1,0,0],
    [1,0,1],
    [1,1,0],
    [1,1,1]])
# The 4 desired result values
y = np.array([0,0,0,1])
# Heaviside function
def heaviside( sum ):
    """Calculation of the decision on the sum value
       Input:  sum
       Output: 1, if sum >= 0,
               0 else
    """
    if sum >= 0:
        return 1
    else:
        return 0
# Perceptron calculation (forward path)
def perceptron_eval(X,y):
    """ Perceptron calculation
    Input:  X, input vector
            y, the desired output
    Output: The total error, i.e. sum of the amount of the difference
            of calculated and desired output
    """
    # The total error
    total_error = 0;
    # Select the weights so that the robot problem can be solved
    w = np.array([-2,1,1])
    # Index i and element x Determination of array X
    for i, x in enumerate(X):
        # x = Use line by line
        # Inner product between x and w
```

```
        sum = np.dot(w,x)
        result = heaviside(sum)
        # Error
        error = np.abs(result - y[i])
        # Total error
        total_error += error
        # Output
        print("sensor L = {}, sensor R = {}, desired result = {}, calculated result = {}, error = {}".
              format(x[1], x[2], y[i], result, error))
    # Total error per epoch over the entire training dataset
    return total_error
#-----------------------------
# Core function for analyzing the input
total_error = perceptron_eval(X,y)
print("Total error = %1d" % (total_error))
# Output:
sensor L = 0, sensor R = 0, desired result = 0, calculated result = 0, error = 0
sensor L = 0, sensor R = 1, desired result = 0, calculated result = 0, error = 0
sensor L = 1, sensor R = 0, desired result = 0, calculated result = 0, error = 0
sensor L = 1, sensor R = 1, desired result = 1, calculated result = 1, error = 0
Total error = 0
```

Listing 3.7 The Perceptron Control System for the Robot

3.13 Summary

The perceptron is now all yours. You've carried out the theoretical and practical setup and calculations in the perceptron, which included such important things as the weighted sum, threshold value calculations, graphs, and graphic analyses. Admittedly, we've so far only implemented analyses and haven't yet considered the learning of the perceptron. We'll make up for this immediately in the following chapter.

3.14 Further Reading

- "The Perceptron: A Probabilistic Model for Information Storage and Organization in the Brain" by F. Rosenblatt (1958, *www.ling.upenn.edu/courses/cogs501/Rosenblatt1958.pdf*)
- UC Irvine Machine Learning Repository: Iris dataset (*https://archive.ics.uci.edu/dataset/53/iris*)
- *Flatland. A Romance of Many Dimensions* by Edwin Abbott under the pseudonym A. Square (1884, *www.gutenberg.org/ebooks/201*)

Chapter 4
Learning in a Simple Network

What does learning in a neural network mean? We'll describe learning in this context and what that looks like in concrete terms. You'll also see that you can learn in more than one way!

This chapter takes us back to the beginnings of learning for artificial neural networks (ANNs). In 1949, Donald Hebb discovered the effect that the connection between two neurons is strengthened if they fire simultaneously, which is the basic mechanism for learning and memory. Motivated by these discoveries, a directional learning strategy emerged called *perceptron learning*, which explained learning for datasets that are cleanly separable. Encouraged by the success of this learning approach and aware of its shortcomings, the next strategy was developed—*Adaline*—which, among other things, forms the theoretical basis for current learning strategies.

4.1 Background: Plans Are Being Made

Our starting position is relatively clear: we know that we want a perceptron that delivers a desired result based on certain specifications (see Table 4.1).

P1	P2	→	E
0	0	→	0
0	1	→	1
1	0	→	1
1	1	→	1

Table 4.1 The Well-Known Planning Problem

We pretty much picked apart the analysis by the network in Chapter 3 and looked at every aspect, no matter how small. Our findings are summarized in Figure 4.1. Focus on the forward path in the diagram, and briefly think about the green dots again. What happens at those times?

4 Learning in a Simple Network

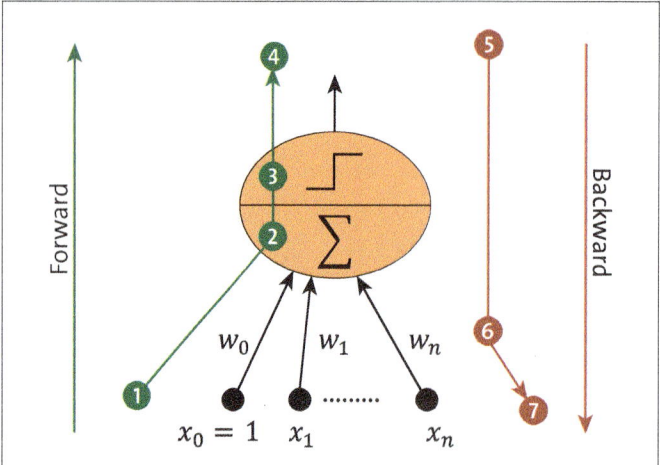

Figure 4.1 Going Forward and Backward in the Perceptron

The main steps were as follows:

1. Provide the input x_i ❶.
2. Calculate the weighted sum x_i ❷.
3. Use the step function x_i ❸, and finally, admire the output x_i ❹.

Very good! From now on, the backward path is the path of interest for us because this path concerns learning in the ANN. Based on the determined error in ❺, the weights are adjusted ❻, and the input is also included in the calculation of the error ❼.

We'll now take a look at these steps again, slowly and in detail. For this purpose, we'll start with greenfield implementations and use a *scikit-learn* implementation template as a guide. This template contains a method that adapts the ANN to the problem (fit method) and a method that enables you to use the ANN for analyses (predict method).

4.2 Learning in Python Code

You'll now get to know different variants of perceptron learning: firstly, those that work great if the dataset is *linearly separable*, but which will fail completely if that isn't the case. We'll also discuss another approach that can deal with the second situation, the *Adaline* approach mentioned earlier.

4.3 Perceptron Learning

Imagine the tables with the 0s and 1s again. We now take a row and want the ANN to return a specific result for the input. Of course, this won't always work, and the network will produce different results, which means that something needs to be changed in the network. If we look again at the calculation, we can consider which parameters can be adjusted:

$$s = \sum_{i=0}^{n} w_i \cdot x_i = \vec{w}^t \cdot \vec{x}$$

In other words, this means that the input values x_i from the predecessor neurons are multiplied by a weight w_i and then added together. For example, the equation for the weight values $\vec{w}^t = (0.1, 0.1, 0.1)$ would look like this:

$s = 0.1 \cdot x_0 + 0.1 \cdot x_1 + 0.1 \cdot x_2$

In addition, the input $\vec{x} = (1.0, 0.0, 0.0)$ results in the following:

$s = 0.1 \cdot 1.0 + 0.1 \cdot 0.0 + 0.1 \cdot 0.0 = 0.1$

Now, it's your turn!

> **Task**
>
> What does the equation for the weight values $\vec{w}^t = (-0.9, 0.1, 0.1)$ look like?
>
> What is the value for $\vec{x} = (1.0, 0.0, 0.0)$?

Solution:

$s = -0.9 \cdot 1.0 + 0.1 \cdot 0.0 + 0.1 \cdot 0.0 = -0.9$

What have you discovered that can be changed? Because the inputs x_i are predetermined by the question, only the weights can be adjusted in this equation. This makes it clear that when we talk about learning, it must be about x_i adjusting the weightings.

You may now be asking yourself how you should adjust the weights. Here's the answer:

1. Present a learning example to the network.
2. Have it carry out the calculations.
3. Compare the *calculated* value $\hat{y}$ with the *desired* value y of the learning example.
4. Adjust the weights based on the comparison result.

The ANN should learn the examples better and better and deliver the desired results. In this way, each pass should provide a reduction in the calculation error. This behavior is referred to as *convergence*, and the rule according to which this is achieved is called the *learning algorithm*. Such a learning algorithm could look as follows:

1. Initialize the weights and thus also the threshold value. The weights can assume arbitrarily small values, usually in the interval (-1,1). There are many different strategies for initialization; we want to keep it simple and start with a random selection in the (-1,1) interval.

2. If the calculated output of a neuron is 1 (or 0) and the desired value is 1 (or 0), then the weights aren't changed. And why should they be? The network already determines the correct values!

3. If the output is 0, but the desired value is 1, then all weights must be increased. This is done for the simple reason that the calculated value is too low, and a change must be made so that the calculated result is higher. The amount of change is relative to the error.
4. If the output is 1, but the desired value is 0, the weights get reduced.

This process can of course be translated into the language of mathematics, and this gives us the following representation of the learning algorithm:

$$w_i^{new} = w_i^{old} + \Delta w_i$$

where

$$\Delta w_i = (y - \hat{y}) \cdot x_i$$

w_i^{old} indicates the weight with the current value, for example, -0.3.

w_i^{new} is the weight after a change to the old weight, that is, if, for example, 0.1 is added to the old weight. For our example, this would be -0.3 + 0.1 = -0.2.

Δw_i is called the delta w_i. The small triangle is therefore referred to as a *delta* and is intended to express that a change is taking place. This means that each weight is changed with each step, either by -1, 0, or +1.

> **Task**
>
> Where does the certainty come from that the weight change per step can only be -1, 0, or +1? You can see the individual steps to the solution in Figure 4.2.

In the learning algorithm, the different elements mean the following:

- w_i is the *i*-th weight of the weight vector, where w_0 is the negative threshold value $-\theta$—and we already know that.
- Δw_i is the change in weight w_i.
- y is the desired output of the neuron for the learning example.
- $\hat{y}$ is the actual calculated output for the learning example.
- x_i is the input for the neuron of the learning example, where $x_0 = 1$. We've defined this in such a way that the threshold value $-\theta$ can be incorporated into the beautifully uniform presentation of the weighted sum. This procedure is sometimes referred to as a *bias trick*.

When we talk about *old weight* and *new weight*, this implies that time passes in between, or at least a time step. In Figure 4.2, we show you one step. We start at t, which is, for example, the 150th learning step, and switch to step $t + 1$, which is the 151st learning step.

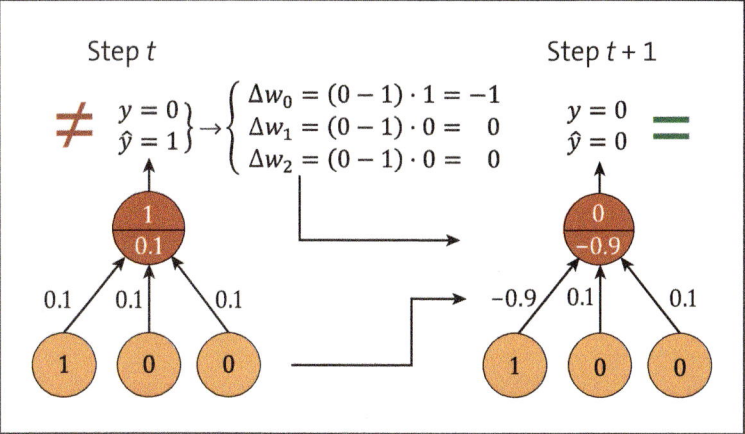

Figure 4.2 A Perceptron Learning Step

Solution:

A weight adjustment from step t to step $t+1$ is determined based on the difference between the desired output and the calculated output, as shown in Figure 4.2. The starting point is the following formula:

$\Delta w_i = (y - \hat{y}) \cdot x_i$

The output can only be 0 or 1 due to the step function, and the desired output can also only be 0 or 1; this results in the situation shown in Table 4.2.

y_i	$\hat{y}_i$	$(y - \hat{y})$	$(y - \hat{y}) \cdot x_i$
0	0	0	0
0	1	-1	$-x_i$
1	0	1	x_i
1	1	0	0

Table 4.2 Possible Errors in the Perceptron and the Corresponding Change in Weight

With this knowledge, you can now go back to the formula $w_i^{new} = w_i^{old} + \Delta w_i$ and carry out replacements:

- w_i^{new} is the value in step $t+1$, so we can write $w_i(t+1)$.
- w_i^{old} is the value in step t, so we can write $w_i(t)$.

These substitutions result in the following calculation steps for changing the weights:

- $w_i(t+1) = w_i(t)$ with correct output
- $w_i(t+1) = w_i(t) + x_i$ for output 0 and desired output 1
- $w_i(t+1) = w_i(t) - x_i$ for output 1 and desired output 0

4 Learning in a Simple Network

4.4 Separating Line for a Learning Step

In Figure 4.3, we've drawn the separating line that separates the 0 and 1 regions from each other. These are the two desired results, which are also referred to as *classes* or *categories*. If you're interested in the math behind this, then check out Appendix B where we show you how to get from the weights to a dividing line.

Determination of the Separating Line for Step t + 1

We start from the weighted sum:

$x_0 \cdot w_0 + x_1 \cdot w_1 + x_2 \cdot w_2 = 0$

Then, we enter the values:

$1 \cdot (-0.9) + x_1 \cdot 0.1 + x_2 \cdot 0.1 = 0$

A rewriting for x_2 provides the following result:

$x_2 = -x_1 + 9$

Point 1: *If* $x_1 = 0$, $x_2 = 9$, that is, the point (0,9) is located on the separating line.

Point 2: *If* $x_1 = 9$, *then* $x_2 = 0$, that is, the point (9,0) is located on the separating line.

In step *t*, the point (0,0) is still in the wrong region (we've highlighted the point in red in Figure 4.3)—it's classified as 1, while it should actually be 0.

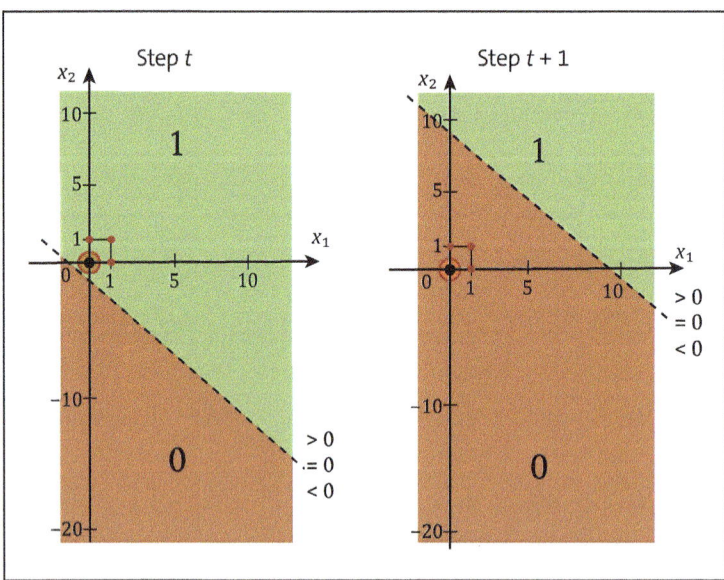

Figure 4.3 Changing the Position of the Separating Line via a Learning Step with Adjusted Weights

The learning algorithm now changes the weights in step *t* + 1, and you can see that the separating line shifts as a result. This in turn means that the 0 and 1 regions have shifted, just as desired for our individual example from the *training dataset*. During the learning step, the training datasets are run through, and the corresponding changes are made to the weights.

The changes to the weights are carried out immediately in the algorithm. In the *convergence theorem*, Frank Rosenblatt was able to prove that the preceding learning method can be used to learn all solutions a perceptron can represent, that is, all linearly separable solutions. On the other hand, if the input set isn't linearly separable, the algorithm doesn't stop.

Time for coding: let's take a look at the details of the implementation of the perceptron learning algorithm using detailed Python code. Again, the planning task serves as a simple example. If you don't remember the table, take another look at Table 4.1.

4.5 Perceptron Learning Algorithm

As a first step, we recommend that you create a new notebook called Chapter 4. Then, create a Markdown cell with a heading, for example, "Perceptron learning algorithm", and a code cell for the code to make it easier to read.

The core of the code is the `fit` function, in which learning takes place by means of weight adjustment, as shown in Listing 4.1. The `iterations` parameter determines how many steps are used to learn the `training_data_set` data.

```python
# Plot
import matplotlib.pyplot as plt
# Random number generator
from random import choice
# For the mathematical operations
from numpy import array, dot, zeros, random
# Very important, otherwise the plot will not be displayed
%matplotlib inline

# The Heaviside step function as a lambda function
heaviside = lambda x: 0 if x < 0 else 1

# Training
def fit(iterations, training_data_set,w):
    """ Learning in the perceptron
    
    iterations: A forward and backward run of all training examples
    trainings_data_set: The training examples
    w: The weights to start with
    
    """
```

```python
        errors = []
        weights = []
        for i in range(iterations):
            # random selection of a training example random.choice
            training_data = choice(training_data_set)
            x = training_data[0]
            y = training_data[1]
            # Determine the calculated output: Weighted sum with
            # downstream step function
            y_hat = heaviside(dot(w, x))
            # Calculate the error as the difference between the desired and
            # current output
            error = y - y_hat
            # Collect errors for output
            errors.append(error)
            # Collect weights for later output
            weights.append(w)
            # Weight adjustment = Learning... x_i is either 0 or 1
            w += error * x
        # Return of errors and weights
        return errors, weights

def main():
    """ Main program
    The individual components are put together here
    """
    # Training data
    # Per line: the binary input data and the desired binary output
    # in a list of tuples
    # At index position 0 of the input vector, the bias neuron
    training_data_set = [
        (array([1,0,0]), 0),
        (array([1,0,1]), 1),
        (array([1,1,0]), 1),
        (array([1,1,1]), 1),
    ]

    # Initial initialization of the random generator due to
    # reproducibility of the results
    random.seed( 12 ) # any value

    # Initialize array of length 3 with 0
    w = zeros(3)
```

```python
    # The number of passes. Experience through trial and error
    iterations = 30

    # Training using fit()
    # We collect the errors/weights in each step for the graphical output
    errors, weights = fit(iterations, training_data_set,w)
    # Output the last weight vector
    w = weights[iterations-1]
    print("Weight vector at the end of the training:")
    print(w)

    # Analysis after training
    print("Analysis at the end of the training:")
    for x, y in training_data_set:
        y_hat = heaviside(dot(x, w))
        print("{}: {} -> {}".format(x, y, y_hat))

    # Graphic for errors per learning example :-)
    # Get current axes to set the visibility of the grid
    ax = plt.gca()
    # Control visibility
    ax.grid(True)
    # Figure Numbers Start
    fignr = 1
    # Print size in inches
    plt.figure(fignr,figsize=(10,10))
    # Output error as plot
    plt.plot(errors)
    # Grid
    plt.style.use('seaborn-v0_8-whitegrid')
    # Labels
    plt.xlabel('Iteration')
    # Label the y-axis using LaTeX
    plt.ylabel(r"$(y - \hat y)$")
    # Show plot
    plt.show()

# Main program
main()
# Output: Perfect!
Weight vector at the end of the training:
[-1.  1.  1.]
Analysis at the end of the training:
[1 0 0]: 0 -> 0
```

```
[1 0 1]: 1 -> 1
[1 1 0]: 1 -> 1
[1 1 1]: 1 -> 1
```

Listing 4.1 Perceptron Learning Algorithm

Figure 4.4 shows the graphical output.

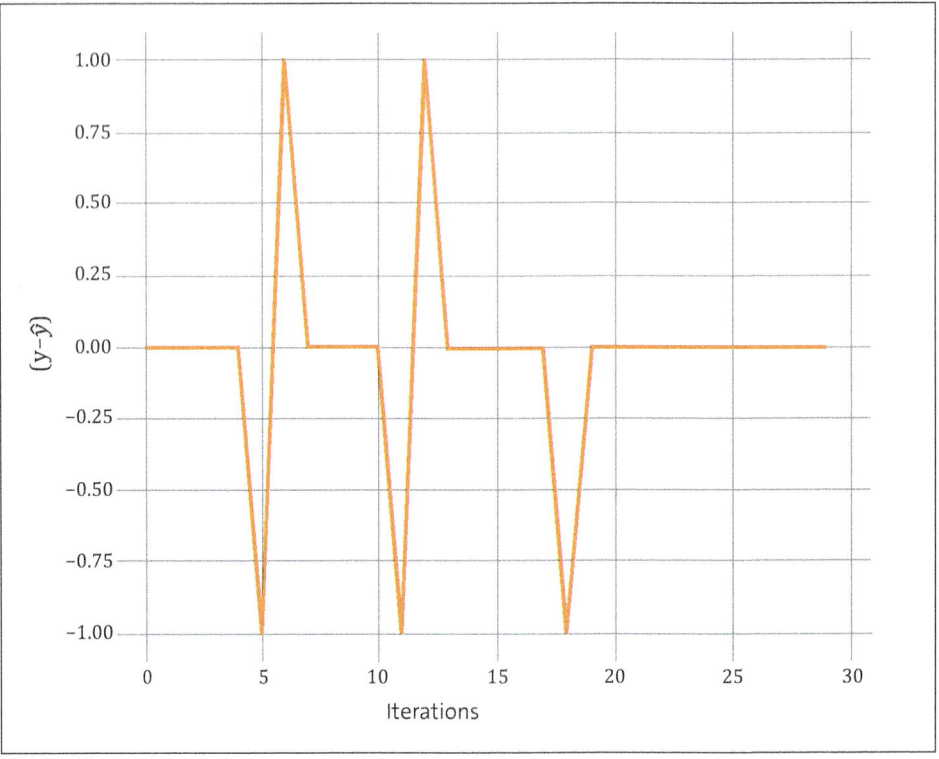

Figure 4.4 Difference between the Desired and Calculated Output $(y - \hat{y})$ during Learning per Randomly Selected Training Example

The most important thing we need at the start is the training data, which consists of the inputs and the desired output. The inputs are the vectors, which computer scientists refer to as an *array* or *list*, for example.

The output is a simple value, that is, a *scalar*. In the code, we've designated the set of training data as `training_data_set`. If you look closely, you'll see that several examples are stored in *tuples* in the array. This type of storage is rather arbitrary and was created on a whim. It could just as well be kept as a two-dimensional list or completely different in the program (e.g., `numpy ndarray`).

In the code, the output function is defined using a *lambda* construct: a step function named after the mathematician and physicist Oliver Heaviside. This suddenly jumps to the output value 1 at point 0. The weights are then initialized (as already mentioned, also with 0), and the number of `iterations` is defined.

In the case of the perceptron, this isn't so easy. If the problem can't be solved from the perceptron's point of view, it won't stop learning and will frantically try to find a solution. One way to complete the learning in any case is to set a fixed number of learning steps, sometimes referred to as *iterations*.

The connection with other useful terms is available to you in Table 4.3. We use 25 iterations with batch size 1 in our program.

Term	Description
Epoch	A forward and backward run of *all* training examples
Batch size	The number of training examples in a forward and backward run
Iterations	The number of runs, where each run runs with the batch size number of examples

Table 4.3 Epoch, Batch Size, and Iterations

Then, the function that carries out the learning is already in the program: `fit`. The name is derived from the `fit` from scikit-learn, as we'll see in a moment. Learning takes place online because we present the perceptron with one learning example after the other and let it learn each time—that is, adjust weights. In this case, we speak of a batch size = 1, as exactly one example is used for learning. The learning examples are selected at random. The weighted sum (`dot`) and the output function (`heaviside`) are then calculated, and a weight adjustment is carried out, which can also be omitted if everything has already been recognized correctly. For each step, we note the difference between the desired output and the calculated output so that we can draw a graph.

You'll be watching this graph with interest in the future. Will the network learn the task or not? This graph is one of your most important analysis tools. As you can see in , the differences ($y - \hat{y}$) per randomly selected training example jump wildly back and forth between -1, 0, and 1 at the beginning, before leveling off at 0 after a few iterations. This means the task was learned for the training examples that were selected using `choice()`.

4.6 The Separating Lines or Hyperplanes for the Example

The linear separability, that is, the division of the input examples, can also be nicely illustrated for low dimensions, which you'll see in this section.

Let's take a look at the example from before, where we've already visualized the error of the perceptron learning algorithm. Now let's look at how the perceptron learning algorithm adjusts the weights iteration by iteration to arrive at the correct classification.

4 Learning in a Simple Network

But first, we have a little math problem for you, in which you first adjust and correct the weights manually.

> **Task: Correction of the Weights**
>
> Let's assume that the weight vector has the values (-0.28, 0.02, 0.05). The vector (1,0,1) is transferred as input to the ANN with the desired output 1. What do the new weights look like after the correction? We ask for your calculations!

Solution:

1. First, calculate $\hat{y}$:

   $-0.28 \cdot 1 + 0.02 \cdot 0 + 0.05 \cdot 1 = -0.23 < 0 \rightarrow 0$

2. Then, calculate Δw:

   $(1 - 0) \cdot x = 1 \cdot (1,0,1) = (1,0,1)$

3. Grand finale—calculate the new weight:

   $(-0.28, 0.02, 0.05) + (1,0,1) = (0.72, 0.02, 1.05)$

Ta-da! Fortunately, the result matches line 1 in Table 4.4, which we'll now construct in the following, as shown in Figure 4.5, Figure 4.6, and Figure 4.7.

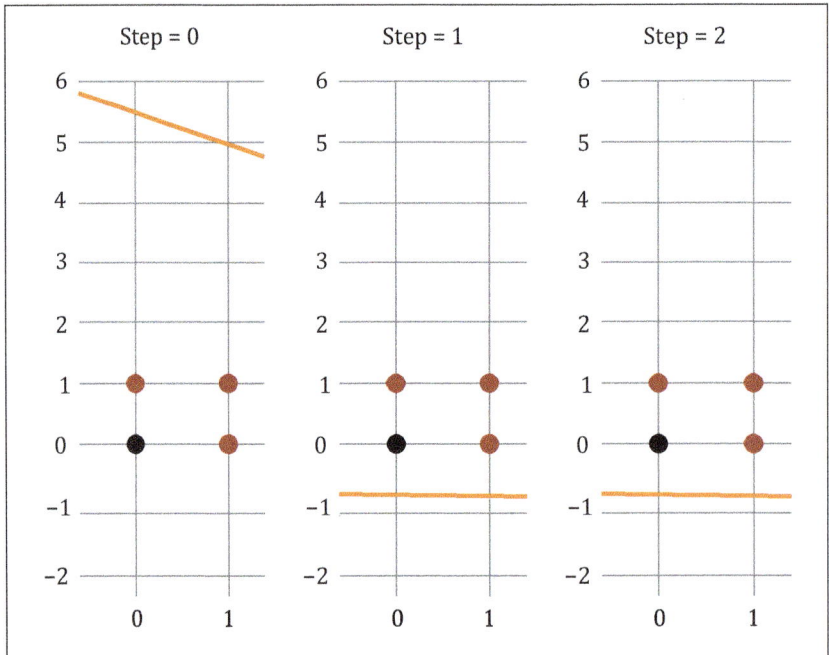

Figure 4.5 Step 0, Step 1, and Step 2

4.6 The Separating Lines or Hyperplanes for the Example

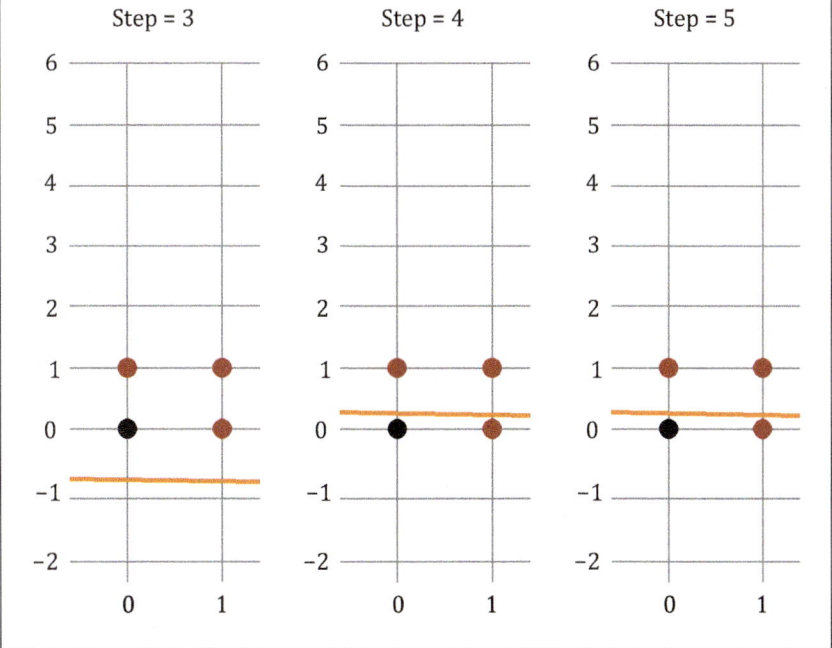

Figure 4.6 Step 3, Step 4, and Step 5

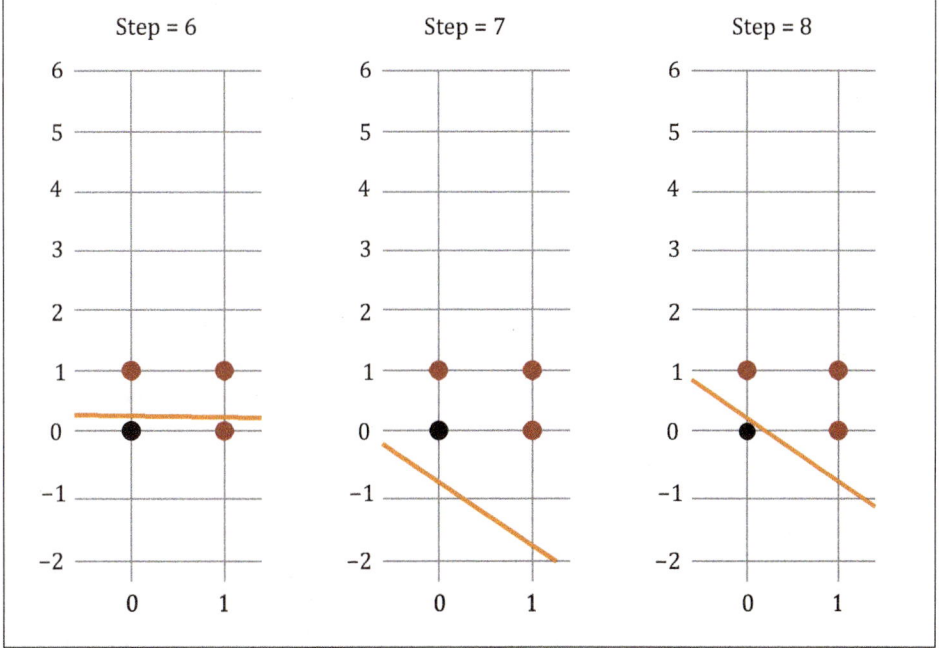

Figure 4.7 Step 6, Step 7, and Step 8

4 Learning in a Simple Network

The separating lines we've drawn in the training dataset change from step to step. The learning algorithm adjusts the weights, which in turn changes the position of the separating line. In our example, the perceptron succeeds in finding a solution to the task within nine steps.

We've summarized the steps in Table 4.4. We start with a random initialization of the weight vector. The Step column indicates the *running index*, which starts with 0 and is incremented by 1 per step. The column on the far right contains the figure that belongs to the step.

Step	x	y	$\hat{y}$	Error $(y - \hat{y})$	Correction $(y - \hat{y}) \cdot x$	w	Mapping
						(-0.28, 0.02, 0.05)	
0	(1,0,0)	0	0	0	(0,0,0)	(-0.28, 0.02, 0.05)	Figure 4.5
1	(1,0,1)	1	0	1	(1,0,1)	(0.72, 0.02, 1.05)	Figure 4.5
2	(1,0,1)	1	1	0	(0,0,0)	(0.72, 0.02, 1.05)	Figure 4.5
3	(1,1,0)	1	1	0	(0,0,0)	(0.72, 0.02, 1.05)	Figure 4.6
4	(1,0,0)	0	1	-1	(-1,0,0)	(-0.28,0.02,1.05)	Figure 4.6
5	(1,0,0)	0	0	0	(0,0,0)	(-0.28,0.02,1.05)	Figure 4.6
6	(1,1,1)	1	1	0	(0,0,0)	(-0.28,0.02,1.05)	Figure 4.7
7	(1,1,0)	1	0	1	(1,1,0)	(0.72,1.02,1.05)	Figure 4.7
8	(1,0,0)	0	1	-1	(-1,0,0)	(-0.28,1.02,1.05)	Figure 4.7

Table 4.4 Learning Steps and Adjustments

In the first line, only the start weight is shown, which was initialized randomly with (-0.28, 0.02, 0.05). In Figure 4.5, you can see the corresponding line or the resulting areas for classification. That's not very uplifting yet. For this reason, a correction is necessary!

Now follows the use of the scikit-learn template for the implementation of ANNs, referred to in technical jargon as the *estimator*. After all, we're all about programming!

If you wish, you can read about object orientation in Appendix A before we discuss the implementation. After reading this, you'll understand the following lines without any difficulty.

4.7 scikit-learn Compatible Estimator

With regard to reuse, it makes sense to store the developed code in an implementation that is compatible with scikit-learn. As a matter of fact, there are templates that can be used and that we've included here. The following objects are the basic components of the implementation, such as the *estimator* object for learning and the *predictor* object for performing the classification:

- **Estimator**

    An estimator is an object that learns from data, for example, using a classification, regression, or clustering algorithm. The base object `sklearn.base.BaseEstimator` implements the `fit` method to learn from the data:

    `estimator = Perceptron.fit(data,targets)`

    `Perceptron` is the name of an object we'll instantiate using the `PerceptronEstimator` estimator. When the `fit` method is called, the reference to the object itself gets returned, so the object reference `estimator` can be assigned the returned object reference. We'll develop the `PerceptronEstimator` estimator together in a moment.

- **Predictor**

    For example, the method for classification for supervised learning is implemented in the predictor object:

    `prediction = Perceptron.predict(data)`

    We'll also implement the predictor method in the `PerceptronEstimator` later.

- **Transformer**

    We could implement data filtering or data changes in the transformer object, but we don't do that.

    `new_data = obj.transform(data)`

- **Model**

    In the model object, we can implement the quality according to the goodness of fit measure, which we also omit.

    `score = obj.score(data)`

We'll now break down the explanation for the implementation of our estimator into separate parts, although you should think of the code in one piece without all the explanations. In addition, you should be able to understand the code easily because you've already developed it. We only distribute the coding passages appropriately in the methods.

Create another code cell in your Jupyter Notebook and transfer the different coding passages one below the other into the cell, starting from here. Let's start with the declaration:

- The new estimator we're building inherits from the `BaseEstimator` and `ClassifierMixin` classes, as shown in Listing 4.2.

    ```
    # Numpy helps us with the arrays
    import numpy as np
    # Graphical display
    ```

```python
import matplotlib.pyplot as plt
# These are our basic classes
from sklearn.base import BaseEstimator, ClassifierMixin
# Check routines for the consistency of the data, etc.
from sklearn.utils.validation import check_X_y, check_is_fitted, check_random_state
# Buffering the different target values
from sklearn.utils.multiclass import unique_labels
# Very important, otherwise the plot will not be displayed
%matplotlib inline


# Our estimator, appropriately labeled, and the base classes
class PerceptronEstimator(BaseEstimator, ClassifierMixin):
```

Listing 4.2 Declaration Part and Class Inheritance

- The classifier has three methods: initialize (`__init__`), learn (`fit`), and analyze (predict).
- The `__init__` method shouldn't accept any training data; it should rather be passed to the `fit` method.
- It should also be possible to instantiate the estimator without parameters. This means that all parameters of the `__init__` method require a default value.
- All parameters of the `__init__` method must be stored in object attributes with the same name, as shown in Listing 4.3.

```python
# Initialization
def __init__(self, n_iterations=20, random_state=None):
    """ Initialization of the objects
    n_iterations: Number of iterations for learning
    random_state_seed: In order to guarantee repeatability, a
                numpy.random.RandomState object can be constructed,
                which was initialized via random_state_seed-Seed
    """
    # The number of iterations
    self.n_iterations = n_iterations
    # The seed for the random generator
    self.random_state = random_state
    # Buffer the errors in the learning process for the plot
    self.errors = []
```

Listing 4.3 The __init__ Method

- We write the step function as before, but we write it as a method, not a lambda (see Listing 4.4).

```python
# A step function named after the mathematician and physicist
# Oliver Heaviside
def heaviside(self, x):
    """ A step function

    x: The value for which the step function is analyzed

    """
    if x < 0:
        result = 0
    else:
        result = 1
    return result
```
Listing 4.4 The Step Function as a Method

- If a random number generator (RNG) is used in the code, which is the case in our example, then `numpy.random.random()` shouldn't be used; instead, use `numpy.random.RandomState`. For reasons of repeatability, especially if the `fit` method is called multiple times, the RNG should be generated in the `fit` method, and this is how it works: The `__init__` method requires a parameter called `random_state` and should default this to `None`. In addition, the method should save the `random_state` unchanged in an attribute.

- The `fit` method uses `check_random_state` to generate an RNG and saves it in an attribute called `random_state_`, as shown in Listing 4.5.

```python
# Learn
def fit(self, X=None, y=None ):
    """ Train
    X: Array-like structure with [N,D], where
        N = rows = number of learning examples and
        D = columns = number of features
    y: Array with [N], with N as above
    """
    # Generation of the random number generator (RNG)
    random_state = check_random_state(self.random_state)
    # Initialization of the weights
    # np.size(.,1) = number of columns
    self.w = random_state.random_sample(np.size(X,1))
    # Check whether X and y have the correct shape: X.shape[0] = y.shape[0]
    X, y = check_X_y(X, y)
    # Save the unique target values
    self.classes_ = unique_labels(y)
    # Save learning data for later testing in predict method
    self.X_ = X
```

```python
        self.y_ = y
        # Learn
        for i in range(self.n_iterations):
            # random mix, for batch size = 1
            # np.size(.,0) = number of rows
            rand_index = random_state.randint(0,np.size(X,0))
            # A random input vector
            x_ = X[rand_index]
            # A matching output
            y_ = y[rand_index]
            # Determine the calculated output:
            # Weighted sum with downstream step function
            y_hat = self.heaviside(np.dot(self.w, x_))
            # Calculate the error as the difference between the desired and
            # current output
            error = y_ - y_hat
            # Collect errors for output
            self.errors.append(error)
            # Weight adjustment = learning
            self.w += error * x_
            # Return of the estimator for linked calls
        return self
```

Listing 4.5 The "fit" Method

We should say a few more words about the `fit` method. This method receives the training data in the two-dimensional array X and the desired values in the array y. A new RNG with the *initialization value* (*seed*) `random_state` is then generated. The weights are initialized depending on the size of the input vector x. The `check_X_y` method checks whether the vectors match. The learning data is stored in object attributes, and the familiar learning iterations take place. That's it for the learning method.

Now follows the last method, `predict`, as shown in Listing 4.6. The analysis can then take place. Here, too, scikit-learn provides support methods, such as checking whether learning has already taken place.

```python
# Analyze
def predict(self, x):
    """ Analyzing a vector

    x: A test input vector

    """
    # Check whether fit has already been called
    # The data was set in the fit method
    check_is_fitted(self, ['X_', 'y_'])
```

```python
# Analyze, forward path
y_hat = self.heaviside(np.dot(self.w,x))

return y_hat
```

Listing 4.6 The "predict" Method

Of course, we can also implement methods in the estimator class that can be useful in addition to the standard methods, for example, a method for outputting the error curve. We've called this method `plot` (see Listing 4.7).

```python
# Plot
def plot(self):
    """ Output of the error

    Output the errors stored in the error array as a graphic

    """
    # Figure numbers start
    fignr = 1
    # Print size in inches
    plt.figure(fignr,figsize=(10,10))
    # Output error as plot
    plt.plot(self.errors)
    # Grid
    plt.style.use('seaborn-v0_8-whitegrid')
    # Labels
    plt.xlabel('Iteration')
    plt.ylabel(r"$(y - \hat y)$")
```

Listing 4.7 The "plot" Method

This completes the estimator class, and we're now ready to use it. First, the training data is set up in the `main()` function, which we only create to better structure the script; then our perceptron estimator gets instantiated. We've set the number of iterations to 30 and selected 10 as the initial value for the RNG. The `fit` method receives the training data for learning, and our estimator can then be used for evaluation, as shown in Listing 4.8.

```python
def main():
    # Training data
    X = np.array([[1,0,0], [1,0,1], [1,1,0],[1,1,1]])
    y = np.array([0,1,1,1])
    # Learn
    Perceptron = PerceptronEstimator(30,10)
    Perceptron.fit(X,y)
```

```
    # Test data
    x = np.array([1,0,0])
    # Analysis after training
    print("Analysis at the end of the training:")
    for index, x in enumerate(X):
        p = Perceptron.predict(x)
        print("{}: {} -> {}".format(x, y[index],p))
    # Output graph
    Perceptron.plot()

# Main program
main()
```

Listing 4.8 Testing the Custom Estimator

The predict method analyzes the training data correctly, and the plot also shows what it should show (see Listing 4.9): Success!

```
# Output:
Analysis at the end of the training:
[1 0 0]: 0 -> 0
[1 0 1]: 1 -> 1
[1 1 0]: 1 -> 1
[1 1 1]: 1 -> 1
```

Listing 4.9 Output for the "predict" Method

Figure 4.8 shows the graphical output.

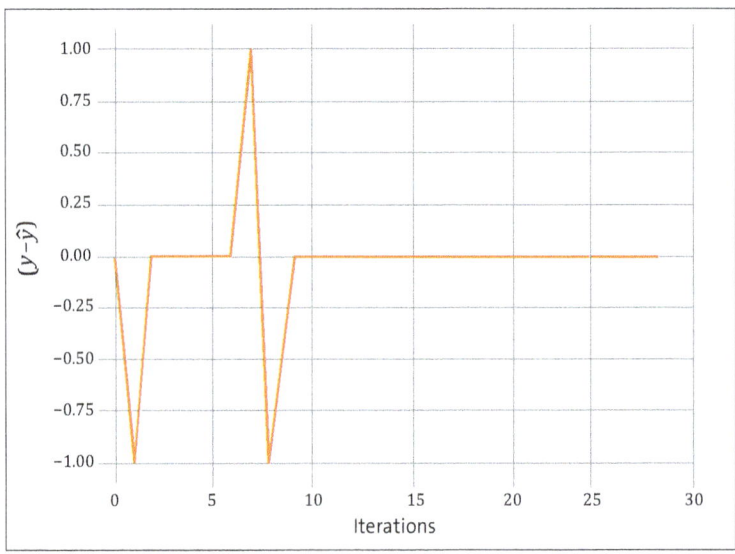

Figure 4.8 Differences in the Output of the Perceptron Estimator

That was quite a bit of work. Congratulations!

But wait a minute, is it possible that an industrious developer has already done the work and programmed a perceptron in scikit-learn? Let's take a look!

4.8 scikit-learn Perceptron Estimator

After searching for "Perceptron" and doing a little research on *https://scikit-learn.org/stable/*, it's clear that an implementation is already available in scikit-learn. But the time spent on our own implementation wasn't wasted because practice is invaluable, and you'll understand the following use of the perceptron without much explanation.

In this section, we'll use another feature of scikit-learn: the creation of datasets for learning and analysis purposes. It's best to create a new code cell in your Jupyter Notebook and start the implementation. We first import the required classes from sklearn to create the datasets and use the perceptron. This is followed by data generation, learning, and graphical analysis, as shown in Listing 4.10.

```python
# Mathematics
import numpy as np
# Graphical output for data blobs and classification
import matplotlib.pyplot as plt
# Data generation
from sklearn import datasets
# Ladies and Gentlemen - the Perceptron
from sklearn.linear_model import Perceptron

# Create data for learning with the help of scikit-learn
# We create two point clusters with two categories that are linearly separable
# n_samples = number of data points per category
# n_features = number of categories
# centers = number of point clusters
# random_state = seed for random generator
X, y = datasets.make_blobs(n_samples=100, n_features=2, centers=2,\
                           random_state=3)
# Classifications
# Set up a grid to analyze and draw
s = 0.02 # Step size in the grid
# Determine the 1-D arrays that represent the coordinates in the grid
# Slices are explained in Appendix A
x_min, x_max = X[:, 0].min() - 1, X[:, 0].max() + 1 # first coordinat
y_min, y_max = X[:, 1].min() - 1, X[:, 1].max() + 1 # second coordinate
```

```python
# np.arange returns an ndarray with evenly distributed values
# np.meshgrid returns coordinate matrixes of coordinate vectors
xx, yy = np.meshgrid(np.arange(x_min, x_max, s),
                     np.arange(y_min, y_max, s))
# Instantiate the perceptron
# max_iter = maximum number of iterations
# tol = termination criterion
Perceptron = Perceptron(random_state=42,max_iter=1000)
# Learn, please
Perceptron.fit(X,y)
# Analysis for all grid points; an array of grid points is created for this
purpose
# ravel() creates a 1-D array
# np.c_ creates a pair of point arrays for each grid point,
# which serve as input for the perceptron
Prediction = Perceptron.predict(np.c_[xx.ravel(), yy.ravel()])
# Display data in a plot
# Plot the point cluster first
plt.plot(X[:, 0][y == 0], X[:, 1][y == 0], 'b^') # blue triangles
plt.plot(X[:, 0][y == 1], X[:, 1][y == 1], 'ys') # yellow squares
# Conversion of the 1-D array into the grid dimensions
#                          [x_min, x_max] mal [y_min, y_max]
Prediction = Prediction.reshape(xx.shape)
# Plot the predictions
plt.contourf(xx, yy, Prediction, cmap=plt.cm.Paired)
# and show all
plt.show()
```

Listing 4.10 The scikit-learn Implementation of the Perceptron with Dataset Generation and Visualization of the Categories

First, we import numpy again for the mathematical calculations and matplotlib.pyplot for the graphical output, in which we'll draw the categories this time. Then, we get the datasets class from sklearn to be able to generate the data, and the object of desire, the *perceptron* estimator.

The dataset gets created using make_blobs(), which consists of two categories with 100 data points each. We've made sure that the data clusters can be separated linearly.

We then prepare the data for the plot and create a meshgrid, which is used to carry out the analyses of the perceptron and can subsequently be visualized.

This is immediately followed by the instantiation of the perceptron estimator, with random number seed and the number of iterations—1,000 in our example. The line after instantiation calls the fit() function to learn the perceptron, followed immediately by predict() to predict the meshgrid points. The predicted values of the perceptron are then available for the plot, which is created in the following lines of the program and which you can see in Figure 4.9.

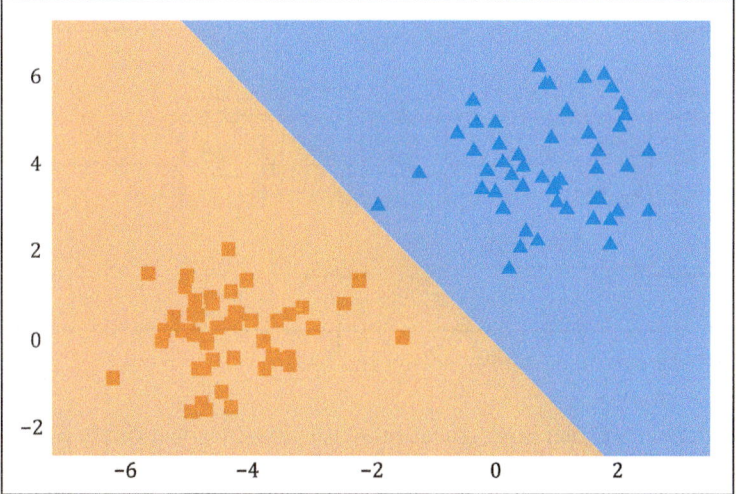

Figure 4.9 scikit-learn Perceptron Analysis and Data Points

4.9 Adaline

The problem with the perceptron is that it doesn't provide a stable solution if the problem isn't linearly separable. And to be honest, most problems in real life simply aren't linearly separable. Thus, an ANN is needed that at least remains stable, even if it can't solve a task perfectly. In this case, stability means that the ANN doesn't immediately forget everything it has already learned. Shortly after the development of the perceptron, Bernard Widrow and Marcian Hoff proposed such an ANN they called *Adaline*, named after the terms *adaptive linear*. (You may also see these versions of the name: *ADAptive LInear Neuron* or *ADALINE*, or *ADAptive LInear Element*.)

Adaline is based on the perceptron, but it uses a different type of error calculation called *root mean square* (RMS).

The error correction still determines how much the weights need to be changed, that is, Δw_i. Instead of the binary output ŷ, Widrow and Hoff use *s*, that is, continuous values (see Figure 4.10).

4 Learning in a Simple Network

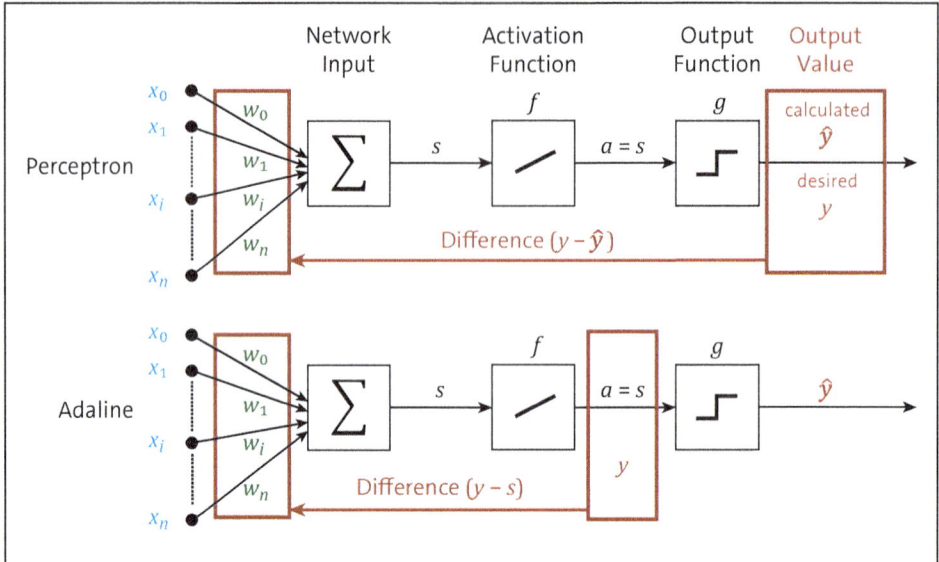

Figure 4.10 Difference Determination with Perceptron and Adaline

This allows them to achieve a much finer gradation. With perceptron learning, you've seen in Table 4.4 that only three error values are available for the correction of the weights: -1, 0, 1. In the case of Adaline learning, we can instead continuously approach the ideal weights.

Thus,

$\Delta w_i = \eta \cdot (y - \hat{y}) \cdot x_i$, or

$\Delta w_i = (y - \hat{y}) \cdot x_i$, for $\eta = 1$

becomes

$\Delta w_i = \eta \cdot (y - s) \cdot x_i$

The implementation of Adaline is similar to that of the perceptron, but the *learning rate* η (Greek letter referred to as eta) is also introduced, which defines how strong the influence of the error values is on the new weights.

Learning Rate

The value of the learning rate is generally selected from the (0,1] interval. The smaller the value of the learning rate, the smaller the effect of the error correction; the larger the value, the greater the effect. Figure 4.11 illustrates the effects of selecting the value of the learning rate.

The error is represented using a quadratic function that has an error value depending on the weight selected, for example, the error 9 for weight -3. Learning is about correcting the weight, and depending on how the learning rate is selected, there may be an optimal change toward the error value 0 or other changes with error values not equal to 0. We've illustrated the effects of the choice of learning rate in Figure 4.11. Further information can be found in Chapter 6.

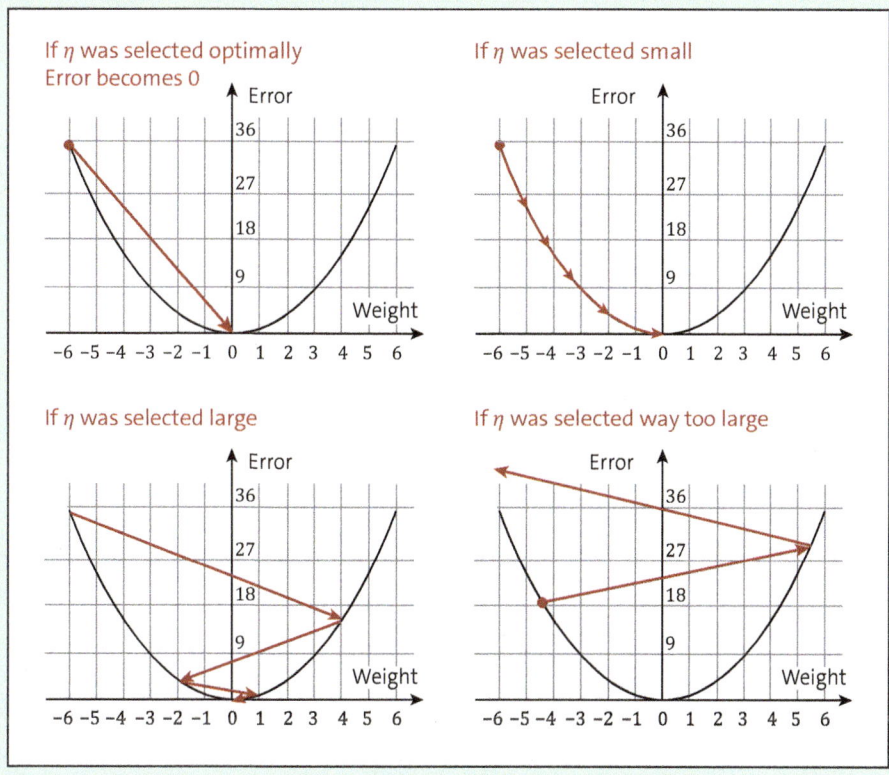

Figure 4.11 Effects of the Choice of Learning Rate on Error Minimization

The error calculation is based on *s*; everything else can remain as with the perceptron.

Let's take a look at the change in the code. This is again implemented according to the scikit-learn estimator pattern, which means that the `__init__`, `fit`, and `predict` methods are available. The code is a little longer, but you already know parts of the individual passages.

As an application example, we create a random dataset with two classes in the `main()` section (see Listing 4.11), where we can control how well the classes are linearly separable.

```python
# Numpy helps us with the arrays
import numpy as np
# Graphical display
import matplotlib.pyplot as plt
# These are our basic classes
from sklearn.base import BaseEstimator, ClassifierMixin
```

```python
# Check routines
from sklearn.utils.validation import check_X_y, check_array, check_is_fitted,
check_random_state
# Saving
from sklearn.utils.multiclass import unique_labels
# Very important, otherwise the plot will not be displayed
%matplotlib inline

# Our estimator, appropriately labeled, and the base classes
class AdalineEstimator(BaseEstimator, ClassifierMixin):
    # Initialization
    def __init__(self, eta=.001, n_iterations=500, random_state=None):
        """ Initialization of the objects
        eta:         Learning Rate
        n_iterations: Number of iterations for learning
        random_state: In order to guarantee repeatability, a
                      numpy.random.RandomState object can be constructed,
                      which was initialized via random_state_seed-Seed
        """
        # The number of iterations
        self.n_iterations = n_iterations
        # The seed for the random generator
        self.random_state = random_state
        # Buffer the errors in the learning process for the plot
        self.errors = []
        # The learning rate
        self.eta = eta
        # Weights for the calculation in the ANN
        self.w = []
        # All weights for plot, for drawing the separating line
        self.wAll = []
```

Listing 4.11 Declaration of the AdalineEstimator Using the __init__ Method

In contrast to the `PerceptronEstimator`, the `AdalineEstimator` also requires the learning rate, which is passed to the `__init__` method. Furthermore, we buffer the data of the weights to draw the separating lines later in the plot and thus visualize their change in position during the training, as shown in Listing 4.12.

```python
# The weighted input
def net_i(self, x):
    """ Calculate the weighted input w*x
    x: A vector
    """
    return np.dot(x, self.w)
```

```python
# Activation function
def activation(self, x):
    """ Linear activation function
    """
    return self.net_i(x)


# Output function, where the output can be 1 and -1,
# in contrast to the perceptron, where 1 and 0 are output
def output(self, x):
    """ Output function
    """
    if self.activation(x) >= 0.0:
        return 1
    else:
        return -1
```

Listing 4.12 Weighted Input, Activation, and Output as Separate Methods

In this implementation, we've mapped the calculation of the weighted input, activation, and output as separate methods, as shown in Listing 4.13.

```python
# Learn
def fit(self, X=None, y=None):
    """ Train
    X: Array-like structure with [N,D], where
        N = rows = number of learning examples and
        D = columns = number of features
    y: Array with [N], with N as above
    """
    # Generation of the random number generator (RNG)
    random_state = check_random_state(self.random_state)
    # Initialization of the weights
    # np.size(.,1) = number of columns
    self.w = random_state.random_sample(np.size(X,1))
    # Check whether X and y have the correct shape: X.shape[0] = y.shape[0]
    X, y = check_X_y(X, y)
    # Store learning data for later use
    self.X_ = X
    self.y_ = y
    # Learning with gradient descent
    for i in range(self.n_iterations):
        # random mix, for batch size = 1
        # np.size(.,0) = number of rows
        rand_index = random_state.randint(0,np.size(X,0))
```

```python
        # A random input vector
        x_ = X[rand_index]
        # A matching output (+1,-1)
        y_ = y[rand_index]
        # calculate net input s
        s = np.dot(x_, self.w)
        # Calculate error as the square of the difference between
        # desired output and net input
        error = (y_ - s)**2
        self.errors.append(error)
        # Online Adaline learning, as described
        self.w += self.eta * x_ * (y_ - s)
        # .copy() copies the list
        self.wAll.append(self.w.copy())
```

Listing 4.13 Learning in Adaline

In contrast to perceptron learning, in Adaline learning, the correction is based on the difference between the desired output and the input, as can be seen in the line after the comment # Online Adaline learning in Listing 4.14 and as we've previously discussed and shown in Figure 4.10.

```python
    # Analyze
    def predict(self,x):
        """ Analyzing a vector
        x: A test input vector
        """
        # Check whether fit has been called
        # The data was set in the fit method
        check_is_fitted(self, ['X_', 'y_'])
        # Analyze, forward path
        y_hat = self.output(x)

        return y_hat

    # Plot
    def plot(self):
        """ Output of the error and the learning curve
        Output the errors stored in the error array as a graphic
        Output the separating lines from the stored weights
        """
        x1 = []
        x2 = []
        colors = []
```

```python
        for i in range(self.X_.shape[0]):
            x1.append(self.X_[i][1])
            x2.append(self.X_[i][2])
            y = self.y_[i]
            if y == 1:
                colors.append('r') # rot
            else:
                colors.append('b') # blue
        # Grid
        plt.style.use('seaborn-v0_8-whitegrid')
        # Errors
        plt.plot(self.errors)
        # Learning Curve
        plt.figure(1)
        plt.show()
        # Scatter
        plt.figure(2)
        plt.scatter(x1, x2,c=colors)
        # Result Line
        x1Line = np.linspace(0.0, 1.0, 2)
        x2Line = lambda x1, w0, w1, w2: (-x1*w1 - w0) / w2;
        alpha = 0.0
        for idx, weight in enumerate(self.wAll):
            # alpha = transparency, the closer to the target, the darker
            if(idx % 100 == 0):
                alpha = 1.0 #( idx / len(self.wAll) )
                plt.plot(x1Line, x2Line(x1Line,weight[0],weight[1],weight[2]),
alpha=alpha,linestyle='solid',label=str(idx),linewidth=1.5)
        # Result line
        plt.plot(x1Line, x2Line(x1Line,weight[0],weight[1],weight[2]),alpha=
alpha,linestyle='solid',label=str(idx),linewidth=2.0)
        plt.legend(loc='best', shadow=True)
```

Listing 4.14 The "predict" and "plot" Methods for Adaline

The predict method is largely the same as in the perceptron. However, the plot method differs in that, in addition to the data points, the error curve and the separating lines during the training process are also shown. You can therefore see where the separating line is after 0, 100, 200, and 299 iterations (see Listing 4.15).

```python
def main():
    # Generation of the random number generator (RNG)
    random_state = check_random_state(1)
    # Initialization of the datasets
    I = []
    o = []
```

```
        # Create datasets for two categories
        # This time without scikit-learn
        for x in random_state.random_sample(20):
            y = random_state.random_sample()
            I.append([1, x, y+1.0]) # If +0.0, then overlapping categories
            o.append(1)

        for x in random_state.random_sample(20):
            y = random_state.random_sample()
            I.append([1, x, y-1.0]) # If +0.0, then overlapping categories
            o.append(-1)

        # Training data
        X = np.array(I)
        y = np.array(o)
        # Do your duty, Estimator
        Adaline = AdalineEstimator(eta=0.01,n_iterations=300, random_state=10)
        # Learn
        Adaline.fit(X,y)
        # Output graphs
        Adaline.plot()

# Main program
main()
```

Listing 4.15 The "main" Function of Adaline

The main function again summarizes the steps required to use the estimator and supply it with data. We've created our own datasets where we can control the degree of overlap.

The output generated by calling main() is shown in Figure 4.12 and Figure 4.13.

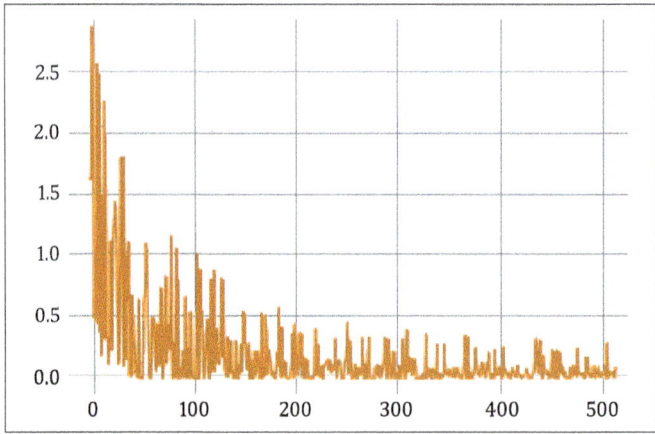

Figure 4.12 Adaline Learning Curve

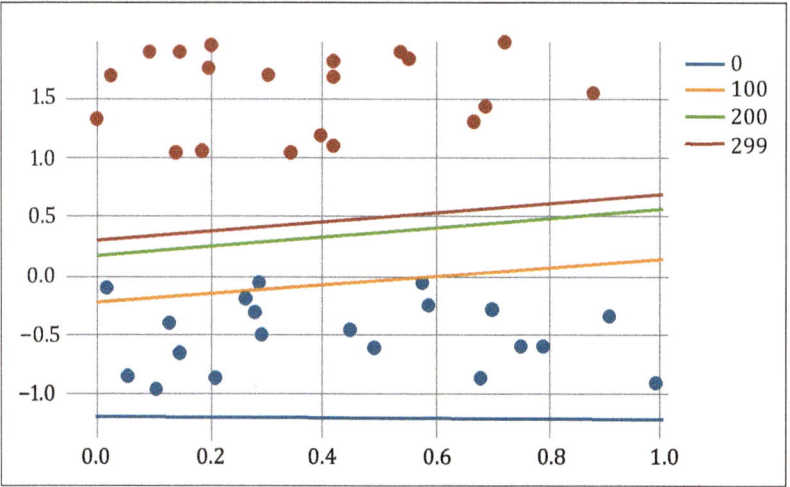

Figure 4.13 Adaline Separating Lines in Different Steps

We need to highlight two places in the code: first, let's discuss the implementation of the `fit` method. The error is now no longer determined on the basis of the outputs of the ANN as in the perceptron, but on the basis of the net input, *s*. There is a deep mathematical reason for this, which is discussed in more detail in Chapter 6 and in Appendix B. At the latest with the *backpropagation algorithm*, the hour of the *least mean square* (LMS) error strikes. Here and now we simply call it the *squared error* because the difference between the net input and the desired output gets squared:

`self.w += self.eta * x_ * (y_ - s)`

The new weight is calculated in a similar way to the perceptron, but here, the difference between the desired output `y_` and the net input `s` is decisive. You can also see that the `self.eta` learning rate is used to change the size of the contribution (refer to Figure 4.11). With a typical value of 0.01, the learning rate ensures that the weights change slowly. With a large value, for example, 0.5, the weights change abruptly, which can cause problems, for example, if a potentially better solution is virtually skipped and therefore not found. First, you should gain some practical experience with the parameters, which is why we've prepared the following tasks for you.

> **Task: Learning Rate**
>
> Change the number of iterations from 500 to 1,500 and the learning rate from 0.1 to 0.001. Observe the development of the learning curve and the position of the separating line.

Solution:

In Figure 4.14, you can see three possible results.

4 Learning in a Simple Network

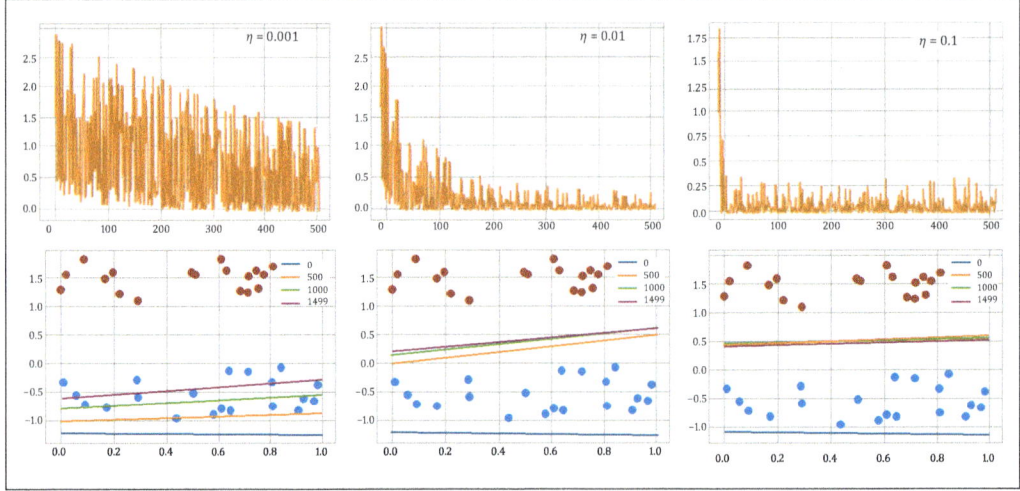

Figure 4.14 Learning Rate Comparison

Task: Change the Overlap Factor

With the value $y + 1.0$ and $y - 1.0$, you have an optimal separation, so no problem for a perceptron. Change the separating value and observe the separating lines with regard to the convergence of the position of the separating lines.

Solution:

Figure 4.15 shows possible solutions.

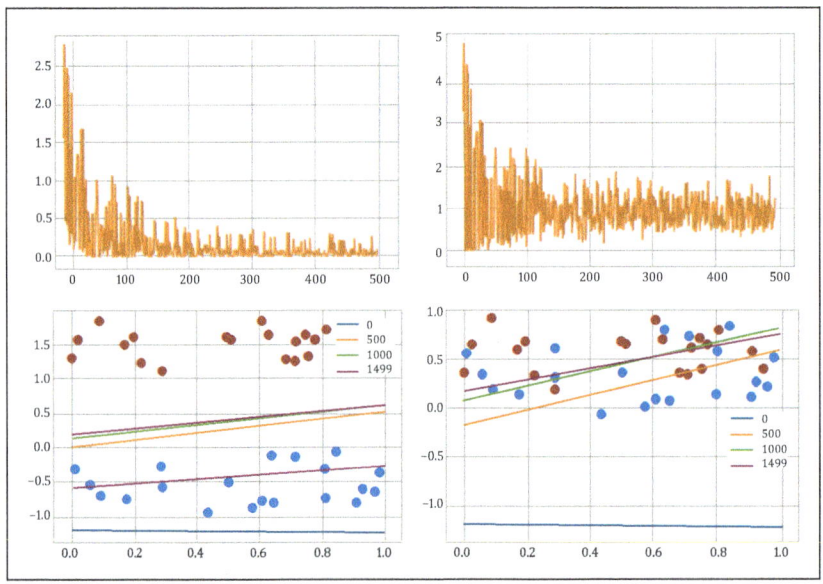

Figure 4.15 Overlap Factor of the Two Classes

4.10 Summary

In this chapter, we looked at Frank Rosenblatt's perceptron in greater detail. We hope that we've clearly shown you that a simple learning rule can achieve initial success and that perceptrons learn to learn:

$w_i^{new} = w_i^{old} + \Delta w_i$, where

$\Delta w_i = (y - \hat{y}) \cdot x_i$

Just as in the ANN history, we followed the path to Adaline, which was subsequently created and motivated by the perceptron. The main difference between the two ANNs is that in Adaline the net input is used for error calculation, in contrast to the step function in the perceptron. This made it possible for Widrow and Hoff to derive learning analytically and to describe the delta rule.

Following is Rosenblatt's delta rule calculation:

$\Delta w_i = \eta \cdot (y - \hat{y}) \cdot x_i$, or

$\Delta w_i = (y - \hat{y}) \cdot x_i$, for $\eta = 1$

Following is Widrow-Hoff's delta rule calculation:

$\Delta w_i = \eta \cdot (y - s) \cdot x_i$

These additions now allow us to further refine the abstract neuron model. Take a look at Figure 4.16 in this context.

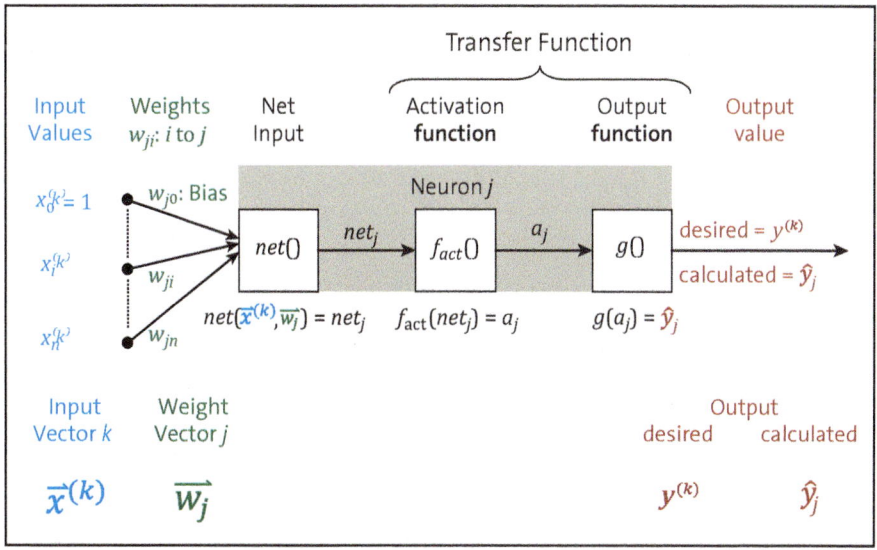

Figure 4.16 Generalized Neuron Model with Details from This Chapter

We've generally designated the calculation of the net input with the net() function, and we've highlighted the neuron in gray and designated it as **Neuron j**. The superscript (k) for input and output indicates that there can be any number of training examples.

The activation function is to be used as described in Chapter 1. In the next chapter, we'll take a look at special activation functions for multilayer ANNs.

4.11 Further Reading

- "Adaptive Switching Circuits" by Bernand Widrow and Marcian E. Hoff (1960, *https://isl.stanford.edu/~widrow/papers/c1960adaptiveswitching.pdf*
- Estimator in scikit-learn: *http://scikit-learn.org/stable/developers/contributing.html#rolling-your-own-estimator*

Chapter 5
Multilayer Neural Networks

As powerful as the perceptron and Adaline are, there are tasks they can't solve. For this reason, we'll take a look at how the network architecture can be changed so that other problem categories can also be solved.

Let's start again from the beginning. So far, we have input values that are weighted and sent to a neuron, which then carries out its calculations. Our only use case up to this point has been the classification of data into two classes with the data separated linearly.

However, here comes the killer problem that killed the perceptron, at least for a short period of time. We'll describe it again in the form of the planning problem.

5.1 A Real Problem

The task is as follows: We're in Mrs. Apple's store. Two employees, Mrs. Carrot and Mr. Leek, are available to assist each day, and their deployment must be planned. There should be *exactly one* employee in the store, as it has turned out that the store is too small for two employees, and the costs per employee aren't negligible. Let's take a look at the planning in the form of a table (Table 5.1). P1 is the first person, P2 is the second, and E is the desired result.

P1	P2	→	E
0	0	→	0
1	0	→	1
0	1	→	1
1	1	→	0

Table 5.1 A Hidden XOR Problem

At first glance, Table 5.1 looks innocent, but if you take a closer look, this innocence disappears very quickly. Figure 5.1 shows why.

5 Multilayer Neural Networks

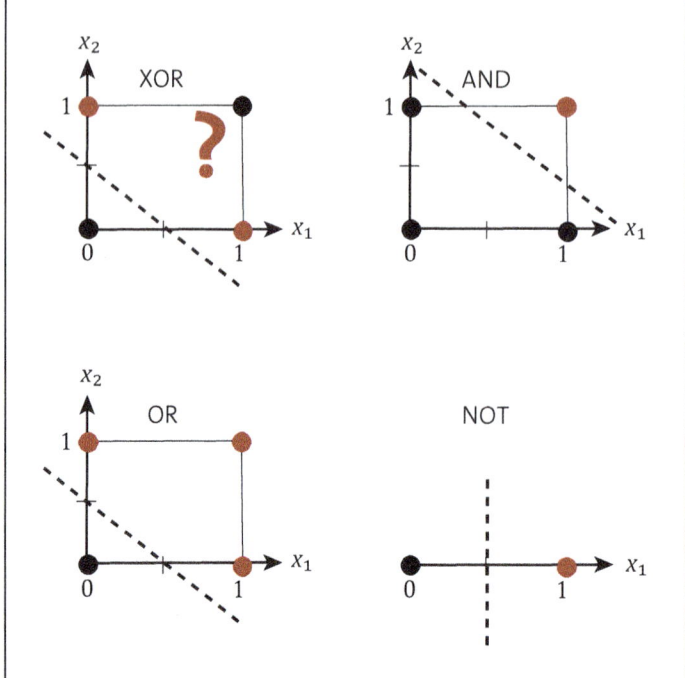

Figure 5.1 Planning Task to Be Solved

How would you solve the XOR task using a ruler? You have exactly one line available, as a perceptron represents exactly one line!

Nice to know: Truth Tables and Boolean Algebra

We've drawn four tables in Figure 5.1 showing the different combinations of zeros and ones (0,0), (0,1), (1,0), and (1,1). x_1 and x_2 can have the values 0 and 1. This is indicated by a point. Each logical operator is displayed in a separate panel. If the point is black, the value *false* (or 0) gets assigned. The red point represents the value *true* (or 1).

For example, with AND, only the combination (1,1) is assigned the value 1 (red point); otherwise, the value is 0 (black point). This brings us to *Boolean algebra*, which forms the basis of electronics and is used in all modern programming languages.

Task: Logical XOR

If you want, you can try to represent XOR using NOT, AND, and OR (see Table 5.2).

Solution:

(NOT x_1 AND x_2) OR (x_1 AND NOT x_2)

x_1	x_2	NOT x_1	NOT x_2	(NOT x_1 AND x_2)	(x_1 AND NOT x_2)	OR
0	0	1	1	0	0	0
0	1	1	0	1	0	1
1	0	0	1	0	1	1
1	1	0	0	0	0	0

Table 5.2 Truth Table for the XOR Problem

You're right, unfortunately, the problem can't be solved that way, and that hasn't helped the perceptron's popularity. It actually led to a fundamental loss of confidence and research funding, which had a negative impact on research in the field of ANN for a long time. However, there's good news: the XOR problem can be solved.

5.2 Solving XOR

Step by step, we'll work out a solution to the XOR problem and thus arrive at a completely new type of ANN. Consider the following idea: If the XOR task can't be separated into two groups by a line, then, in any case, it can be separated by an additional second line, as outlined in Figure 5.2. The two lines allow us to form two groups, which is just what we wanted. Be creative and draw your own solutions for two lines that also perform the desired grouping.

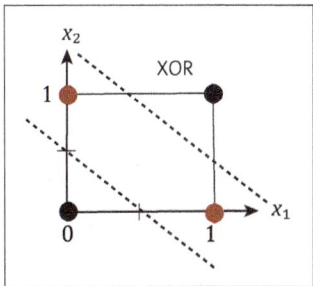

Figure 5.2 Solved XOR Problem

But how can we use our knowledge of the perceptron for this consideration? Let's try the following: we're now using not one but two perceptrons, which are fed from the same inputs, as shown in Figure 5.3.

The two perceptrons—Perceptron 1 and Perceptron 2 in the figure—should now provide the two categories 0 and 1 as the desired outputs. Figure 5.4 shows what we expect from the neurons. The left-hand neuron should learn OR, while the right-hand neuron should learn NOT AND.

5 Multilayer Neural Networks

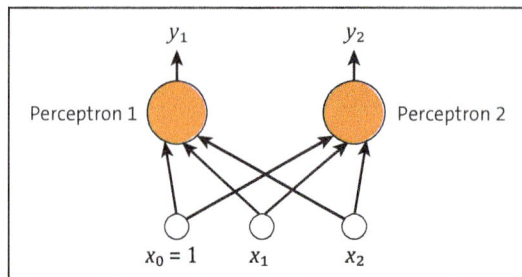

Figure 5.3 Two Perceptrons

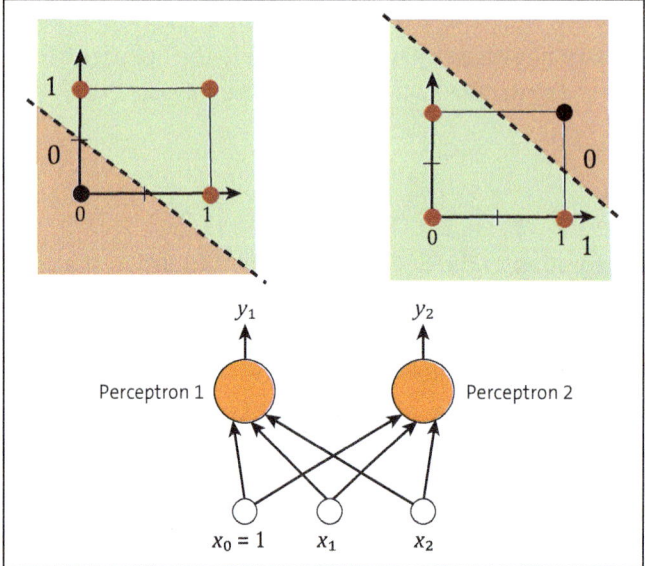

Figure 5.4 Learning Tasks for Perceptron 1 and Perceptron 2

With these requirements, the truth table for the learning content of the perceptrons looks like Table 5.3.

x_1	x_2	Perceptron 1 (y_1)	Perceptron 2 (y_2)
0	0	0	1
1	0	1	1
0	1	1	1
1	1	1	0

Table 5.3 Perceptron 1 and Perceptron 2

To reach an overall decision, we need an additional perceptron that combines the outputs of the two perceptrons (see Table 5.4). This means that Perceptron 3 receives the outputs as inputs and learns to map them.

x_1	x_2	Perceptron 1 (y_1)	Perceptron 2 (y_2)	Perceptron 3 ($\hat{y}$)
0	0	0	1	0
1	0	1	1	1
0	1	1	1	1
1	1	1	0	0

Table 5.4 Perceptron 3

Of course, we create another drawing (see Figure 5.5) so that we can better see what Perceptron 3 has to learn.

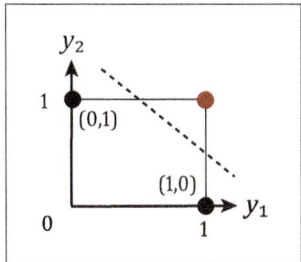

Figure 5.5 Learning by Perceptron 3

In any case, if you take a close look at the Perceptron 1 and Perceptron 2 columns and look carefully at the table rows, you'll already notice one thing. Here's a hint: Are all value combinations of 0 and 1 available?

Correct, the point (0,0) was eliminated by selecting perceptron 1 and 2. The best way to recognize this is to write down the possible input vectors (Perceptron 1, Perceptron 2):

- Row 1: (0,0)
- Row, 2, 3: (1,0)
- Row 4: (1,1)

With this insight, we can finalize the network. Figure 5.6 shows a possibility for the weight values. Such *multilayer* networks are called *multilayer perceptrons* (MLPs). A layer of neurons passes the data on to the next layer.

For example, the following results are obtained for the input $(x_1, x_2) = (1,0)$:

$net_{P_1} = -10 + 20 \cdot 1 + 20 \cdot 0 = 10 \rightarrow 1$
$net_{P_2} = 30 + (-20) \cdot 1 + (-20) \cdot 0 = 10 \rightarrow 1$
$net_{P_3} = -30 + 20 \cdot 1 + 20 \cdot 1 = 10 \rightarrow 1$

5 Multilayer Neural Networks

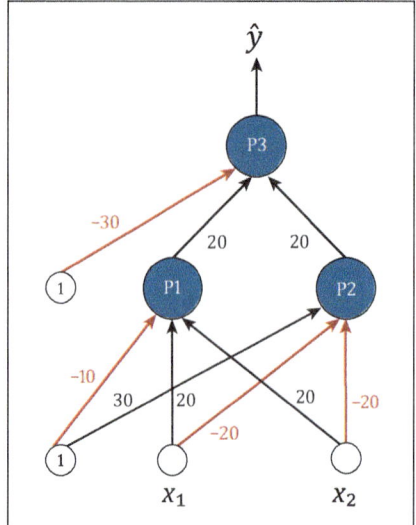

Figure 5.6 The Network for the XOR Problem

For all inputs, this results in the calculations shown in Table 5.5.

Input $(1, x_1, x_2)$	Output P1	Output P2	Output P3 ($\hat{y}$)
(1,0,0)	$-10 \cdot 1 + 20 \cdot 0$ $+20 \cdot 0 = -10$ $\to 0$	$30 \cdot 1 - 20 \cdot 0$ $-20 \cdot 0 = 30$ $\to 1$	$-30 \cdot 1 + 20 \cdot 0$ $+20 \cdot 1 = -10$ $\to 0$
(1,0,1)	$-10 \cdot 1 + 20 \cdot 0$ $+20 \cdot 1 = 10$ $\to 1$	$30 \cdot 1 - 20 \cdot 0$ $-20 \cdot 1 = 10$ $\to 1$	$-30 \cdot 1 + 20 \cdot 1$ $+20 \cdot 1 = 10$ $\to 1$
(1,1,0)	$-10 \cdot 1 + 20 \cdot 1$ $+20 \cdot 0 = 10$ $\to 1$	$30 \cdot 1 - 20 \cdot 1$ $-20 \cdot 0 = 10$ $\to 1$	$-30 \cdot 1 + 20 \cdot 1$ $+20 \cdot 1 = 10$ $\to 1$
(1,1,1)	$-10 \cdot 1 + 20 \cdot 1$ $+20 \cdot 1 = 30$ $\to 1$	$30 \cdot 1 - 20 \cdot 1$ $-20 \cdot 1 = 10$ $\to 0$	$-30 \cdot 1 + 20 \cdot 1$ $+20 \cdot 0 = -10$ $\to 0$

Table 5.5 Determining the Outputs of Different Neurons

Task: Find Your Own Weights for the MLP

Find other weight assignments to solve the XOR problem. Using paper and pencil or Microsoft Excel here is helpful.

You could use Figure 5.4 as a solution strategy for the task at hand. Any separating line that separates the input space, as for Perceptron 1 and 2, can be used as a solution, and the weights can be determined.

Appendix B, Section B.1.6, explains the calculation steps in more detail.

Solution:

As before, Figure 5.7 represents a MLP with three perceptron units, labeled as P1, P2, and P3. The only difference to before is the range of weights. Table 5.6 shows the calculation of the output based on the given input vector.

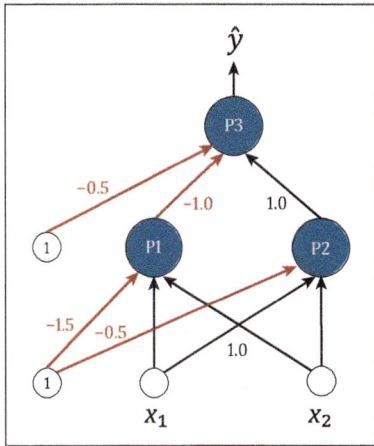

Figure 5.7 Proposed Solution for the Weights in the MLP

Input $(1, x_1, x_2)$	Output P1	Output P2	Output P3 ($\hat{y}$)
(1,0,0)	$-1.5 \cdot 1 + 1.0 \cdot 0$ $+1.0 \cdot 0 = -1.5$ $\rightarrow 0$	$-0.5 \cdot 1 + 1.0 \cdot 0$ $+1.0 \cdot 0 = -0.5$ $\rightarrow 0$	$-0.5 \cdot 1 - 1.0 \cdot 0$ $+1.0 \cdot 0 = -0.5$ $\rightarrow 0$
(1,0,1)	$-1.5 \cdot 1 + 1.0 \cdot 0$ $+1.0 \cdot 1 = -0.5$ $\rightarrow 0$	$-0.5 \cdot 1 + 1.0 \cdot 0$ $+1.0 \cdot 1 = 0.5$ $\rightarrow 1$	$-0.5 \cdot 1 - 1.0 \cdot 0$ $+1.0 \cdot 1 = 0.5$ $\rightarrow 1$
(1,1,0)	$-1.5 \cdot 1 + 1.0 \cdot 1$ $+1.0 \cdot 0 = -1.5$ $\rightarrow 0$	$-0.5 \cdot 1 + 1.0 \cdot 1$ $+1.0 \cdot 0 = 0.5$ $\rightarrow 1$	$-0.5 \cdot 1 - 1.0 \cdot 0$ $+1.0 \cdot 1 = 0.5$ $\rightarrow 1$
(1,1,1)	$-1.5 \cdot 1 + 1.0 \cdot 1 + 1.0 \cdot 1 = 0.5$ $\rightarrow 1$	$-0.5 \cdot 1 + 1.0 \cdot 1$ $+1.0 \cdot 1 = 1.5$ $\rightarrow 1$	$-0.5 \cdot 1 - 1.0 \cdot 1 + 1.0 \cdot 1 = -0.5$ $\rightarrow 0$

Table 5.6 Proposed Solution for the Weighting in the MLP

Note and Outlook

Many layers are used in the more complex networks. These networks are then referred to as *deep neural networks*. You can find out more about *convolutional neural networks* (CNNs) in Chapter 7.

5.3 Preparations for the Launch

Great, we have a solution. To be able to carry out the calculation efficiently, we'll now consider how we can implement the network in Python. To do this, we rotate the network so that the flow of calculation is not from the bottom to the top, as before, but from the left to the right. So, in Figure 5.8, data no longer flows through the network from bottom to top, but from left to right. There's a simple reason for this: we want to present the calculations as simply as possible, and this works better in the horizontal position.

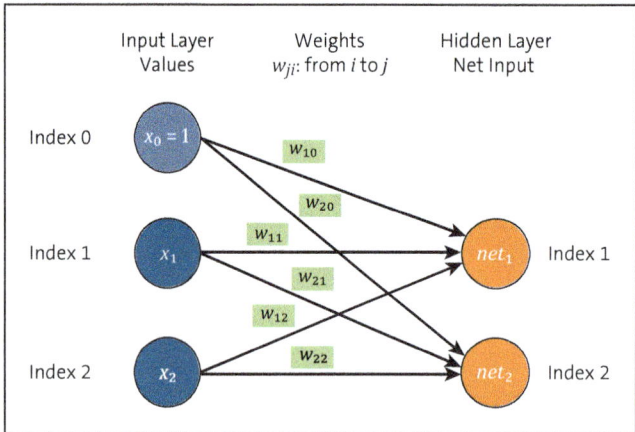

Figure 5.8 The Calculation in the Multilayer Network

In Figure 5.8, you can see the tilted network, or at least the transition from the input to the hidden layer. We want to start with the input layer and input values on the far left. The representation already looks like a *vector*, where x_0 always has the value 1, which we discussed in Chapter 4. Then you see the green boxes, which are the *weights* w_{ji}, above the connections from the input neuron to the hidden neuron, where i is the index of the input neuron and j is the index of the hidden neuron. We've used the term net_j for the hidden neuron to show that it's the calculation of the *weighted sum*, that is, the net input. For example, net_1 is calculated from $w_{10} \cdot x_0 + w_{11} \cdot x_1 + w_{12} \cdot x_2$. We can write this a little more elegantly for all calculations, as shown in Figure 5.9.

Figure 5.10 shows a somewhat more structured representation of the calculation than is possible with the graphical representation of the network. But it doesn't actually contain any new information. However, this representation is an important step toward an efficient calculation algorithm for analysis in the network.

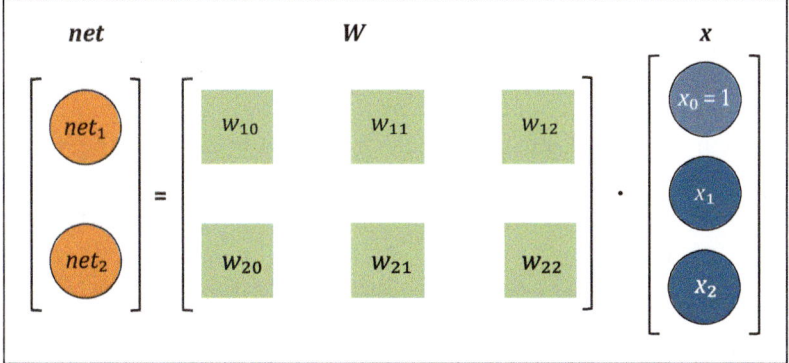

Figure 5.9 On the Way to Matrix Multiplication

Figure 5.10 Matrix Multiplication: Great for NumPy

Finally, this results in the NumPy calculation that we need for the algorithm:

net = numpy.dot(W,x)

We've already used the `numpy.dot()` function with the perceptron. There we've multiplied two vectors with each other. And `numpy.dot()` is so flexible that you can multiply several vectors at once. By the way, mathematicians refer to the weight vectors of Figure 5.10 in the illustration as a *matrix*. (You can find more details about them in Appendix B.) We had to mention the term because it occurs very often and will continue to do so, and because it's also reminiscent of an iconic movie trilogy.

5.4 The Plan for Implementation

The diagram in Figure 5.11 serves as a plan for the implementation. The letters I, H, and O denote the *input*, *hidden*, and *output* layers. W_HI are the weights from the input layer (2 + 1 neurons) to the hidden layer (2 + 1 neurons). W_OH stands for the weights from the hidden layer (2 + 1 neurons) to the output layer (1 + 1 neuron). The 1 in the number of neurons always comes from the bias neuron at index position 0.

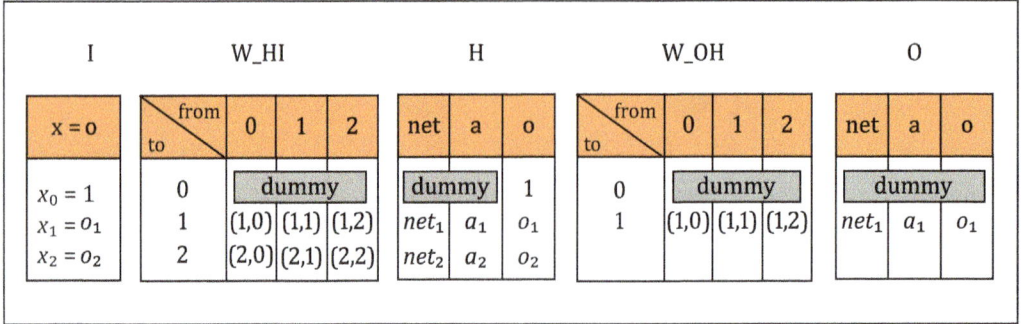

Figure 5.11 Network Components

The numbers denote the indexes for the subsequent arithmetic operations. The weights in the matrix are arranged in exactly the way we described earlier. The "from" and "to" in the matrix rows and columns mean that a weight applies to a connection that leads from a neuron to a neuron; for example, the weight w_{10} denotes the weight of the connection from neuron 0 to neuron 1. To avoid another index, we don't write separately whether the weight is in the matrix W_HI or W_OH. This should be clear from the context.

In addition, the **dummy** in the figure means that it doesn't matter what values are in it, whereby we always use the value 0. Why's that? Think about it: the **dummy** is in the row with index **0**, which means there is a connection from any neuron x in the input layer to neuron 0 in the hidden layer. Neuron 0 is always the bias neuron with the fixed value 1. This means that the weight of such a connection is irrelevant.

The **net**, **a**, and **o** columns denote the net input values to the neuron, the activation after applying the *activation function*, and the output after the output calculation, respectively. The output of the network therefore appears in the last layer, the output layer (in the **o** column). To build on previous representations, we've shown the input x in the input layer, which is identical to the output o of the input layer. You could also say that input x is simply passed through without performing the calculations that take place in the other layers.

5.5 The Setup ("class")

We'll now implement this blueprint in Python using object orientation. The ANN is supposed to be implemented as a Python class, and because it's a multilayer neural network, we'll call the class MLP. We keep the overhead very low and define the MLP class as a shell for the most important methods: `__init__` and `predict`. As a goodie, we also implement the `print` method, which prints out the network architecture, as you've already seen in Figure 5.11. We're going to explain the implementation in chunks as we don't want to overwhelm you with the complete code.

For this purpose, you need to create a new notebook called Chapter 5 in Jupyter and use the tried-and-tested combination of Markdown and code cell again. Enter "Multilayer Perceptron" as the heading, and we start with the code in the code cell from Listing 5.1.

```python
# For the mathematical operations
import numpy as np

# The identical function
def func_id(x):
    return x

# A world-famous activation function: The sigmoid
def func_sigmoid(x):
    # Important: Not math.exp, but np.exp because of array
    # Use operations
    return 1.0 / (1.0 + np.exp(-x))
```

Listing 5.1 The Import and Function Definitions

Of course, we need NumPy, but this time for real. The dot products are calculated again between vectors and—note, this is new—also with matrixes.

Then, we've defined two functions that we'll use for the output function (`func_id`) and for the activation function (`func_sigmoid`) in the class. The sigmoid, which is often referred to as the *gooseneck function*, is one of the classic activation functions. It used to be really popular, and we'll look at it in more detail in Chapter 6 when we talk about learning. Today, however, the `rectifier` function is more popular, as it enables faster learning, for example.

> **Good to Know**
>
> Neurons in which the rectifier function is used are referred to as *rectified linear unit* (ReLU).

The `rectifier` function is defined as

$$f_{ReLU}(x) = \max(0, x)$$

In the following task, you'll use the `rectifier` function, as it will be of benefit to you later with the more complex networks (see Listing 5.2).

5 Multilayer Neural Networks

> **Task: Implement a New Activation Function**
> - Visualize the `rectifier` function using a graph.
> - Create the `func_relu(x)` function, which returns the value of the `rectifier` function. Make sure that you pass an array as a parameter, as the activation function is to be applied to several elements.
> - For example, place the code after the previously discussed function definitions but always before the subsequent class definition so that we can use the function later.

Solution:

The graph looks like that shown in Figure 5.12.

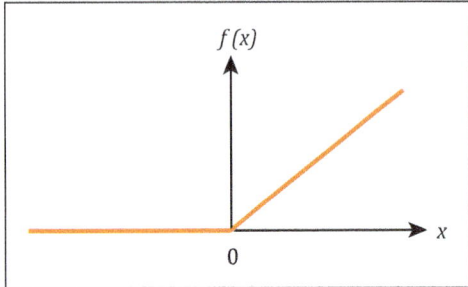

Figure 5.12 The Rectifier Function

```
def func_relu(x):
    return np.maximum(0,x)
```

Listing 5.2 Rectifier Function

5.6 The Initialization ("__init__")

Let's take another look at the code: we define the `MLP` class, which is home to the methods for building, analyzing (`predict`), and printing (`print`) the network architecture (see Listing 5.3).

In the `MLP` class, the `__init__` method is also used to initialize objects, as in all classes. We've defined some self-explanatory parameters to use to influence the appearance of the network, but the basic architecture is currently hard-wired with an input layer, a hidden layer, and an output layer.

```
# Define the network class
class MLP(object):
    """ The multilayer perceptron, MLP """
    def __init__(self,
```

```python
            n_input_neurons=2,n_hidden_neurons=2,n_output_neurons=1,
        weights=None,
        *args, **kwargs):
""" Initialization of the network
We use a fixed I-H-O structure for the start:
(Input-Hidden-Output)
The number of neurons is flexible
It is also possible to initialize the network with weights:
[W_HI,W_OH]
"""
# Activation and output function
self.f_act = func_sigmoid
self.g_out = func_id
# Number of neurons per layer
self.n_input_neurons=n_input_neurons
self.n_hidden_neurons=n_hidden_neurons
self.n_output_neurons=n_output_neurons
# Weight initialization
self.weights = weights
W_HI=[]
W_OH=[]

# All data for the network calculation is stored here
self.network=[]
# Input layer + bias neuron: Columns = o_i
self.inputLayer = np.zeros((self.n_input_neurons+1,1))
# Bias neuron output is always +1
self.inputLayer[0] = 1.0
# Add the input layer to the network
self.network.append(self.inputLayer)
# Weights from the input layer to the hidden W_HI layer
# Neuron: Row x Columns: Rows = # Hidden, Columns = # Input
# Only initialize if weights are actually present
# Simple existence check
if weights:
    W_HI = self.weights[0]
else:
    W_HI = np.zeros((self.n_hidden_neurons+1,self.n_input_neurons+1))
self.network.append(W_HI)
# Hidden layer + bias neuron: Columns = net_i,a_i,o_i
self.hiddenLayer = np.zeros((self.n_hidden_neurons+1,3))
# Bias neuron output is always +1
self.hiddenLayer[0] = 1.0
```

```python
# Add the hidden layer to the network
self.network.append(self.hiddenLayer)
# Weights from the hidden layer to the output layer W_OH
# Neuron: Row x Columns: Rows = # Output, Columns = # Hidden
if weights:
    W_OH = self.weights[1]
else:
    W_OH = np.zeros((self.n_output_neurons+1,self.n_hidden_neurons+1))
self.network.append(W_OH)
# Output layer + bias neuron: Columns = net_i,a_i,o_i
self.outputLayer = np.zeros((self.n_output_neurons+1,3))
# Bias neuron output = 0, as not relevant
# Only available due to standardized indexing
self.outputLayer[0] = 0.0
# Add the output layer to the network
self.network.append(self.outputLayer)
```

Listing 5.3 Class Definition and the __init__ Method

In the __init__ method, the network structure is set up, the weights are initialized, and the activation function (self.f_act) and output function (self.g_out) are defined by assigning functions—in our case func_sigmoid and func_id. The structure of the network corresponds to Figure 5.11. The number of neurons can be varied, whereby we suggest a 2-2-1 network as default, that is, 2 input neurons, 2 hidden neurons, and 1 output neuron, which are completely connected to each other. The layers and weights are implemented using matrixes, as the performance-optimized operations from NumPy can be used for these. The dimensions, that is, the rows and columns, depend on the number of neurons. The matrixes are inserted into the network array one after the other and can therefore be easily read later in the calculation by index.

The bias neurons at index position 0 in the I (input) and H (hidden) layer are always initialized with 1.0. The weights are either supplied with the initial weights from the parameter or initialized with zeros, and the initialization work is over.

Task: Determine the Dimensions of the Matrixes for the Initial Values

How many rows and columns are in each of the following:
- Input layer
- Hidden layer
- Output layer
- Weight matrix W_HI
- Weight matrix W_OH

Solution (compare the information with Figure 5.11):

- 3,1 (2 input neurons + 1 bias neuron, 1 output neuron)
- 3,3 (2 hidden neurons + 1 bias neuron, 1 net input + 1 activation and 1 output neuron)
- 2,3 (1 output neuron + 1 bias neuron, 1 net input + 1 activation and 1 output neuron)
- 3,3 (from 3 input to 3 hidden neurons)
- 2,3 (from 3 hidden to 2 output neurons)

5.7 Something for In-Between ("print")

To loosen things up, let's write a method that outputs the matrixes in sequence, that is, from the input layer to the output layer. In Python, the print function handles the formatting for NumPy matrixes, so we don't have to do much. The only setting we make causes three decimal places to be printed to obtain an output aligned in columns, as shown in Listing 5.4.

```python
def print(self):
    print('Multilayer Perceptron - Network Architecture')
    # 7 digits in total, output with three decimal places
    np.set_printoptions(
        formatter={'float': lambda x: "{0:7.3f}".format(x)})
    for nn_part in self.network:
        print(nn_part)
        print('----------v----------')
```

Listing 5.4 Output of the Network Architecture

You'll see the result in the next section. Although not spectacular, it is very useful for error analysis. By the way, if you want it to be more "Pythonic," you can define the def __str__(self) method, which returns the string to be output. The call is then simply print(mlp).

5.8 The Analysis ("predict")

We've done all the hard work so far for the analysis or, as we say, the predict method, as shown in Listing 5.5. Only one vector gets passed to the predict method as a parameter, which is to be analyzed and represents the input to the network. "Analyzed" here means that the calculations are carried out from layer to layer. Of course, it would be great if the dimension of the input vector matched the dimension of the input layer. Otherwise, it crashes because we haven't built in any special tests.

```python
def predict(self,x):
    """ The output y_hat is calculated for the input x
    A prediction is calculated for vector x and
    the matrix values of the layers (not the weights) are adjusted
    """
    # Input layer
    # Set the input values: All rows, column 0
    self.network[0][:,0] = x
    # Hidden layer
    # Start of line 1 due to bias neuron at index position 0
    # net_i
    self.network[2][1:,0] = np.dot(self.network[1][1:,:],
                                    self.network[0][:,0])
    # a_i
    self.network[2][1:,1] = self.f_act(self.network[2][1:,0])
    # o_i
    self.network[2][1:,2] = self.g_out(self.network[2][1:,1])
    # Output layer
    # Start of line 1 due to bias neuron to 0
    # net_i
    self.network[4][1:,0] = np.dot(self.network[3][1:,:],
                                    self.network[2][:,2])
    # a_i
    self.network[4][1:,1] = self.f_act(self.network[4][1:,0])
    # o_i
    self.network[4][1:,2] = self.g_out(self.network[4][1:,1])
    # The return is the output vector
    return self.network[4][1:,2]
```

Listing 5.5 The "predict" Method

A technique that is very often used in Listing 5.1 earlier is the *slicing* of arrays and column access, as in the following lines:

```
# Set the input values: All rows, column 0
    self.network[0][:,0] = x
```

We use the input layer, that is, the matrix, at index position 0 in the network array. All rows and one column are affected. If you want to restrict the rows or columns, you can do this by specifying an index before and after the colon (start:end). For example, if we're only interested in the output values in the input layer without a bias neuron, then that would be self.network[0][1:,0]. You can find out more about slicing in Appendix A.

In the implementation of the method, the *net input* for the hidden layer gets calculated first. It's calculated by multiplying the weights W_HI by the output of the hidden layer. We've been through this several times before. (Good news, by the way: It probably doesn't get any more complicated than that.)

Once the net input has been determined, the activity values are calculated using the *activation function*. We use the sigmoid. Finally, the output value gets calculated from the activity, whereby in our case, the output value always matches the activity value, as the *identical function* is used for the calculation.

The procedure is repeated for the next layer. In our 2-2-1 network, this is already the output layer, but it could be the next hidden layer in more complex networks. As a result, the predict method then returns the array with the output values to the caller using return.

> **Task: Dynamize the Activation Function**
>
> Find the places where the func_sigmoid function is assigned to the activation function variable self.f_act, and replace the func_sigmoid function with the func_relu function.

Solution:

```
self.f_act = func_relu
```

Once you've changed the activation function and started the program, you'll notice that the calculated output no longer corresponds to the desired output. This example clearly shows that the selected or learned weights must match the activation function. We'll take a closer look at the adjustment of the weights in the MLP in the next chapter.

OK, now we have everything we need for the evaluation, including setting up a new MLP. On to the next step: using our masterpiece.

5.9 The Usage

This is very easy! First, the network gets initialized, and then the predict method is called (see Listing 5.6). We implemented the initialization with weights that we had previously determined as a possible solution to the XOR problem. It's important to note that the weights apply to the sigmoid activation function, not to the relu!

```
def main():
    # Initialization of the weights
    W_HI = np.matrix([[0.0,0.0,0.0],[-10,20.0,20.0],[30,-20.0,-20.0]])
    W_OH = np.matrix([[0.0,0.0,0.0],[-30,20.0,20.0]])
```

5 Multilayer Neural Networks

```
    weights=[]
    weights.append(W_HI)
    weights.append(W_OH)
    nn = MLP(weights=weights)
    # Output network
    nn.print()
    # Test
    X=np.array([[1.0,1.0,1.0],[1.0,0,1.0],[1.0,1.0,0],[1.0,0,0]])
    y=np.array([0,1.0,1.0,0])
    print('Predict:')
    for idx,x in enumerate(X):
        print('{} {} -> {}'.format(x,y[idx],nn.predict(x)))

# Main program
main()
```

Listing 5.6 Using the MLP Class

If you execute the preceding code, the network structure will be output with the matrixes, and the prediction for our XOR learning examples will be provided, as shown in Listing 5.7.

```
# Output:
Multilayer Perceptron - Network Architecture
[[ 1.000]
 [ 0.000]
 [ 0.000]]
----------v----------
[[  0.000   0.000   0.000]
 [-10.000  20.000  20.000]
 [ 30.000 -20.000 -20.000]]
----------v----------
[[ 1.000  1.000  1.000]
 [ 0.000  0.000  0.000]
 [ 0.000  0.000  0.000]]
----------v----------
[[  0.000   0.000   0.000]
 [-30.000  20.000  20.000]]
----------v----------
[[ 0.000  0.000  0.000]
 [ 0.000  0.000  0.000]]
----------v----------
Predict:
[ 1.000  1.000  1.000] 0.0 -> [ 0.000]
```

```
[   1.000    0.000    1.000] 1.0 -> [   1.000]
[   1.000    1.000    0.000] 1.0 -> [   1.000]
[   1.000    0.000    0.000] 0.0 -> [   0.000]
```

Listing 5.7 Output of the Architecture with Weights and the Prediction

There you go!

5.10 Summary

This chapter has pushed you a huge step forward. Starting from one perceptron, we built a network of multiple perceptrons that was able to solve the common XOR problem. Not only have you built up this network, you've also developed your own activation function. By the way, hopefully some exciting NumPy aspects have stuck. At the very least, you've already had intensive contact with vector and matrix multiplications.

Thinking about activation functions led us to build a feed-forward network that used the sigmoid as an activation function. As mentioned, this use will be of great benefit to us in the next chapter, as we can use it to implement learning in the multilayer network.

Chapter 6
Learning in a Multilayer Network

A lifetime of learning, over and over again, is almost like the backpropagation algorithm.

Understanding neural networks requires some programmable math, which is provided in this chapter: gradient descent, feed-forward calculation, backpropagation up to the output layer, backpropagation up to the hidden layer, and correction of weights. Whether it's backpropagation at a snail's pace or as fast as a rocket, a little math won't hurt.

6.1 How Do You Measure an Error?

"Error" is a word that doesn't have many positive connotations. But in our task of teaching the artificial neural network (ANN) to learn, it's great because it gives us the opportunity to show the network how to learn. The error here is the measure, which determines how the weights must be adjusted. We'll answer the following question in the upcoming pages: How can a measured error be used to program learning for an ANN?

Let's first consider how to measure an error or a deviation from a target value. So far, we've seen from the output layer of our networks that there is a desired output y and an actual output $\hat{y}$. In Chapter 4, we measured the error of the perceptron using $E = (y - \hat{y})$, that is, the desired output of the perceptron minus the calculated output, whereby the calculated output was determined using the step function. Using Adaline, we measured the error differently, namely using $E = (y - s)$, where s is the net input. Any calculation could be used to calculate the error; there is no reason why the error shouldn't be described as

$$E = (y - \hat{y})^2$$

with the ulterior motive of keeping the error between the desired and calculated value as small as possible. If we compare the two error measurement methods, that is, $(y - \hat{y})$ and $(y - \hat{y})^2$, the graph looks like the one shown in Figure 6.1.

With the perceptron learning rule, the behavior is actually quite logical. Our goal is for the error to be 0. If the difference between the desired and calculated value is < 0, then the calculated value is greater than the desired value. This means that we have to reduce the weights so that the difference is smaller, that is, greater in terms of value. The reverse situation applies as well. So, we either climb up the line or slide down it.

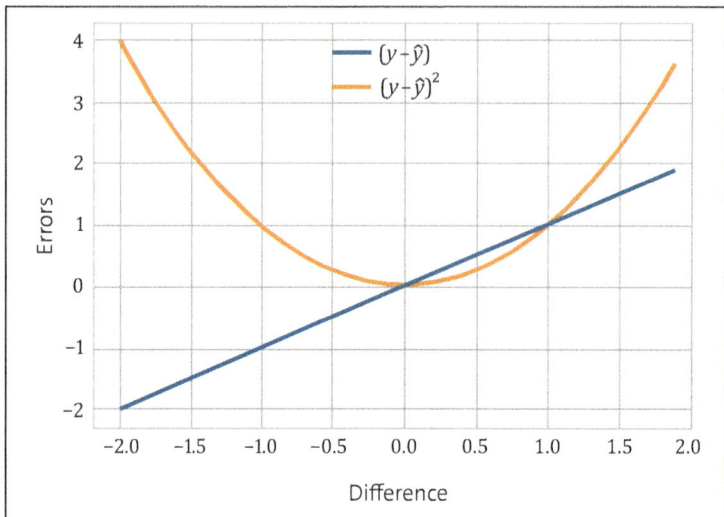

Figure 6.1 Error Curves

If we look at the square of the error measurement, which plays a central role in the derivation of the Adaline learning rules, we discover a completely different situation, which is caused by the square of the error (experts refer to this as the *squared error* [SE]). How can we proceed here so that the error becomes 0? If we were a sphere, things would be easy: We would simply switch off our brains and let gravity and friction take over. It would go down on one side and maybe up again on the other, but each time back and forth, we would get closer to 0—friction and gravity would take care of that.

Unfortunately, gravity and friction aren't available to us to determine in which direction it goes down, but we have another means, *gradient descent*. "Descent" is relatively clear, like in mountaineering, but what about *gradient*? Perhaps an alternative name for the gradient descent would help: the *method of steepest descent*. This mathematical process helps us find the direction in which there are fewer errors. Don't panic because of the word "mathematical"; we'll simply present the results from the mathematical foundations and accept them as given. For the brave and intrepid among you, we've provided a brief excursion into the mathematical principles of gradient descent in Section 6.2.

In short, the learning method based on gradient descent is the *backpropagation algorithm* (*backprop* for short), which is a generalization of the *delta rule*. This algorithm ensures that the weights are adjusted in such a way that the overall error E gets minimized. This means that if we want to use the backprop, we also need to know how to calculate the descent (i.e., the gradients). We'll make this information available to you in the next section.

6.2 Gradient Descent: An Example

For the frightened among you, we'll show you the gradient descent in pictures in this section. This approach plays too central a role for us to simply acknowledge it by saying "that's the way it is."

6.2.1 Gradient Descent: The Concept

The basic concept of gradient descent is simple: find the direction in which the error decreases. Our aim is to present the concept and the algorithm of the gradient descent graphically to develop a picture for you and let it arise in your mind. This will then help you to develop an intuitive feel for the algorithms in the more complex topics.

Take, for example, the concept of the gradient, which is used permanently. What does this word mean, and what is its underlying concept? As always, it helps to look at the original meaning of the word. In the case of *gradient*, the origin lies in the Latin *gradiens*, meaning "rise", "fall," or "slope." So, if I'm a mountaineer climbing up a mountain, I'm more interested in the ascent. If I'm a driver with a huge trailer on my way to the ocean and I'm in the Appalachians, just before a descent, then I'm more interested in the gradient aspect of the slope, which might be 15%.

Take a look at Figure 6.2 where we've drawn three arrows, *t*, *u*, and *v*, which point downward and start from the same starting point. Remember, we want to minimize the error, so we're interested in the gradient. You probably remember from your driving test that the gradient is calculated as the ratio of height to distance.

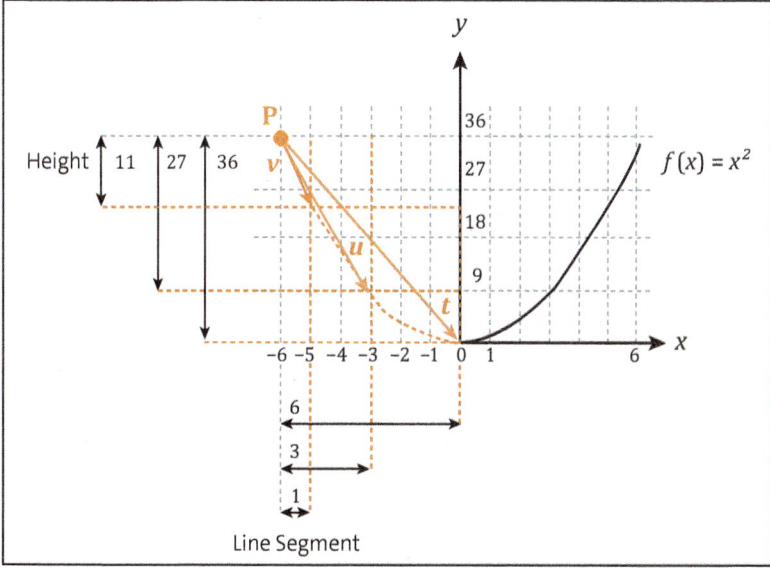

Figure 6.2 The Gradient

6 Learning in a Multilayer Network

If we compile the gradient for t, u, and v in a table, you'll make an interesting observation. Take a look at the Gradient column in Table 6.1.

Vector	Height	Distance	Gradient (Height/Distance)
t	33	6	6
u	27	3	9
v	13	1	13
	5.75	0.5	11.5
	1.19	0.1	11.9
	Smaller and smaller	Smaller and smaller	12

Table 6.1 The Approximation: Step by Step Toward the Gradient

As you can see, the gradient approximates to the value 12, so we can say that the gradient at the point -6 has the value 12. As mathematicians, we would say that the value of the gradient or the *first derivative* of $f(x) = x^2$ at the point -6 has the value -12. We've thus intuitively discovered how to determine the value of the steepest descent.

Task

Create the gradient table for the function $f(x) = x^2$, starting at the point $x = 0$ and calculating the gradient for the distances 6, 3, 1, and 0.1.

Solution:

We leave the table to you, but the gradient should approximate 0.

In the previous task, you determined the gradient for a point at which the minimum of the function is located, and this is 0. This finding therefore also represents our defined goal for determining the minimum point: At which point of the error function is the value of the gradient equal to 0?

6.2.2 Algorithm for the Gradient Descent

Of course, we have an answer to this question. Let's take another look at the algorithm using a diagram (see Figure 6.3). Compared to before, we're now changing the names so that we move in the direction of neural networks.

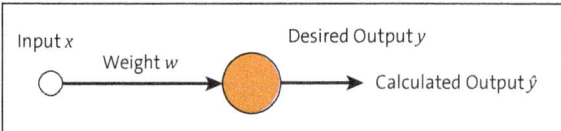

Figure 6.3 Simple ANN

Let's start with the ANN: It's really simple, there is only one input, one neuron, and the identical function $f(x) = x$ as the activity function. And let's assume that the network is presented with $x = 0.2$ as input, and we want to have $y = 0.2$ as output. Figure 6.4 shows the error caused by the changes in the weight w:

$$E(w) = \frac{1}{2} \cdot (y - \hat{y})^2$$

$E(w)$ is the *cost function* (or *loss function*), which we want to minimize by changing the weight w. This in turn means that we want to find the w for which $E(w)$ is minimal. The factor $\frac{1}{2}$ is added to simplify the mathematical calculation. The details of this and the following explanations can be found in Appendix B, Section B.2.4.

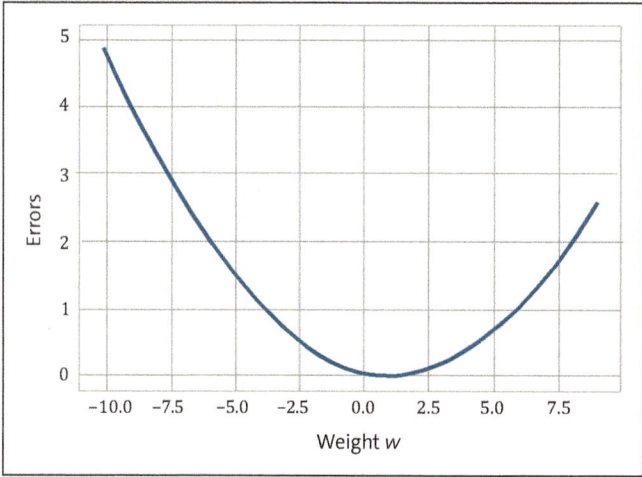

Figure 6.4 Error Depending on Weight w

From the graph in Figure 6.4, you can see that the error is 0 for the weight $w = 1$. We now want to determine the value using the *gradient descent method*, which can be summarized very simply as follows:

1. Start with an arbitrary value for the weight w.
2. Determine the negative gradient for this weight. As described previously, it points to the lowest point.
3. Move along the negative gradient toward the minimum, but only for a certain step length (as connectionists, we call this the *learning rate*).
4. At the new point that we've determined, everything starts again from the beginning (go to step 2), unless we've reached an error that is sufficient for us, for example, less than 0.001, or if the error can no longer be reduced.

We still need to divulge the formula for determining the gradient at point w. For the following function,

$$E = \frac{1}{2} \cdot (y - \hat{y})^2 = \frac{1}{2} \cdot (y - w \cdot x)^2$$

the gradient is

$$\nabla E(w) = \frac{1}{2} \cdot 2 \cdot (y - w \cdot x) \cdot (-1) \cdot x = (-1) \cdot x \cdot (y - \hat{y})$$

The ∇ symbol indicates that this is the gradient. It has an interesting name: *nabla*. For our special case with a weight and an input value, you probably know a different name and a different symbol for the gradient. First, the gradient is referred to as the *first derivative*, and second, an apostrophe is used instead of the nabla symbol. Then, the formula looks like this:

$$E'(w) = (-1) \cdot x \cdot (y - \hat{y})$$

Because we don't want to go in the direction of the strongest rise, but in the opposite direction, we still have to multiply the gradient $E'(w)$ by (-1):

$$(-1) \cdot E'(w) = (-1) \cdot (-1) \cdot x \cdot (y - \hat{y}) = x \cdot (y - \hat{y})$$

The background stories on trailer driving from the previous section are only of limited relevance for the implementation of the algorithm for gradient descent. Basically, it's sufficient to know that the weight correction is made by adding $x \cdot (y - \hat{y})$.

Listing 6.1 shows the implementation of the gradient descent for the simple ANN. The output of the program is provided in Table 6.2 with the intermediate results and a plot with the change in error and the change in weight in individual steps.

```python
# Graphical display
import matplotlib.pyplot as plt
# Very important, otherwise the plot will not be displayed
%matplotlib inline

# Identical function
def func_id(x):
        return x

# Initializations
x = 0.2
y = x
# Start weight
weight = -10.0
# For the plot
weights = []
errors = []
w_deltas = []

# Create table
# Print heading
```

```python
print("Input x = {:.6f}, Desired output y = {:.2f}".format(x,y))
print("{}\t{}\t{}\t{}\t{}\t{}\t{}\t{}\t{}"
        .format('Iter', 'x','w','net i',
                'a','y_hat','y','E',"E'",'w delta'))
# Fixed 120 steps
for step in range(120):
    # Calculate net input
    net_i = weight * x
    # Activation (identical function)
    activation = func_id(net_i)
    # Calculated output
    y_hat = activation
    # Squared error: desired - calculated output
    error = 0.5*(y - y_hat)**2
    # Gradient
    derivative = (-1.0)*x*(y - y_hat)
    # Delta for weight adjustment
    w_delta = (-1)*derivative
    # Data for the plot (weight,error)
    weights.append(weight)
    errors.append(error)
    w_deltas.append(w_delta)
    # Output the changes every 10 steps
    if step % 10 == 0:
        print("{}\t{}\t{:6.2f}\t{:5.2f}\t{:5.2f}"
              "\t{:5.2f}\t{:.2f}\t{:.2f}\t{:.2f}\t{:.2f}"
           .format(step, x,weight,net_i,activation,y_hat,
                   y,error,derivative,w_delta))
    # That's why we do all this: Weight adjustment = learning
    weight += w_delta

# Create plot
# Figure and subplot
fig, ax1 = plt.subplots()
ax1.plot(weights, errors, label="Error")
ax1.plot(weights,w_deltas, label="w Deltas")
# Title
ax1.set_title('Gradient Descent')
# Legend
legend = ax1.legend(loc='best', fancybox=True, framealpha=0.5)
# Grid
plt.style.use('seaborn-v0_8-whitegrid')
```

```
# Label
plt.xlabel('Weight w')
plt.show()
```

Listing 6.1 Gradient Descent for the Simple ANN

The program consists of the initialization of the required variables, followed by the loop with the iterations in which the variables of the network are calculated, and the output of the collected data. Let's take a look at Table 6.2 to see the output of the program that is collected in the iterations. The data calculated in this iteration is displayed per line. We haven't included all iterations in the table, but only the 10 iteration steps. You should already be familiar with the first columns. It gets interesting from column E onward, where the SE is displayed. The gradient follows in the column next to it, and finally the change in weight (w delta) is displayed. Before the subsequent iteration starts, the weight is adjusted using `w delta`, as you can also read in the code.

Iteration	x	w	Net Input	a	y_hat	y	E	E'	w Delta
0	0.2	-10.00	-2.00	-2.00	-2.00	0.20	2.42	-0.44	0.44
10	0.2	-6.31	-1.26	-1.26	-1.26	0.20	1.07	-0.29	0.29
20	0.2	-3.86	-0.77	-0.77	-0.77	0.20	0.47	-0.19	0.19
30	0.2	-2.23	-0.45	-0.45	-0.45	0.20	0.21	-0.13	0.13
40	0.2	-1.15	-0.23	-0.23	-0.23	0.20	0.09	-0.09	0.09
50	0.2	-0.43	-0.09	-0.09	-0.09	0.20	0.04	-0.06	0.06
60	0.2	0.05	0.01	0.01	0.01	0.20	0.02	-0.04	0.04
70	0.2	0.37	0.07	0.07	0.07	0.20	0.01	-0.03	0.03
80	0.2	0.58	0.12	0.12	0.12	0.20	0.00	-0.02	0.02
90	0.2	0.72	0.14	0.14	0.14	0.20	0.00	-0.01	0.01
100	0.2	0.81	0.16	0.16	0.16	0.20	0.00	-0.01	0.01
110	0.2	0.88	0.18	0.18	0.18	0.20	0.00	-0.00	0.00
120	0.2	0.92	0.18	0.18	0.18	0.20	0.00	-0.00	0.00

Table 6.2 Gradient Descent for the Simple ANN

If you recall from Chapter 4, you'll notice that the formula we derived earlier is almost equivalent to the Widrow-Hoff rule except for the learning rate η: we've set this rate to 1.0, so it can be omitted.

> **Task**
>
> Extend the program in such a way that a learning rate can be set and then experiment with it. Test your program for learning rates 2.0, 1.5, 1.0, and 0.01:
>
> - You may need to adjust the number of iterations.
> - You may also need to adjust the output accuracy of the E column ({:.6f}).

Solution:

```
# Initializations
x = 0.2
y = x
eta = 0.1
```

and

```
# Delta for weight adjustment
    w_delta = (-1)*derivative*eta
```

In Figure 6.5, the reduction of the error is shown depending on the change in weight. At each iteration step, we collect the data on the weights, the SE, and the calculated step length for the next weight change. You can clearly see that the rate of weight change adapts to the residual error: The greater the error, the greater the change in weight, and the smaller the error, the smaller the change in weight.

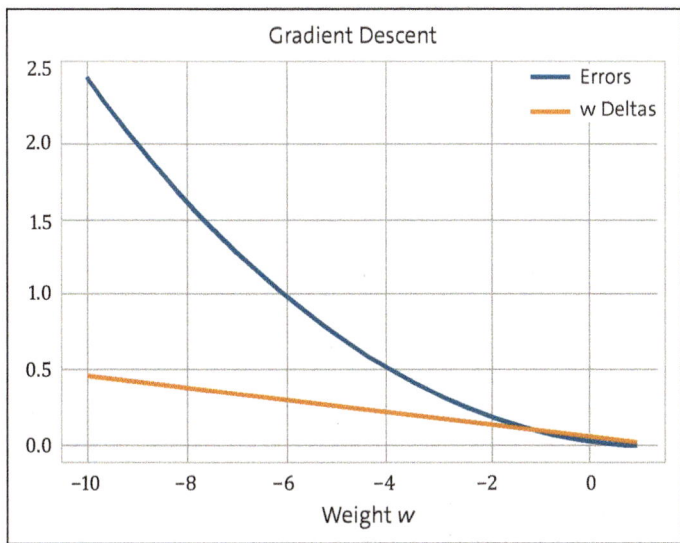

Figure 6.5 Error in Gradient Descent and Step Length during Weight Adjustment

6 Learning in a Multilayer Network

So far, we've only used one linear activation function in our program for the gradient descent, namely the *identical function*. This means that only linearly separable functions can be learned, regardless of how many neurons we later connect together in a complex network. For this reason, your next task is to insert the sigmoid function into the gradient descent.

> **Task**
>
> Change the gradient descent in such a way that the sigmoid activation function gets used.

Following are some hints for the solution (of course, whether you try it yourself first or look at the first hint is up to you):

- Hint 1: Define the sigmoids (also import `numpy`) as in the previous chapter:

$$f(x) = \frac{1}{1 + e^{-x}}$$

- Hint 2: Calculate the activation as in the previous chapter.
- Hint 3: Recalculate the derivative. This is the most difficult part, as you need to know the derivative of the sigmoid function (see Figure 6.6):

sigmoid' = *sigmoid* · (1 - *sigmoid*)

```
# Gradient
#    derivative = (-1.0)*x*(y - y_hat)
     derivative = (-1.0)*activation*(1.0-activation)*(y - y_hat)
```

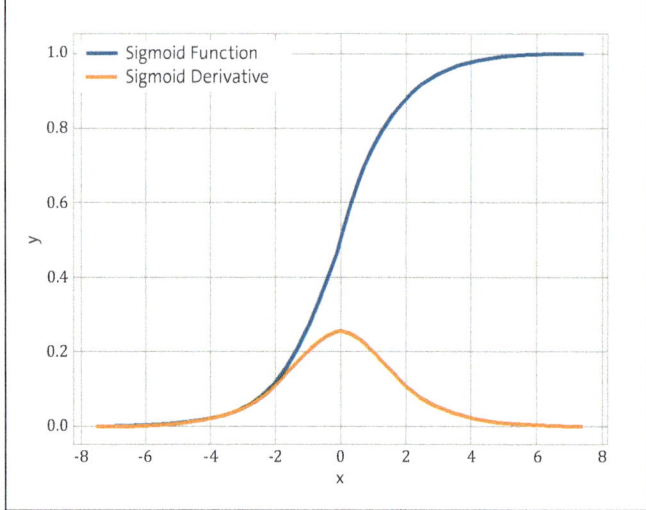

Figure 6.6 The Sigmoid Function and Its Derivative

- Hint 4: You may need to adjust the number of iterations and the learning rate.

The last exercise was a tough one. But as you'll see, this was the optimal preparation for the backpropagation algorithm, which we'll discuss in the next section.

6.3 A Network of Sigmoid Neurons

For the next steps, we assume that all neurons except the input neurons use a *sigmoid function* as activation function. The big trick is that we represent the learning problem using a small error formula, as we did previously with the gradient descent.

The output of the ANN is a desired output value (e.g., 0.5) and a calculated value (e.g., 0.8). The *difference* between the two values is of interest here. As the difference can also be negative, for example, 0.5 - 0.8 = -0.3, we need a means to obtain only positive values; otherwise, the gradient descent wouldn't work. (Where should you go if the value can become arbitrarily small?) In the previous discussion, we recognized the square of the difference (aka squared error, or SE), or *mean squared error* (MSE), as a solution method. Of course, we still have to calculate the error defined in this way for all output neurons. If we put this together as a compact formula, then we understand why mathematicians like to speak in symbols so much. One paragraph fits in one line—now that's an argument for math:

$$E = \frac{1}{2} \cdot \sum_{i=1}^{n} (y_i - \hat{y}_i)^2$$

We hope you've already taken a look at the gradient descent in the previous section because you'll recognize that we've used the sum to extend the error formula from one neuron to a large number of neurons.

The real search is only starting now. We want to change the weights so that the desired output is achieved as well as possible. This means that we want to keep the E in the preceding formula as small as possible. But where are the weights? They were hiding, but not well enough for us. If we now unpack a lot of math, we come to the basic rules of how to adjust the weights. This procedure is known as *error backpropagation*. It was rediscovered by David Rumelhart and his team in 1985 and has since become an indispensable basic tool for a seasoned connectionist. Let's get started.

As mentioned, our focus is on changing the weights. The algorithm provides the previously discussed calculation rule:

$w_{ji}^{new} = w_{ji}^{old} + \Delta w_{ji}$, with

$\Delta w_{ji} = -\eta \cdot \delta_j \cdot o_i$

The *eta*, that is, η, is again the learning rate, o_i is the output of the neuron i, and small *delta j*, that is, δ_j, is the value that was calculated proportionally for the weight w_{ji}. This means that every weight gets a small slice of the error pie. And now follows the question of all questions: How large is the proportion for each individual weight? The algorithm also provides a calculation rule for that, which we've already adapted for the use of the sigmoid activation function, as defined earlier (see also Figure 6.8):

$$\delta_j = \begin{cases} o_j \cdot (1 - o_j) \cdot (y_j - \hat{y}_j), & \text{if } j \text{ output neuron} \\ o_j \cdot (1 - o_j) \cdot \sum_k \delta_k \cdot w_{jk}, & \text{if } j \text{ hidden neuron} \end{cases}$$

Secret

We've adapted the calculation rule for δ_j for the sigmoid function. Of course, other activation functions could also be used, in which case, the $o_j \cdot (1 - o_j)$ would have to be replaced in the calculation rule. The details can be found again, for the fearless ones among you, in Appendix B.

If you've already actively worked through the section on gradient descent, the backpropagation algorithm will probably be pretty familiar to you. Let's take a look at the whole thing using an implementation. We'll use matrix and vector operations in NumPy to achieve the necessary speed for the calculations.

6.4 The Cool Algorithm with Forward Delta and Backpropagation

Let's take a look at the algorithm step by step. As always, we have some examples for classification: X is the set of input vectors x_i, each of which comes with a desired result y_i, for example, an image of a traffic sign as input and 1 if it contains a stop sign. Here are the algorithm's steps:

1. The forward calculation, that is, the `predict` method, is performed first. We've already implemented this method in the previous chapter.
2. Then, we follow the backward steps:
   - Determining the error at the output layer
   - Determining the errors in the hidden layers
3. Correct the weights.

There's not much more to say about forward calculation, is there? If we take a closer look at the backpropagation algorithm from earlier, the calculation of δ_j appears to be somewhat more complex. A closer look reveals that o_j must already be calculated to determine the output. This means that we can easily calculate the value for the derivative $o_j \cdot (1 - o_j)$ in forward gear and store it temporarily for each neuron.

6.4.1 The __init__ Method

Before we can start the implementation, we need additional parameter declarations and attributes in the `__init__` method:

6.4 The Cool Algorithm with Forward Delta and Backpropagation

- **eta**

  For the learning rate, we need the `eta` parameter and the `self.eta` attribute, which takes the value from the parameters.

- **n_iterations**

  The number of iterations is also given as a parameter and saved in the `self.n_iterations` attribute.

- **random_state parameter and self.random_state attribute**

  To specify the seed value for initializing the random generator, we need the `random_state` parameter and, of course, the `self.random_state` attribute.

- **self.errors**

  To collect the errors, `self.errors` should be defined in the attributes so that we can draw an error curve later.

We've incorporated the changes into the code from Chapter 5 and marked the changed sections with `# NEW`.

After the changes, the `__init__` method looks as shown in Listing 6.2.

```python
def __init__(self,
            n_input_neurons=2,
            n_hidden_neurons=2,
            n_output_neurons=1,
            weights=None,
            # NEW
            eta=0.01,n_iterations=10,random_state=2,
            *args, **kwargs):
    """ Initialization of the network
    We use a fixed I-H-O structure for the start
        (Input-Hidden-Output)
    The number of neurons is flexible
    It is also possible to initialize the network
        with weights [W_HI,W_OH]
    """
    # Activation and output function
    self.f_act = func_sigmoid
    self.g_out = func_id
    # Number of neurons per layer
    self.n_input_neurons=n_input_neurons
    self.n_hidden_neurons=n_hidden_neurons
    self.n_output_neurons=n_output_neurons
    # Weight initialization
    self.weights = weights
    W_HI=[]
    W_OH=[]
```

```python
# NEW Learning rate
self.eta = eta
# Iterations
self.n_iterations=n_iterations
# NEW Random generator
self.random_state = random_state
# NEW Generation of the random number generator (RNG)
random_state = check_random_state(self.random_state)
# NEW Errors during fit
self.errors=[]
# All data for the network calculation is stored here
self.network=[]
# Input layer + bias neuron: Columns = n
#   et_i, a_i, o_i,d_i,delta_i
self.inputLayer = np.zeros((self.n_input_neurons+1,5))
# Bias neuron output is always +1
self.inputLayer[0] = 1.0
# Add the input layer to the network
self.network.append(self.inputLayer)
# Weights from the input layer to the hidden W_HI layer
# Neuron: Row x Columns:
#   Rows = # Hidden, Columns = # Input
if weights:
    W_HI = self.weights[0]
else:
    # NEW
    W_HI = 2 * random_state.random_sample(
    (self.n_hidden_neurons+1,self.n_input_neurons+1)) - 1
self.network.append(W_HI)
# NEW Hidden layer + bias neuron:
# Columns = net_i,a_i,o_i,d_i,delta_i
self.hiddenLayer = np.zeros((self.n_hidden_neurons+1,5))
# Bias neuron output is always +1
self.hiddenLayer[0] = 1.0
# Add the hidden layer to the network
self.network.append(self.hiddenLayer)
# Weights from the hidden layer to the output layer W_OH
# Neuron: Row x Columns:
#   Rows = # Output, Columns = # Hidden
if weights:
    W_OH = self.weights[1]
else:
    # NEW
    W_OH = 2 * random_state.random_sample(
```

```
    (self.n_output_neurons+1,self.n_hidden_neurons+1)) - 1
self.network.append(W_OH)
# NEW output layer + bias neuron:
# Columns = net_i,a_i,o_i,d_i,delta_i
self.outputLayer = np.zeros((self.n_output_neurons+1,5))
# Bias neuron output = 0, as not relevant
# Only available due to standardized indexing
self.outputLayer[0] = 0.0
# Add the output layer to the network
self.network.append(self.outputLayer)
```

Listing 6.2 The Adjusted __init__ Method

6.4.2 The "predict" Method

Our task is to adjust the `predict` method from Chapter 5 as we've just shown for the `__init__` method. In the `__init__` method, you must add a column for the derivation to the arrays for `self.inputLayer`, `self.hiddenLayer`, and `self.outputLayer`. This could look as follows for the input layer:

```
# Input layer + bias neuron: Columns = net_i,a_i,o_i,d_i
  self.inputLayer = np.zeros((self.n_input_neurons+1,4))
```

Following are some hints for the solution:

- Hint 1: The new or changed positions are marked with NEW.
- Hint 2: The input, hidden, and output layers have the same structure in the implementation. They are implemented with an array with three columns ("net", "act", "out").
- Hint 3: If you're still using the `round()` function to determine the output, as was required in an example in Chapter 5, then you must extend this function now; otherwise, the algorithm wouldn't work.

Solution:

The predict method is updated to include an additional column for storing the derivative of the sigmoid function in both the hidden and output layers. It follows the same structure as before, but now computes and stores der_j = o_j (1 - o_j) and der_k = o_k (1 - o_k). These modifications ensure that the network can correctly compute gradients.

```
def predict(self,x):
    """ The output y is calculated for the input x.
    """
    ###############
    # Input layer
    # Set the inputs: All rows, column 2
```

6 Learning in a Multilayer Network

```python
        self.network[0][:,2] = x
        ###############
        # Hidden layer
        # Start from line 1 due to bias neuron at index position 0
        # net_j = W_ij . x
        self.network[2][1:,0] = np.dot(self.network[1][1:,:],
                                      self.network[0][:,2])
        # a_j
        self.network[2][1:,1] = self.f_act(
                                      self.network[2][1:,0])
        # o_j
        self.network[2][1:,2] = self.g_out(self.network[2][1:,1])
        # NEW der_j = o_j*(1-o_j) Derivative of sigmoid function
        self.network[2][1:,3] = self.network[2][1:,2] * \
                                ( 1.0 - self.network[2][1:,2])
        ###############
        # Output layer
        # Start from line 1 due to bias neuron at 0
        # net_k = = W_jk . h
        self.network[4][1:,0] = np.dot(self.network[3][1:,:],
                                      self.network[2][:,2])
        # a_k
        self.network[4][1:,1] = self.f_act(
                                      self.network[4][1:,0])
        # o_k
        self.network[4][1:,2] = self.g_out(self.network[4][1:,1])
        # NEW der_k = o_k*(1-o_k) Derivative of sigmoid function
        self.network[4][1:,3] = self.network[4][1:,2] * \
                                ( 1.0 - self.network[4][1:,2])
        # Return of the output vector
        return self.network[4][:,2]
```

Listing 6.3 The "predict" Method Adapted to the Needs of the Backpropagation Algorithm

During the first step backward, the error is calculated at the output. We've already described this in the earlier algorithm. Then, it continues backwards, and it's now very exciting to see how the error is distributed in the hidden layers. You can view the diagram in Figure 6.7.

We've already calculated the layer that follows the hidden layer, that is, the output layer. There, δ_k was determined for each neuron. Thus, the error component for neuron j results from the following calculation:

$$\delta_j = o_j \cdot (1 - o_j) \cdot \sum_k \delta_k \cdot w_{kj}$$

6.4 The Cool Algorithm with Forward Delta and Backpropagation

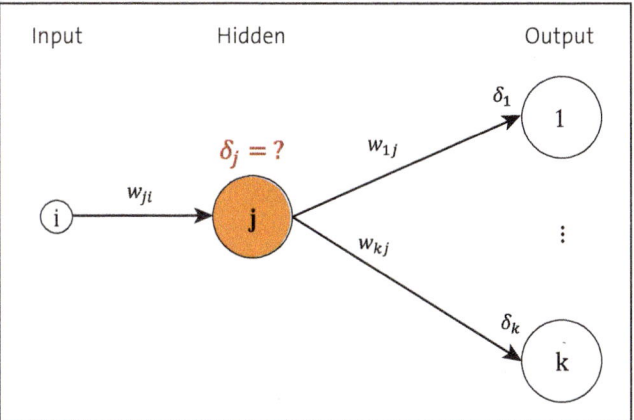

Figure 6.7 From Output to the Hidden Layer

This step can be applied to any number of layers in the manner explained earlier. And, because a picture says more than a thousand formulas, we've once again prepared a diagram.

In Figure 6.8, you can see the multiplications that are necessary to implement the algorithm. The forward path ❶ leads from layer i to j and from j to k. The way back ❷ is also shown. The error e at the output can simply be calculated as the difference between the desired and the actual output. It's best to add a new column for each layer.

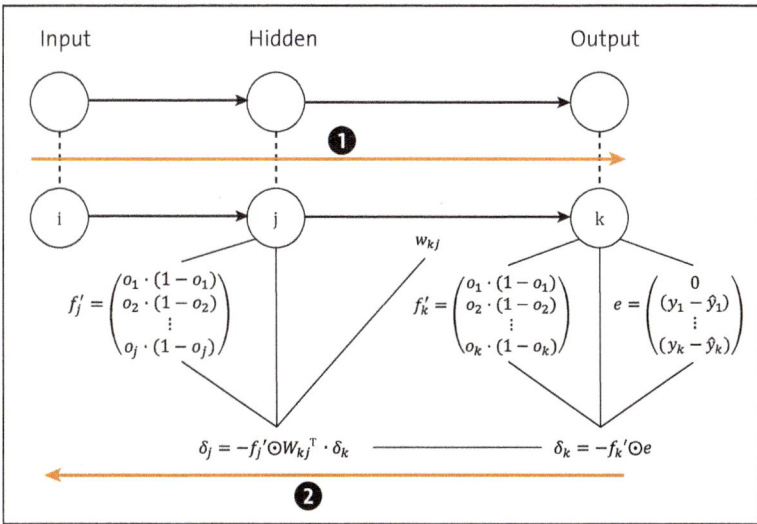

Figure 6.8 Vector and Matrix Multiplications for Programming

Let's now move on to another task: Insert a new column for the error or the delta for the backward path. Use the previous solution or the following for the input layer as a guide:

```
# Input layer + bias neuron: Columns = net_i,a_i,o_i,d_i,delta_i
self.inputLayer = np.zeros((self.n_input_neurons+1,5))
```

Solution:

In the previous hints, we showed how you can extend the input layer so that the delta can be saved. This exact change needs to be implemented for the hidden and output layers. This results in the code change to

```
# Columns = net_i,a_i,o_i,d_i,delta_i
self.hiddenLayer = np.zeros((self.n_hidden_neurons+1,5))
```

and

```
# Columns = net_i,a_i,o_i,d_i,delta_i
self.outputLayer = np.zeros((self.n_output_neurons+1,5))
```

Note the multiplication, for example, in $\delta_k = -f_k' \odot e$. In contrast to the *dot product* (*inner product*), this multiplication multiplies the components with each other, and the vector then has the same dimension, that is, the same number of entries as before. This multiplication is referred to as the *Hadamard product* (see Figure 6.9).

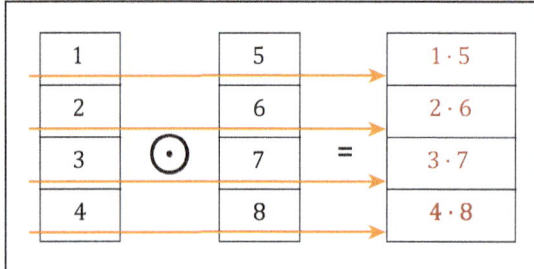

Figure 6.9 Hadamard Product

However, this knowledge isn't absolutely necessary to implement the algorithm because NumPy does the following for us (see Listing 6.4).

```
import numpy as np
a= np.array([1, 2, 3, 4])
b= np.array([5, 6, 7, 8])
print(a*b)
# Output:
[ 5 12 21 32]
```

Listing 6.4 Hadamard Multiplication in NumPy

Another detail concerns the weight matrix W_{jk}^T; the superscript T means that the matrix is transposed. Details can be found in Appendix B. So, we replace the "from" or "to" with the transposition, which allows us to use NumPy's matrix multiplication again.

Finally, the weights need to be adjusted. It's important that the adjustment only takes place once all correction values have been calculated; otherwise, the direction of descent will be distorted. And, to make things really fast, we can use matrix multiplications again:

$$\Delta W_{kj} = \eta \cdot \delta_k \cdot o_j^T$$
$$\Delta W_{ji} = \eta \cdot \delta_j \cdot o_i^T$$

To give you a better idea of what the calculation rule means, Figure 6.10 shows the multiplication $\delta_k \cdot o_j$ or $\delta_j \cdot o_i$ in detail.

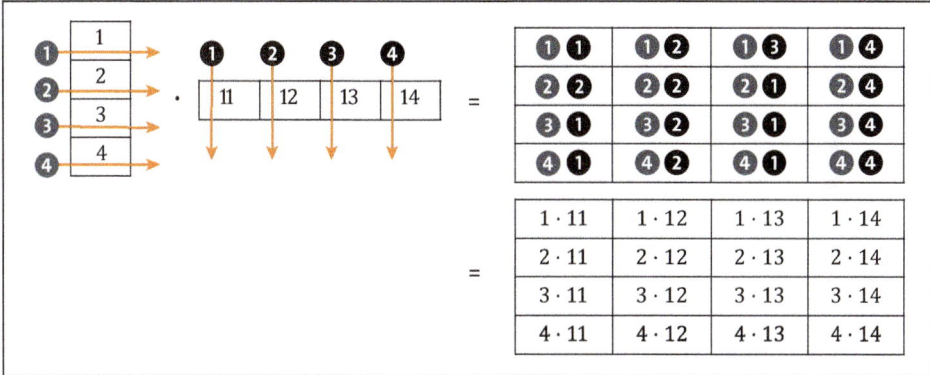

Figure 6.10 Vector Multiplication

The multiplication of two vectors creates a new matrix! But luckily NumPy carries out these calculations for us.

Well, that was an exhausting part with the necessary dose of theory, pared down to the bare essentials, but it was worth it. We've thus created the basis for implementing the backpropagation algorithm, which we'll do in the next section.

6.4.3 The "fit" Method

The implementation of the backward path can be found in the fit method of the MLP method, as shown in Listing 6.5. We need the training examples X and the desired results Y as parameters. In the examples, we've again entered the value 1.0 at index position 0 for the bias neuron. In the case of the result, this is only a dummy value so that we can carry out the calculations consistently.

```
def fit(self,X,Y):
    """ Learn
    """
    # Weight changes
```

```python
        delta_w_jk = []
        delta_w_ij = []
        # Error
        self.errors = []
        # All iterations
        for iteration in range(self.n_iterations):
            # For all training examples
            error = 0.0
            #for xIdx,x in enumerate(X):
            for x,y in zip(X,Y):
                #####################
                # Forward path
                y_hat = self.predict(x)
                # Difference
                diff = y - y_hat
                # Squared error
                error += 0.5 * np.sum(diff * diff)
                # For a more compact notation
                net = self.network
                #####################
                # Output layer
                # delta_k in the output layer = der_k * diff
                net[4][:,4] = net[4][:,3] * diff

                #####################
                # Hidden layer
                # delta_j in the hidden layer =
                #   der_j * dot(W_kj^T,delta_k)
                net[2][:,4] = net[2][:,3] * \
                            np.dot(net[3][:].T,net[4][:,4])
                #####################
                # Weight deltas from W_kj
                # delta_w = eta * delta_k . o_j^T
                delta_w_jk = self.eta * \
                            np.outer(net[4][:,4],net[2][:,2].T )
                # Weight deltas from W_ji
                # delta_w = eta * delta_j . o_i^T
                delta_w_ij = self.eta * \
                            np.outer(net[2][:,4],net[0][:,2].T )
                #####################
                # Adjust weights
                net[1][:,:] += delta_w_ij
                net[3][:,:] += delta_w_jk
```

6.4 The Cool Algorithm with Forward Delta and Backpropagation

```
    # Collecting the error for all examples
    self.errors.append(error)
```

Listing 6.5 The Backpropagation Algorithm in the "fit" Method

We start by determining the result for an example from X using the `predict` method. This is followed by the calculation of the difference `diff` for the output neurons, and the SE is collected for later output using the `plot` method, which you can see in Figure 6.11.

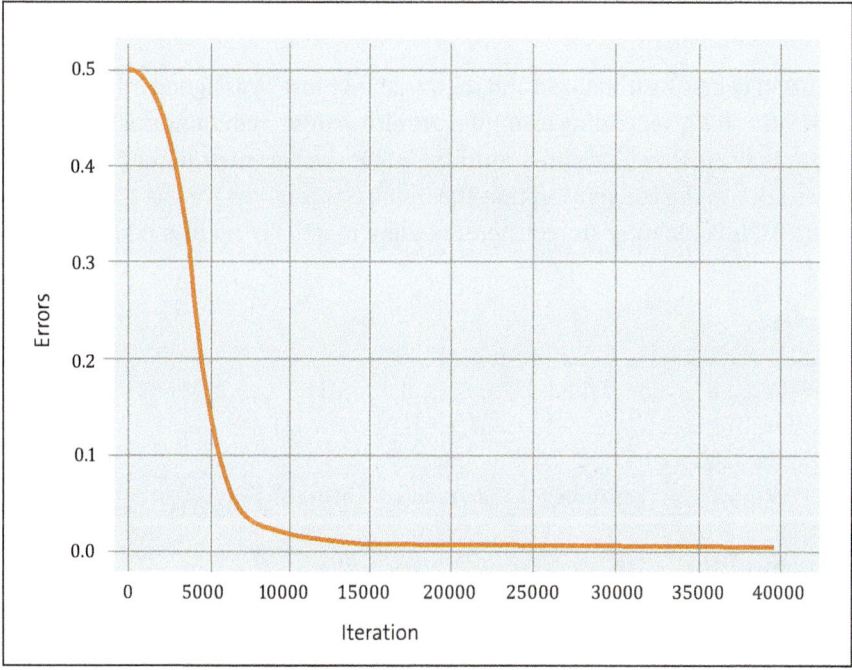

Figure 6.11 A Beautiful Error Curve

6.4.4 The "plot" Method

We output the error curve using a function, as shown in Listing 6.6. There isn't much to say about this, as we've already done this type of output several times.

```
def plot(self):
    """ Output of the error
    Output the errors stored in the error array as a graphic
    """
    # Figure Numbers Start
    fignr = 1
    # Print size in inches
    plt.figure(fignr,figsize=(5,5))
```

```python
        # Output of the error as a plot
        plt.plot(self.errors)
        # Grid
        plt.style.use('seaborn-v0_8-whitegrid')
        # Labels
        plt.xlabel('Iteration')
        plt.ylabel('Error')
```

Listing 6.6 Output of the Error Curve

6.4.5 The Complete Picture

The ANN must of course be initialized and the variables must be assigned initial values, as shown in Listing 6.7. The training examples are also a must. X contains the inputs line by line, and at index 0, there is again a constant 1.0 for the bias neuron output. For the outputs Y, we stick to the convention that the index 0 is reserved for the bias neuron, but the value 0.0 is constantly stored there, as there is actually no bias neuron in the output layer.

```python
def main():
    # Initialization of the training examples
    X=np.array([[1.0,1.0,1.0],[1.0,0,1.0],[1.0,1.0,0],[1.0,0,0]])
    Y=np.array([[0.0,0.0],[0.0,1.0],[0.0,1.0],[0.0,0.0]])
    # Initialize network
    nn = MLP(eta=0.03,n_iterations=40000,random_state=42)

    # Train the network with the fit method and
    # Output after training
    nn.fit(X,Y)
    nn.print()

    # Error output as graph
    nn.plot()

    # Test the prediction of the training dataset
    print('Predict:')
    for x,y in zip(X,Y):
        print('{} {} -> {}'.format(x,y[1],nn.predict(x)[1:2]))

# Main program
# Warning: Takes a little longer to produce an output
main()
```

Listing 6.7 The Program

After the program has run, the network architecture appears with the calculated values for the weights and with the analysis of the vector that was used for the last iteration, as shown in Listing 6.8. In our example, this is (0, 0).

```
# Output:
Multilayer Perceptron - Network Architecture
[[  1.000    1.000    1.000    1.000    1.000]
 [  0.000    0.000    0.000    0.000    0.000]
 [  0.000    0.000    0.000    0.000    0.000]]
----------v----------
[[  7.495   -5.248   -6.321]
 [  2.108   -5.657   -5.653]
 [ -6.578    4.224    4.223]]
----------v----------
[[  1.000    1.000    1.000    1.000   -0.005]
 [  2.108    0.892    0.892    0.097    0.001]
 [ -6.578    0.001    0.001    0.001    0.000]]
----------v----------
[[  0.416   -0.959    0.940]
 [  4.037   -8.164   -8.261]]
----------v----------
[[  0.000    0.000    0.000    0.000    0.000]
 [ -3.253    0.037    0.037    0.036   -0.001]]
----------v----------
Predict:
[  1.000    1.000    1.000]  0.0 -> [  0.042]
[  1.000    0.000    1.000]  1.0 -> [  0.957]
[  1.000    1.000    0.000]  1.0 -> [  0.957]
[  1.000    0.000    0.000]  0.0 -> [  0.037]
```

Listing 6.8 Output of the Program

Let's take a look at the individual sections of the output:

- **Multilayer Perceptron - Network Architecture**
  - It starts with the *input matrix* (five columns for net input, activation, output, derivative, and delta); the third column shows the last input vector (1,0,0), whereby the first dimension is intended for the bias neuron. For this reason, the input is (0,0).
  - Then, it follows a *separator* ----------v----------, which is repeated for each matrix output.
  - After that, there is the *weight matrix* for the connections from the input layer to the hidden layer.

6 Learning in a Multilayer Network

- This is followed by the matrix for the *hidden neurons*, which is structured in the same way as the input matrix.
- The weight matrix from the hidden to the output layer is still missing, and the last matrix represents the *output* for the applied input. The calculated value can again be found in column 3, namely 0.037, that is, approximately 0.

- **Predict**
  - After the output of the matrixes, the application of the network to the training dataset is shown. If you look at the calculated figures in the `Predict` section, you can see that the desired output is already well achieved by the calculated outputs, that is, it was expected:

$$y = \begin{pmatrix} 0 \\ 1 \\ 1 \\ 0 \end{pmatrix} \cdot \hat{y} = \begin{pmatrix} 0.042 \\ 0.957 \\ 0.957 \\ 0.037 \end{pmatrix}$$

In the next section, we'll take a step-by-step look at the path through the algorithm.

6.5 A "fit" Run

The run refers to the network structure shown in Figure 6.12. In this section, we take a look at the change of the network architecture from an **OLD** state to a **NEW** state.

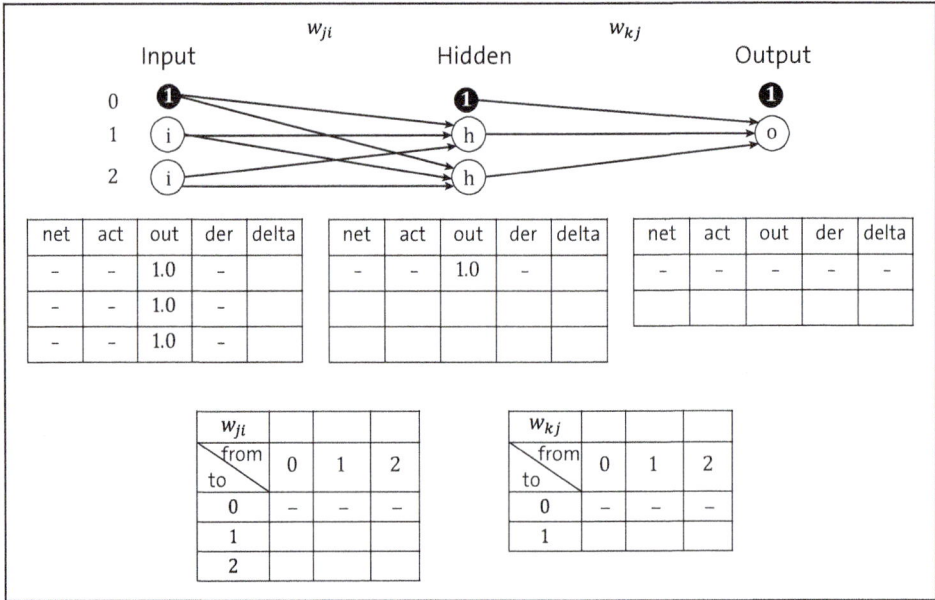

Figure 6.12 Network Structure for the Iteration

The state transition is created by adjusting the weights using the backpropagation algorithm, as shown in Figure 6.13 for the two states and generated by the code in Listing 6.9.

```python
# Mathematics
import numpy as np

def main():
    # Training data
    X=np.array([[1.0,1.0,1.0]])
    Y=np.array([[0.0,0.0]])

    # Initialize the network with an iteration
    nn = MLP(eta=0.03,n_iterations=1,printOn=False,random_state=42)

    # Predict and output the network architecture
    nn.predict(X[0])
    nn.print()

    # Learn an example and output the network architecture
    nn.fit(X,Y)
    nn.print()

# Main program
main()
```

Listing 6.9 A Learning Step in the MLP Method

The **OLD** state on the left-hand side of Figure 6.13 represents the situation of the architecture after the initialization of the multilayer perceptron (MLP) and the prediction based on the input X[0]. On the right-hand side, you can see the **NEW** state, which emerges from the **OLD** state after a learning step. For this purpose, we use the input x = X[0] = [1.0,1.0,1.0] with the desired output y = Y[0] = [0.0, 0.0] for the learning step.

> **Tip**
>
> You can find out more about learning methods in Chapter 12.

The matrixes are stored in a Python list, whereby the access indexes between **OLD** and **NEW** states are visible. For each state description of the network architecture, the matrixes with the values are displayed one below the other. For example, the data for

6 Learning in a Multilayer Network

the input layer can be found line by line under **OLD**. Both the net input (**net**) and activation (**act**) aren't yet relevant in this layer. The output column (**out**) contains the output value of the input neuron, for example, 1.000 for neuron 0 (bias neuron), 1, and 2. The values that are colored gray aren't relevant for the calculations and are only there for technical reasons (dummy) so that the matrix operations can be applied. The value of the gradient calculation is stored temporarily in the **der** column, and **delta** indicates the delta for the weight change, whereby **delta** in **OLD** hasn't been calculated yet.

OLD

Input	net	act	out	der	delta
0	1.000	1.000	1.000	1.000	1.000
1	0.000	0.000	1.000	0.000	0.000
2	0.000	0.000	1.000	0.000	0.000

W_ji

to \ from	0	1	2
0	-0.251	0.901	0.464
1	0.197	-0.688	-0.688
2	-0.884	0.732	0.202

Hidden	net	act	out	der	delta
0	1.000	1.000	1.000	1.000	1.000
1	-1.179	0.235	0.235	0.180	0.000
2	0.051	0.513	0.513	0.250	0.000

W_kj

to \ from	0	1	2
0	0.416	-0.959	0.940
1	0.665	-0.575	-0.636

Output	net	act	out	der	delta
0	0.000	0.000	0.000	0.000	0.000
1	0.203	0.551	0.551	0.247	0.000

NEW

Input	net	act	out	der	delta
0	1.000	1.000	1.000	1.000	1.000
1	0.000	0.000	1.000	0.000	0.000
2	0.000	0.000	1.000	0.000	0.000

W_ji

to \ from	0	1	2
0	-0.254	0.899	0.461
1	0.198	-0.688	-0.688
2	-0.883	0.733	0.203

Hidden	net	act	out	der	delta
0	1.000	1.000	1.000	1.000	-0.091
1	-1.179	0.235	0.235	0.180	0.014
2	0.051	0.513	0.513	0.250	0.022

W_kj

to \ from	0	1	2
0	0.416	-0.959	0.940
1	0.661	-0.576	-0.638

Output	net	act	out	der	delta
0	0.000	0.000	0.000	0.000	0.000
1	0.203	0.551	0.551	0.247	-0.136

Matrix Index: 0, 1, 2, 3, 4

Figure 6.13 From OLD to NEW: An Iteration Step

Below the input layer matrix is the weight matrix from the input layer to the hidden layer, here labeled W_{ji}. The **from/to** cell indicates more precisely how the matrix is to be read. For example, the weight from input neuron 0 to hidden neuron 1 is 0.197.

The result of the calculations from the `predict` method can be seen in the output layer in the **out** column. The output layer is also the layer with which the adjustments begin.

6.5.1 Initialization

The weights are initialized in the `__init__` method, as shown in Listing 6.10. We've selected the weights from the interval [1.0,1.0].

```
*********************************************
Multilayer Perceptron - Network Architecture
*********************************************
[[1.000 1.000 1.000 1.000 1.000]
 [0.000 0.000 1.000 0.000 0.000]
 [0.000 0.000 1.000 0.000 0.000]]
----------v----------
[[-0.251 0.901 0.464]
```

```
 [0.197 -0.688 -0.688]
 [-0.884 0.732 0.202]]
----------v----------
[[1.000 1.000 1.000 1.000 1.000]
 [-1.179 0.235 0.235 0.180 0.000]
 [0.051 0.513 0.513 0.250 0.000]]
----------v----------
[[0.416 -0.959 0.940]
 [0.665 -0.575 -0.636]]
----------v----------
[[0.000 0.000 0.000 0.000 0.000]
 [0.203 0.551 0.551 0.247 0.000]]
----------v----------
```

Listing 6.10 Assignment of the Matrixes after Initialization

6.5.2 Forward

The first step in backpropagation is the analysis of a training example, as shown in Listing 6.11.

```python
def fit(self,X,Y):
    # Weight changes
    delta_w_jk = []
    delta_w_ij = []
    # All iterations
    for iteration in range(self.n_iterations):
        # For all training examples
        error = 0.0
        for x,y in zip(X,Y):
            #####################
            # Forward path
            y_hat = self.predict(x)
```

Listing 6.11 The Relevant Section in the "fit" Method for the Forward Path

Vector x is used as the input and y as the desired output. In the code, y_hat denotes the calculated output, as shown in Listing 6.12.

```
----------- Forward
x = [1.000 1.000 1.000]
y = [0.000 0.000]
Result
[0.000 0.551]
```

Listing 6.12 Input "x" and Desired Output "y" Leading to the Calculated Output: Result

The result of the calculation is 0.551, which isn't quite what we wanted because the plan was 0.0. So, we have the need to learn and the backward path can begin.

6.5.3 Output

The backward path starts at the output layer. The difference between the desired and calculated output gets determined (see Listing 6.13). The SE is then determined for the output of the error curve.

```
######################
# Forward path
y_hat = self.predict(x)
# Difference
diff = y - y_hat
# Squared error
error += 0.5 * np.sum(diff * diff)

######################
# Output layer
# delta_k in the output layer = der_k * diff
self.network[4][:,4] = self.network[4][:,3]*diff
```

Listing 6.13 Determining the Difference, the Squared Error, and delta_k

As a result of these processing steps, you'll see the value of the difference, naturally as a vector, and the calculated `delta_k`. First, the difference `diff` is determined:

y - y_hat = [0.0,0.0] - [0.0,0.551] = [0.0,-0.551]

Then, the SE gets calculated, which is decisive for the representation of the plot (we leave it to you as homework to understand this calculation), and then `delta_k`:

`self.network[4][:,4] = self.network[4][:,3]*diff`

The value is determined for column 4, which is the **delta** column of the matrix at index 4 (i.e., the output matrix). For this purpose, the value from the gradient column **der** in the output matrix is multiplied by the previously determined difference (see Figure 6.14).

Now let's calculate the ***delta**$_k$* for output neuron *k*:

$delta_k = diff \odot der$

$$= (y - out) \odot der$$

$$= \begin{pmatrix} 0.0 \\ -0.551 \end{pmatrix} \odot \begin{pmatrix} 0.0 \\ 0.247 \end{pmatrix}$$

$$= \begin{pmatrix} 0.0 \\ -0.136 \end{pmatrix}$$

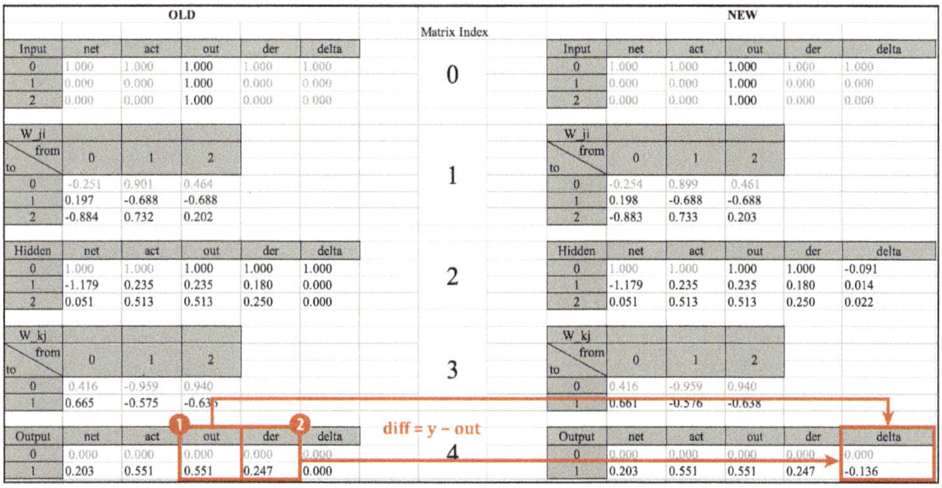

6.5 A "fit" Run

Figure 6.14 Calculation of the Output Layer

OLD

Input	net	act	out	der	delta
0	1.000	1.000	1.000	1.000	1.000
1	0.000	0.000	1.000	0.000	0.000
2	0.000	0.000	1.000	0.000	0.000

W_ji from / to

	0	1	2
0	-0.251	0.901	0.464
1	0.197	-0.688	-0.688
2	-0.884	0.732	0.202

Hidden	net	act	out	der	delta
0	1.000	1.000	1.000	1.000	1.000
1	-1.179	0.235	0.235	0.180	0.000
2	0.051	0.513	0.513	0.250	0.000

W_kj from / to

	0	1	2
0	0.416	-0.959	0.940
1	0.665	-0.575	-0.636

Output	net	act	out	der	delta
0	0.000	0.000	0.000	0.000	0.000
1	0.203	0.551	0.551	0.247	0.000

NEW

Input	net	act	out	der	delta
0	1.000	1.000	1.000	1.000	1.000
1	0.000	0.000	1.000	0.000	0.000
2	0.000	0.000	1.000	0.000	0.000

W_ji from / to

	0	1	2
0	-0.254	0.899	0.461
1	0.198	-0.688	-0.688
2	-0.883	0.733	0.203

Hidden	net	act	out	der	delta
0	1.000	1.000	1.000	1.000	-0.091
1	-1.179	0.235	0.235	0.180	0.014
2	0.051	0.513	0.513	0.250	0.022

W_kj from / to

	0	1	2
0	0.416	-0.959	0.940
1	0.661	-0.576	-0.638

Output	net	act	out	der	delta
0	0.000	0.000	0.000	0.000	0.000
1	0.203	0.551	0.551	0.247	-0.136

Matrix Index: 0, 1, 2, 3, 4 — diff = y − out

6.5.4 Hidden

The deltas are also calculated in the hidden layer. As we've already discussed, the calculation is based on the deltas of the output layer, the weights between the hidden and output layers, and the gradient value in the hidden layer (see Listing 6.14).

```
#####################
# Hidden layer
# delta_j in the hidden layer = der_j * dot(W_jk^T,delta_k)
self.network[2][:,4] = \
self.network[2][:,3] * np.dot(self.network[3][:].T,self.network[4][:,4])
```

Listing 6.14 delta_j Calculation in the Hidden Layer

In the matrix with index 2, which is the matrix of the hidden layer neurons, the column with index 4, which is δ_j, gets calculated. For this purpose, the data as shown in Figure 6.15 is used for the calculation.

If we now insert the figures into the calculation rule, it looks as follows:

$delta_j = 1 \odot 2 \cdot 3$

$$delta_j = \begin{pmatrix} 1.000 \\ 0.180 \\ 0.250 \end{pmatrix} \odot \begin{pmatrix} 0.416 & 0.665 \\ -0.959 & -0.575 \\ 0.940 & -0.636 \end{pmatrix} \cdot \begin{pmatrix} 0.000 \\ -0.136 \end{pmatrix}$$

$$= \begin{pmatrix} 1.000 \\ 0.180 \\ 0.250 \end{pmatrix} \odot \begin{pmatrix} -0.091 \\ 0.078 \\ 0.086 \end{pmatrix}$$

$$= \begin{pmatrix} -0.091 \\ 0.014 \\ 0.022 \end{pmatrix}$$

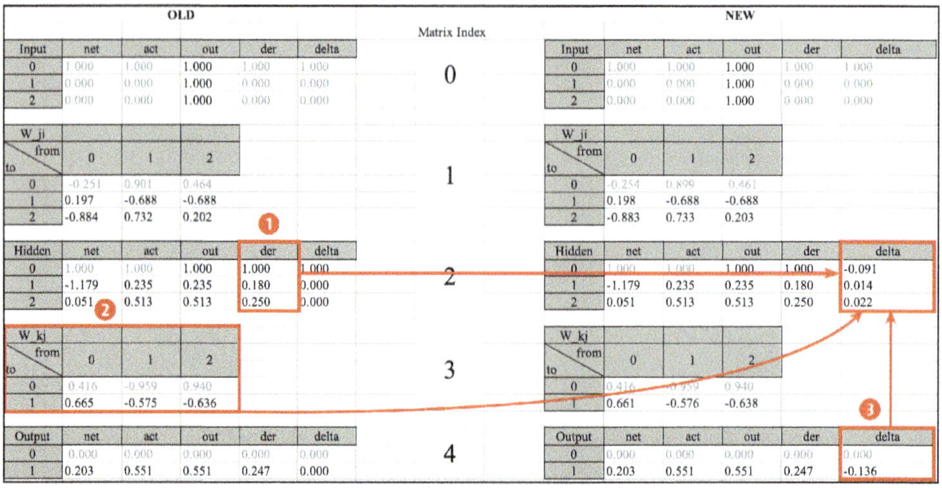

6 Learning in a Multilayer Network

OLD						Matrix Index	NEW					

Figure 6.15 Delta Calculation in the Hidden Layer

There is always a small rounding error here due to the limited number of decimal places.

6.5.5 Delta W_kj

The calculation of the change in weights can be carried out in the next step (see Figure 6.16). Here again, we need the transpose and the NumPy operation `outer` to multiply the two vectors with each other (see Listing 6.15).

```
#####################
# Weight deltas from W_kj
# delta_w = eta * delta_k . o_j^T
delta_w_jk = self.eta * np.outer(self.network[4][:,4],self.network[2][:,2].T )
# Weight deltas of W_ij
# delta_w = eta * delta_j . o_i^T
delta_w_ij = self.eta * np.outer(self.network[2][:,4],self.network[0][:,2].T )
```

Listing 6.15 Calculation of the Weight Deltas

We've used a learning rate of $\eta = 0.03$ for the following calculation:

$$\Delta W_{kj} = \eta \cdot 1 \cdot 2$$

$$\Delta W_{kj} = 0.03 \cdot \begin{pmatrix} 0.000 \\ -0.136 \end{pmatrix} \cdot \begin{pmatrix} 1.000 & 0.235 & 0.513 \end{pmatrix}$$

$$= 0.03 \cdot \begin{pmatrix} 0.000 & 0.000 & 0.000 \\ -0.136 & -0{,}032 & -0{,}070 \end{pmatrix}$$

$$= \begin{pmatrix} 0.000 & 0.000 & 0.000 \\ -0.004 & -0.001 & -0.002 \end{pmatrix}$$

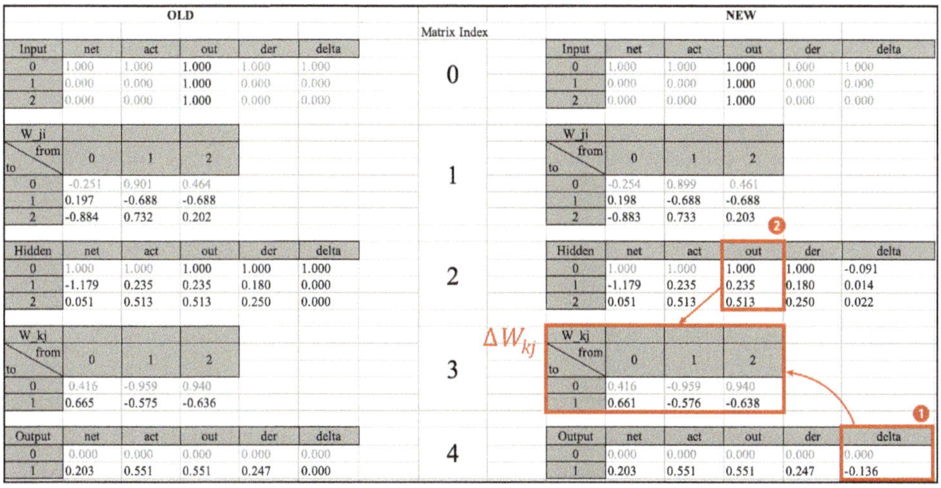

6.5 A "fit" Run

	OLD						NEW				
Input	net	act	out	der	delta	Input	net	act	out	der	delta
0	1.000	1.000	1.000	1.000	1.000	0	1.000	1.000	1.000	1.000	1.000
1	0.000	0.000	1.000	0.000	0.000	1	0.000	0.000	1.000	0.000	0.000
2	0.000	0.000	1.000	0.000	0.000	2	0.000	0.000	1.000	0.000	0.000

Matrix Index: 0

W_ji from \ to	0	1	2			W_ji from \ to	0	1	2		
0	-0.251	0.901	0.464			0	-0.254	0.899	0.461		
1	0.197	-0.688	-0.688			1	0.198	-0.688	-0.688		
2	-0.884	0.732	0.202			2	-0.883	0.733	0.203		

Matrix Index: 1

Hidden	net	act	out	der	delta	Hidden	net	act	out	der	delta
0	1.000	1.000	1.000	1.000	1.000	0	1.000	1.000	1.000	1.000	-0.091
1	-1.179	0.235	0.235	0.180	0.000	1	-1.179	0.235	0.235	0.180	0.014
2	0.051	0.513	0.513	0.250	0.000	2	0.051	0.513	0.513	0.250	0.022

Matrix Index: 2

W_kj from \ to	0	1	2			W_kj from \ to	0	1	2		
0	0.416	-0.959	0.940			0	0.416	-0.959	0.940		
1	0.665	-0.575	-0.636			1	0.661	-0.576	-0.638		

Matrix Index: 3 — ΔW_{kj}

Output	net	act	out	der	delta	Output	net	act	out	der	delta
0	0.000	0.000	0.000	0.000	0.000	0	0.000	0.000	0.000	0.000	0.000
1	0.203	0.551	0.551	0.247	0.000	1	0.203	0.551	0.551	0.247	-0.136

Matrix Index: 4

Figure 6.16 Calculation of the Weight Delta for Wkj

6.5.6 Delta W_ji

Our next task is to calculate the delta for W_{ji}. Use the following program line as the calculation rule:

```
delta_w_ij = self.eta * np.outer(self.network[2][:,4],self.network[0][:,2].T ).
```

Solution:

We can calculate the delta using the following:

$$\Delta W_{ji} = \begin{pmatrix} -0.003 & -0.003 & -0.003 \\ 0.000 & 0.000 & 0.000 \\ 0.001 & 0.001 & 0.001 \end{pmatrix}$$

6.5.7 W_ji

We've now almost reached the end of our example. Once the weight changes for the weight matrix have been calculated, we can simply add them to the original weights:

$$W_{ji}^{NEW} = W_{ji}^{OLD} + \Delta W_{ji}$$

$$= \begin{pmatrix} -0.251 & 0.901 & 0.464 \\ 0.197 & -0.688 & -0.688 \\ -0.884 & 0.732 & 0.202 \end{pmatrix}$$

$$+ \begin{pmatrix} -0.003 & -0.003 & -0.003 \\ 0.000 & 0.000 & 0.000 \\ 0.001 & 0.001 & 0.001 \end{pmatrix}$$

$$= \begin{pmatrix} -0.254 & 0.899 & 0.461 \\ 0.198 & -0.688 & -0.688 \\ -0.883 & 0.733 & 0.203 \end{pmatrix}$$

Again, we ask you to generously overlook rounding errors.

6.5.8 W_kj

Our final task is to calculate the new weight matrix for W_{kj}. Use the procedure from the previous section.

Solution:

We can calculate the new weight matrix as follows:

$$W_{kj}^{NEW} = \begin{pmatrix} 0.416 & -0.959 & 0.940 \\ 0.661 & -0.576 & -0.638 \end{pmatrix}$$

You made it! You've manually performed an iteration for a training example. If you still have energy, you can now tackle the second training example in Listing 6.7.

6.6 Summary

This chapter has taken you deep into the implementation of learning algorithms. You've implemented the error backpropagation algorithm using Python. Congratulations! That was certainly an exhausting chapter, but definitely worthwhile.

We explained the theoretical basics such as gradient descent, and you saw what learning actually means—namely, error minimization. The meticulous process of running a learning iteration has surely made your calculator or smartphone overheat.

From a programming point of view, this was definitely the most challenging part of the book. In the following chapters, we'll focus on frameworks and introduce you to their use, such as TensorFlow in Chapter 8.

6.7 Further Reading

Neural Networks. A Systematic Introduction. by Raul Rojas (Springer, 1995)

Chapter 7
Examples of Deep Neural Networks

Deep experiences are unforgettable knowledge.

In recent years, more and more network architectures have emerged that generate something such as images, text, videos, and so on. The collective term for this type of network is generative AI. Network architectures that differentiate, classify, or segment belong to discriminative AI. This chapter focuses on one representative from each of these two groups, which represent key milestones in the development of neural networks and their success.

The first network architecture is referred to as a convolutional neural network (CNN). Networks of this type are particularly good at processing images and videos. The second network architecture—the transformer neural networks—heralded the start of large language models (LLMs) in 2017. These include, for example, ChatGPT from OpenAI, Llama from Meta, and Gemini from Google.

7.1 Convolutional Neural Networks

A *convolutional neural network* (CNN) is a special multilayer feed-forward network that is mainly used in the machine processing of image and audio files. CNNs belong to the class of *deep neural networks* or deep learning methods. In this chapter, we'll focus on the processing of image data.

What is special about this form of neural network, and why don't we just use the multilayer networks we learned about in the previous chapters? There are several reasons for this: First, those networks work quite well for simple images, but with a growing size and complexity of the images, the number of neurons and connections required increases almost immeasurably. Second, they represent the two-dimensional image as a one-dimensional long vector, which means we lose valuable neighborhood information.

Higher image processing, or *computer vision*, is one of the supreme disciplines of pattern recognition and has a long scientific history behind it, although this has only brought partial successes. Here, too, there have been repeated attempts to develop algorithms that are based on the way the human visual system works.

The image that is recorded from the visual field of our optical apparatus, that is, our eye, is passed on to the visual cortex (see Figure 7.1). This part of the brain is mainly involved in processing visual information.

7 Examples of Deep Neural Networks

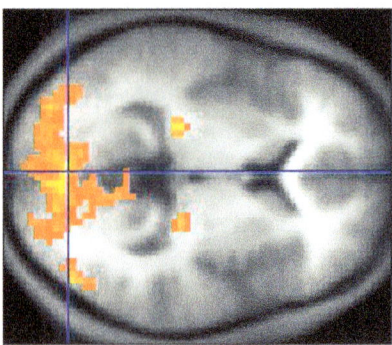

Figure 7.1 A Functional MRI image Showing Activity (Yellow-Red Coloration) in the Visual Cortex during Visual Stimuli

> **Functional Magnetic Resonance Image**
>
> An *fMRI* is an imaging procedure that uses magnetic resonance imaging to depict physiological functions and activities in the body, particularly in the brain. Increased activity in the brain means increased metabolic processes that lead to an increase in the oxygen content (oxygenation) in the active areas, and the difference between less and more oxygenated areas can be recognized and depicted very well by fMRI.

We already know that the visual cortex, as part of the brain, consists of neurons that are connected to each other. But how exactly are these neurons arranged and connected, and what tasks do they perform for image processing?

Fortunately, David H. Hubel and Torsten N. Wiesel provided important information on the structure of the visual cortex as early as 1958 and 1959 in a series of experiments on cats and monkeys (see Section 7.6). They were awarded the Nobel Prize in Physiology and Medicine in 1981 for these groundbreaking findings.

Among other things, Hubel and Wiesel identified the following properties of the neurons in the visual cortex:

- Many neurons only react to a small area of the visual field and therefore only have a local field of perception.
- The perceptual fields of the neurons can overlap.
- Neurons with the same perceptual field react differently. This means that one neuron reacts more to horizontal lines, while the other reacts to punctiform visual stimuli.
- Neurons with a larger perceptual field react to more complex stimuli that are transmitted to them by neurons with a smaller perceptual field.

How can we transfer these findings to a network structure? Because neurons only react to signals from a local perceptual field, and neurons with a larger perceptual field only react to stimuli from a section of upstream neurons, a fully connected network isn't required, but one with far fewer connections. The neurons are arranged in layers, with each subsequent layer reacting only to a local area of the previous layer.

The first work to implement the findings of Hubel and Wiesel's experiments is the *Neocognitron* in 1980. The CNN, whose original form was presented in 1998 in a paper by Yann LeCun (see the "Further Reading" section), developed as a result of technical developments and the availability of more computing power.

7.1.1 The Architecture of Convolutional Networks

The typical architecture of a CNN consists of two sections: the coding block and the prediction block. The *coding block* consists of a sequence of convolutional layer, activation layer, and pooling layer. These layers are used to extract the complex properties of an image. The result of the coding block is, as the name suggests, a coded representation of the image.

This is followed by the *prediction block*, which typically consists of a conventional neural network that classifies the coded representation of the image; usually a fully connected network is used for this, but it could actually be any machine learning (ML) method. Figure 7.2 shows the structure of a typical CNN with coding block and prediction block.

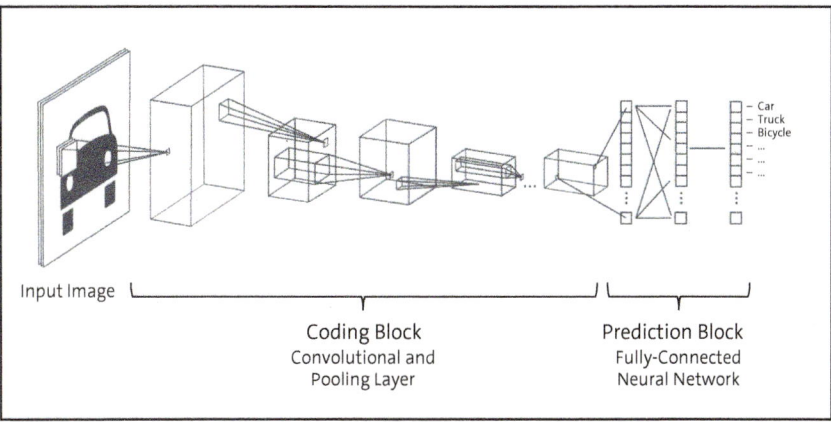

Figure 7.2 The Structure of a CNN with Coding and Prediction Blocks

What Is Convolution?

The mathematical concept of *convolution*—in our case, more precisely *discrete convolution*—can be understood as the product of two functions. In image processing, discrete convolution means "filtering" an image with a 3×3 or 5×5 matrix (see Figure 7.3), whereby there are different filter types (line filter, edge filter, blur filter, etc.). This is exactly what happens in the first layers of the CNN: the input image gets filtered to emphasize certain features (lines, edges, points, corners, etc.).

7 Examples of Deep Neural Networks

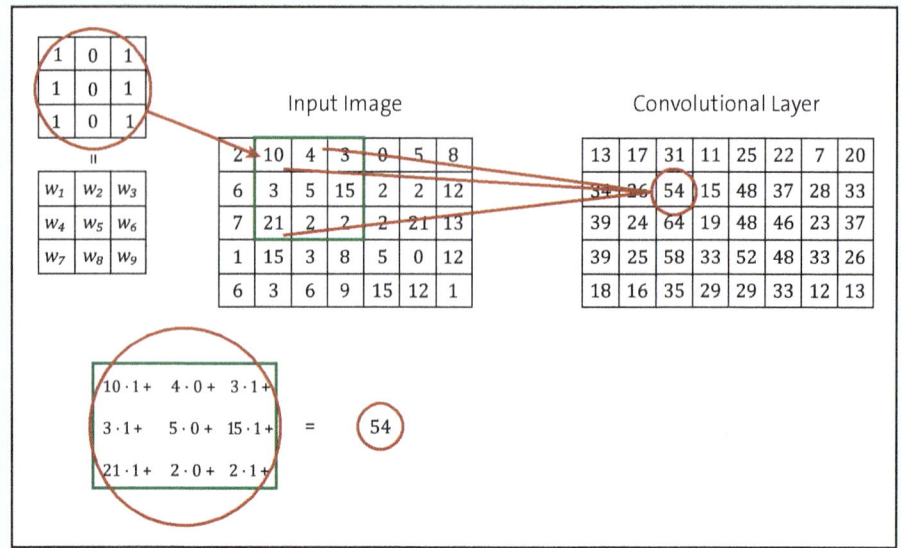

Figure 7.3 A Discrete Convolution with a Vertical Line Filter

7.1.2 The Coding Block

A convolutional module consists of a *convolutional layer, activation function*, and *pooling* or *subsampling layer* and is the essential component based on which the coding block of a CNN is built. This convolutional module can be connected multiple times in series. The underlying concept is that simple structures such as edges, lines, and points are recognized first. In the complex that follows then, structures are recognized that are made up of edges, lines, and points. With the next complex, even more complex structures are recognized, until finally a very large coding vector remains, which describes or codes the image.

The Convolutional Layer

In previous chapters, we've always considered and represented a layer in the neural network as a vector. Now, we consider the input as a two-dimensional plane or matrix. The transition from an image to a matrix is very simple, as you can see in Figure 7.4. This illustration shows a grayscale image or a channel (red, green, or blue) of a color image. The image is the input value for our CNN.

Let's now take a look at the first convolutional layer, which is positioned directly after the input image. This layer consists of a set of neurons, whereby each neuron reacts to a small section of the image, the local perception. The section of the image is typically a 3×3 or 5×5 matrix that runs across the entire image. However, each neuron in a plane only reacts to a certain pattern, which is given by a 3×3 or 5×5 filter. This means, for example, that all neurons in this layer only react to the *horizontal line* or *vertical edge* pattern in a specific 3×3 field of the input image. The advantage here is that the weights from the input image to the corresponding neuron in the convolutional layer are always the same for all filters (see Figure 7.5).

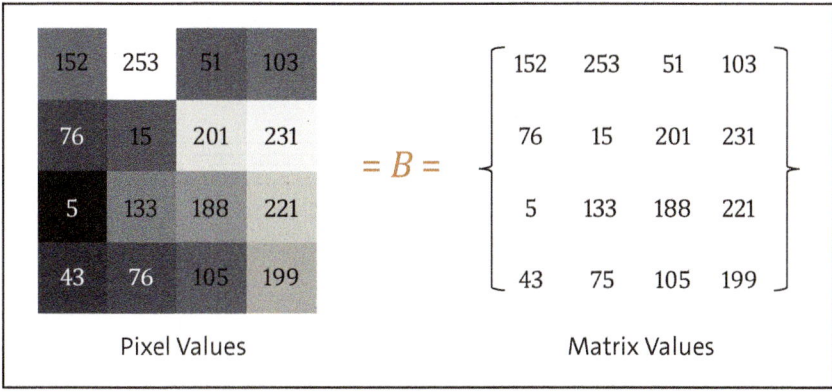

Figure 7.4 From the Image (Pixel Values) to the Matrix

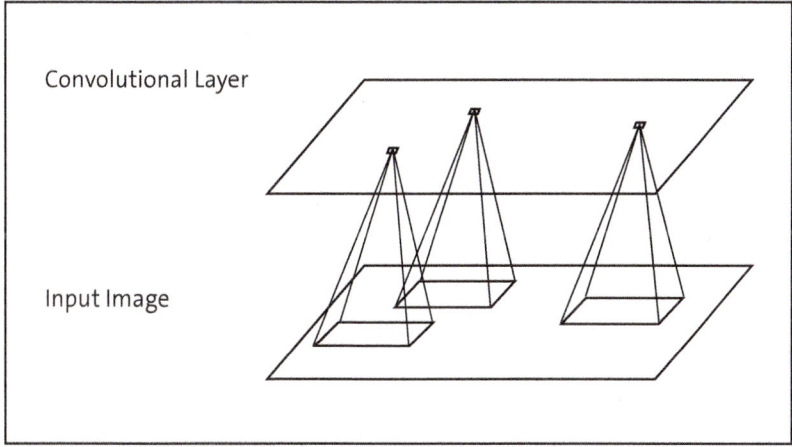

Figure 7.5 The First Convolutional Layer

If this level only reacts to one pattern, then that's not enough for us. This means that the first convolutional layer can consist not only of one layer but also of several layers, with each layer reacting to a different pattern or feature in the input image. The individual levels of a convolutional layer are therefore also referred to as *feature maps* (see Figure 7.6).

This defines the input signal for all neurons $c_{i,j}$ at position i and j of the feature maps in the first convolutional layer with the input image E and the corresponding pixels $e_{l,m}$ and a 3×3 filter $\vec{w}$ as follows:

$$c_{i,j} = w_1 e_{i-1,j-1} + w_2 e_{i-1,j} + w_3 e_{i-1,j+1} + w_4 e_{i,j-1} + w_5 e_{i,j} \\ + w_6 e_{i,j+1} + w_7 e_{i+1,j-1} + w_8 e_{i+1,j} + w_9 e_{i+1,j+1}$$

Note that the weights $\vec{w} = (w_1, \ldots, w_9)$ are the same for all neurons in a layer of the convolutional layer. Only these weights are therefore trained for a feature map during training.

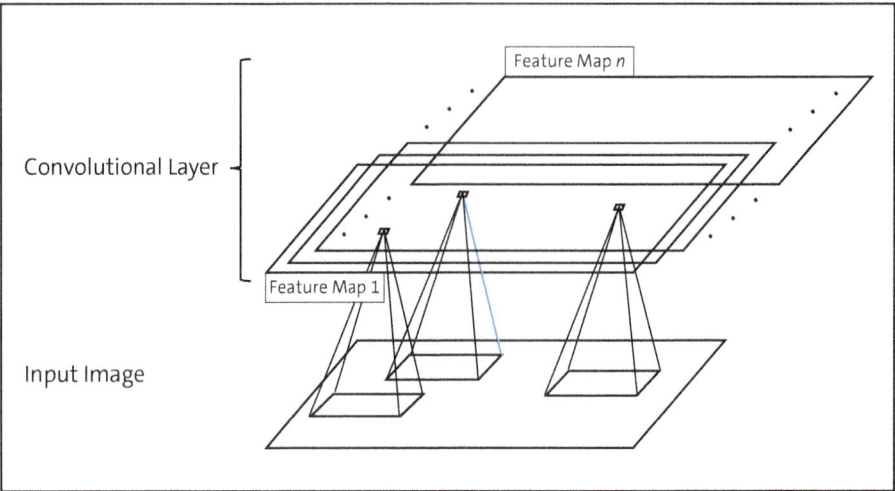

Figure 7.6 A Convolutional Layer Consists of Multiple Feature Maps

As a convolutional layer consists of multiple layers (feature maps), it's usually represented as a cuboid in symbolic representations (refer back to Figure 7.2).

How Many Feature Maps Do You Need?
Finding the right number of feature maps and which pattern in the image a feature map reacts to isn't so easy and is a matter of experience and experimentation. The good news is that the patterns (i.e., the weights represented by the 3×3 filter) are learned independently by the CNN! However, this requires a large number of images in the training phase, but we'll come to that later.

The Activation Function
As we already know, every neuron c also needs an activation function. The activation function has a significant influence on learning in a large neural network with many neurons and connections.

One of the activation functions best suited for CNNs is referred to as the *rectified linear unit* (ReLU) *function* (see Figure 7.7). The function's definition is very simple:

ReLU: $f_{act}(c) = f_{ReLU}(c) = \max(0, c)$

The input signal is c. One of the advantages of this function is the fast calculation, which is very important for large networks.

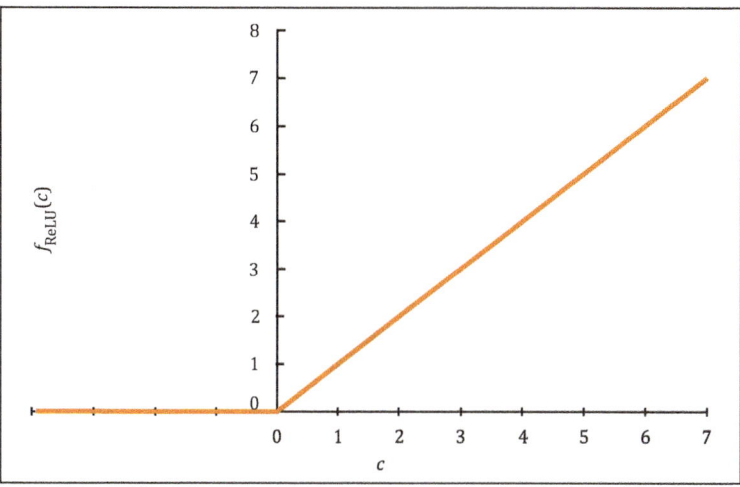

Figure 7.7 The ReLU Activation Function

> **What Is a Hyperparameter?**
>
> Every algorithm needs specific settings. These can be parameters for defining the activation function or even determining which activation function is used. This can also be the determination of which error function is used or very specific parameters that depend on the algorithm. The parameters can be compared to switches, wheels, and screws that have to be set before starting an engine and can have a significant influence on the result. We refer to this type of parameters as *hyperparameters*.
>
> There are, of course, methods that try to find the ideal settings independently. However, these methods are very time-consuming, as new training must be carried out for each change to a hyperparameter. They are therefore only suitable for tasks with a small amount of data or when the corresponding computing capacity is available.

The Pooling Layer

The concept of pooling originates from neurobiology and *lateral inhibition*, which describes an interconnection principle of nerve cells in which an active nerve cell inhibits the activity of neighboring cells.

The aim of pooling is to reduce dimensionality. This filters the layer in front of it and reduces its propagation in the x and y directions. The process is actually the same as for convolutional layers, except that there are no weights involved here. This is usually done by means of *max pooling*. Here, the maximum value is determined within a filter of a small size (2×2, 3×3), and only this value is passed on to the next layer; the other values are laterally inhibited, to put it in neurobiological terms.

There is also an *average pooling*, where the average of all values in the filter is calculated and passed on. Figure 7.8 shows that this is usually done with nonoverlapping filter ranges. In this example, the filter has a dimension of 2×2 and is always continued with a step length of 2—in contrast to the filter areas in the convolutional layer, where overlapping filter areas (i.e., local perception areas) are usually used.

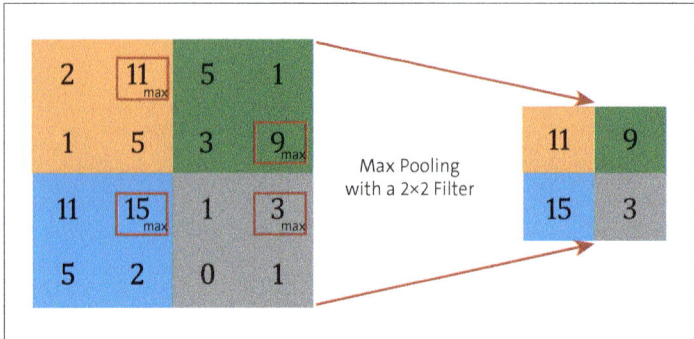

Figure 7.8 Example of Max Pooling with a 2×2 Filter

In addition, the pooling layer significantly reduces the number of weights to be learned, which also has an impact on the computing effort.

Overlapping, Padding, and Step Length

As you've seen, filters play an important role in CNNs. For this reason, you must take a few special features into account, which must also be defined as (hyper) parameters for the individual layers. We'll go into this in more detail here.

Let's start with the filling or *padding* of an image, which becomes necessary when we look at how filters work at the edge of an image. If we run a filter over the edge pixels, a problem arises, as indicated by the question marks in Figure 7.9.

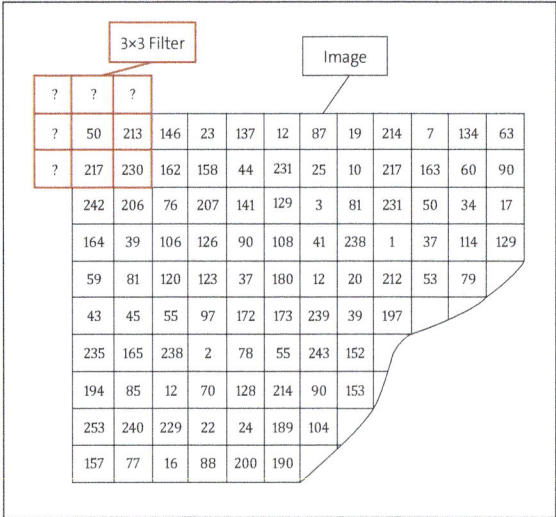

Figure 7.9 Filters: What to Do with Edge Pixels?

Which value should the filter use for multiplication? The solution is to fill in the missing area, which is also known as padding.

There are various ways to fill in the missing values. The Python library TensorFlow provides three options in this regard, which are shown in Figure 7.10: *constant padding*, *symmetric padding*, and *reflect padding*. Of course, you can also program the padding process in Python as you wish. The padding size must be adapted to the filter size. This means that only one additional edge pixel row or column is required for a 3×3 filter, while two edge pixel rows or columns are required for a 5×5 filter. Figure 7.10 shows an example of a 5×5 filter, as the three padding options are more clearly recognizable here. By default, constant padding with the value 0 is used, which is also referred to as *zero padding*.

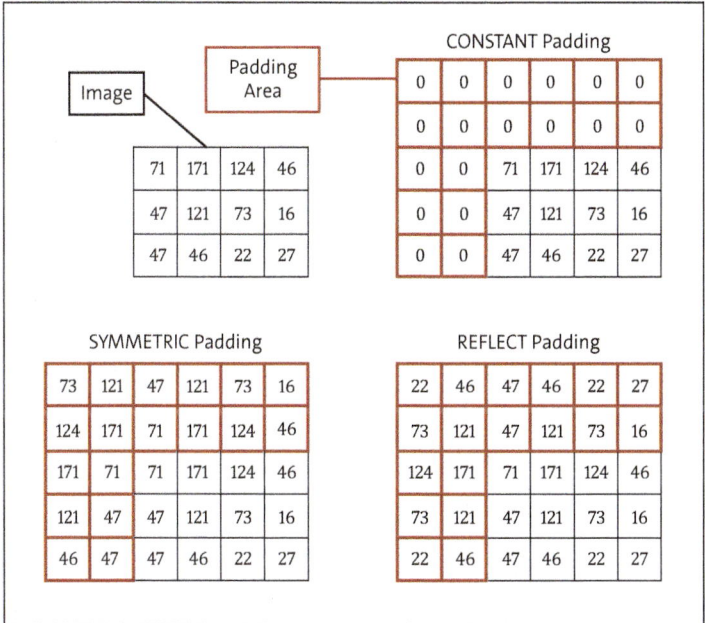

Figure 7.10 Padding Examples for a 5×5 Filter: TensorFlow

So far, we've been talking about a 3×3 or 5×5 filter, partly because filters of this size are used in the majority of CNN applications. To define the filter, you therefore need information on the extension $m.n$, where m defines the horizontal dimensions and n the vertical dimensions. But that's not enough, because we also have to define how the filter moves through the image. The filter is like a flashlight, so to speak, whose light field moves through the image. Of course, this doesn't happen at random, but row by row.

How exactly the light field moves through the image is indicated by the *step length* or *stride*. For the convolutional layer, strides are usually used where the filter areas overlap (see Figure 7.11). In pooling, strides are selected without overlapping.

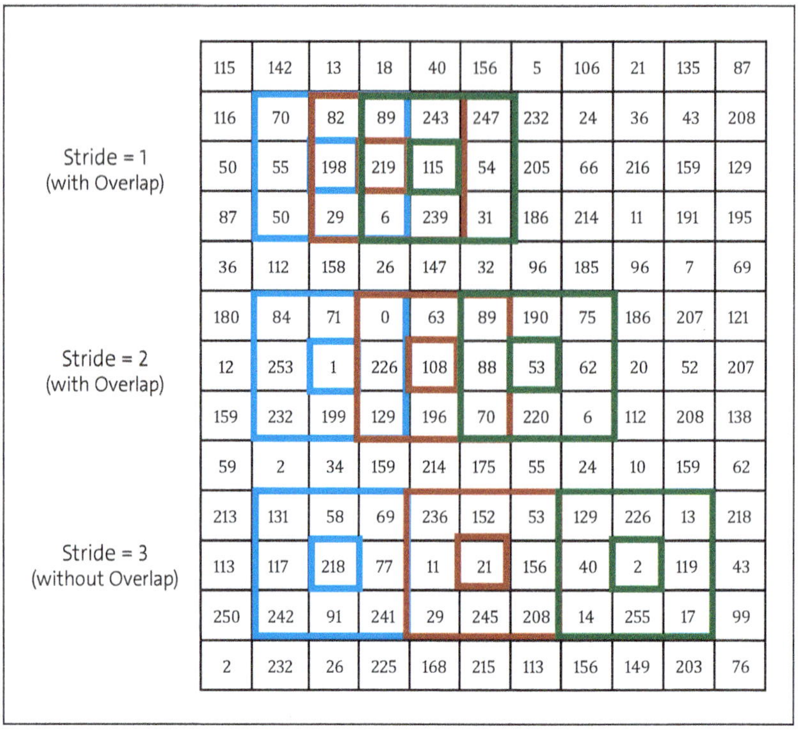

Figure 7.11 The Stride (Step Length) of a Filter

7.1.3 The Prediction Block

Basically, any ML procedure could be appended to the coding block to arrive at the final class. In typical CNNs, however, it's usually a multilayered, fully connected neural network.

Flatten

As shown earlier in Figure 7.2, the last layer of the coding block of a CNN is a pooling layer in which the neurons are arranged three-dimensionally and are also represented as a three-dimensional matrix.

The prediction block consists of a *fully connected neural network* that expects a one-dimensional vector as input. Thus, a flattening of the last pooling block is necessary, which transforms the neurons (or the values of the output signal) into a vector, as shown symbolically in Figure 7.12.

The way in which the coordinates from the pooling layer are arranged in the flattened vector is irrelevant for the subsequent classification. However, they are usually arranged in a specific order from the pooling layer.

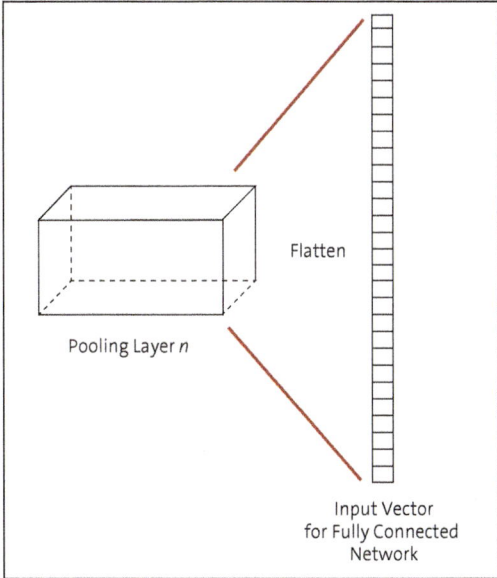

Figure 7.12 Flattening of the Last Block in the Coding Block

Softmax

In the previous chapters, there was usually only a single output neuron, as we often only had a binary task with the answers yes/no, 1/0, or true/false. CNNs are usually used for multi-classification tasks. This means that we specify classes (e.g., car, truck, or bicycle) and now want to assign our images to one of these classes, that is, classify them. More precisely, we want to know the probability of an image belonging to a class for all given classes. This is why the CNNs now always provide output vectors with the same number of elements as the specified classes (refer back to Figure 7.2).

The penultimate layer of the prediction block consists of just such a vector, which has the same number of coordinates as the classes to be recognized. Each coordinate is therefore assigned to a class, and the height of the value correlates with the probability of belonging to a class. This layer is also referred to as the *logits layer*.

If we now want to determine the class, we simply take the class with the highest value. However, if we're interested in the class probability, *softmax* is recommended as an activation function for the logits layer. Softmax transforms the values from the logits layer into a value range of 0 and 1, but in such a way that the sum of all values equals 1.

Softmax can also be regarded as an activation function and is defined as follows for k classes and $i = 1, \ldots, k$ for the class c_j:

Softmax: $f_{\text{act}}(c_j) = f_{\text{softmax}}(c_j) = \dfrac{e^{c_j}}{\sum_{i=1}^{k} e^{c_i}}$

An example is shown in Figure 7.13, where you can see very clearly that the highest logit value also leads to the highest value in the *softmax layer*. These numbers in the softmax layer indicate a probability distribution of the classes.

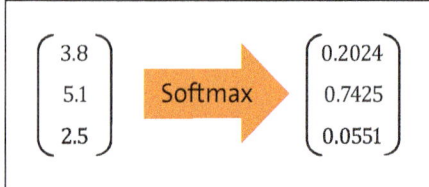

Figure 7.13 Example of Softmax for Three Classes

7.1.4 Training Convolutional Neural Networks

So far, we've mainly looked at the forward direction, that is, from the input image to the classification result. Now we also have to train the network, proceeding as described in Chapter 6. This means that CNNs are also usually trained using backpropagation. The classification error is thus used to change the weights of our network, in a backward direction, starting from the last layer of the network. The training of CNNs is therefore not too different from the training of conventional neural networks.

As you'll see in the following sections, we have to deal with a number of problems during training, such as the problem of exploding/vanishing gradients, overfitting (i.e., overfitting the network to the training dataset), and saturation (i.e., the network improves very little during training).

Many of these challenges result from the size of the CNNs, that is, the number of neurons, weights, or connections. Table 7.1 shows some well-known CNNs and the number of their parameters.

Name of CNN	Number of Parameters
LeNet5 (1998)	60,000
AlexNet (2012)	60,000,000
VGG (2014)	138,000,000
GoogleNet (2014)	4,000,000
ResNet50 (2015)	2,400,000

Table 7.1 Comparison of CNNs Based on Number of Parameters

The Problem of Exploding or Vanishing Gradients

The backpropagation learning algorithm starts from the output layer, works its way layer by layer toward the input layer, and calculates the error gradient in the process. If the error gradient is determined after each layer, the weights of the network can be adjusted based on the error gradients value.. The unpleasant thing is that these gradients become smaller and smaller the closer you get to the input layer, which means that the training doesn't converge (i.e., the weights values change only slightly) to a solution or does so only very slowly. In this case, we speak of a *vanishing gradient*.

Conversely, it can also happen—albeit less frequently and only in certain types of networks—that the error gradient values are big and thus the weights of the layers are changed extremely strongly and the training diverges as a result, which is then also referred to as the problem of *exploding gradients*.

Because of these issues, deep learning networks disappeared into oblivion for a long time. Xavier Glorot and Yoshua Bengio contributed with their work (see Section 7.6) to a better understanding of the causes of this behavior, which can be found in an unfavorable combination of initialization (the weights must have some initial value) and the activation function used (until then, the sigmoid function was very widely used).

Initialization

One basic idea is that the signal must be able to flow as unhindered as possible both in the forward direction for prediction and in the backward direction for passing through the gradients. Glorot and Bengio (Section 7.6) and also more recent work on this topic suggest that this should be done by randomly assigning the weights according to certain specifications.

The random values should be selected in such a way that they correspond to a normal distribution with the mean value 0 and a standard deviation or a uniform distribution between -v and +v. Both the standard deviation and the value v depend on the number of incoming and outgoing connections. Table 7.2 shows the specifications for random initialization depending on the selected activation function and the number of input connections (n_{in}) and output connections (n_{out}). The hyperbolic tangent as an activation function will be new to you, but it's a very common alternative to the sigmoid function, which is why we've included it in our table.

Activation Function	Uniform Distribution $(-v, v)$	Normal Distribution
Sigmoid function	$v = \sqrt{\dfrac{6}{n_{in} + n_{out}}}$	$\sigma = \sqrt{\dfrac{2}{n_{in} + n_{out}}}$
Hyperbolic tangent	$v = 4\sqrt{\dfrac{6}{n_{in} + n_{out}}}$	$\sigma = 4\sqrt{\dfrac{2}{n_{in} + n_{out}}}$
ReLU	$v = \sqrt{2}\sqrt{\dfrac{6}{n_{in} + n_{out}}}$	$\sigma = \sqrt{2}\sqrt{\dfrac{2}{n_{in} + n_{out}}}$

Table 7.2 Uniform Distribution and Normal Distribution for the Random Initialization of the Weights

The initialization types are typically named after the authors who first published them. The initialization for the sigmoid function according to Table 7.2 is therefore called *initialization according to Xavier Glorot* and that according to the ReLU function is called *initialization according to Kaiming He* (see Section 7.6).

But what exactly does that mean? Let's first draw a normal distribution, which is also known as a *Gaussian bell curve* because of its bell shape (see Figure 7.14).

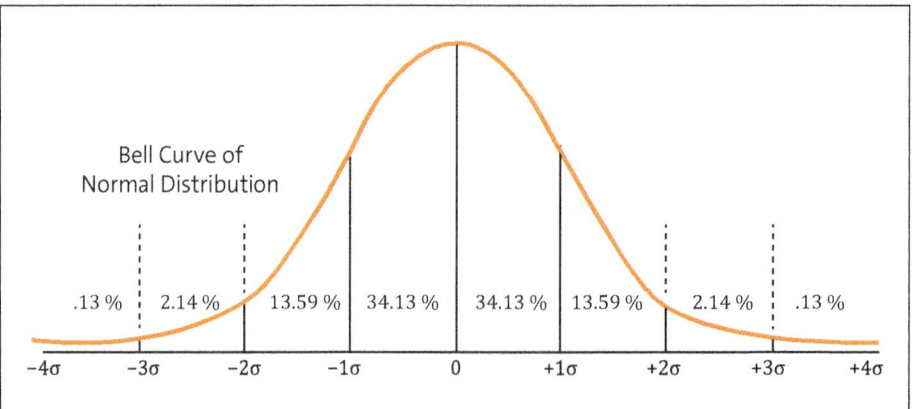

Figure 7.14 Typical Normal Distribution

Every generator of random values must function in such a way that the distribution is maintained. For a normal distribution with a mean value of 0 and the standard deviations defined in Table 7.2, this means that 64.26% of the weight values must lie between -1σ and $+1\sigma$.

Activation Function

The sigmoid function was long considered the ideal choice for the activation function for neural networks. The logistic sigmoid function in Figure 7.15 has the property that the derivative of this function is very small for large positive and negative input values. In a neural network, this also applies to our gradients. Basically, small gradients mean that we only move very slowly toward the optimum in the error area. Because the change in weights is dependent on a sequence of gradients of the individual layers, the total gradient becomes smaller the further you get toward the input layer. This is also called a *saturation* with large input values.

For this reason, activation functions have been considered because they aren't subject to the risk of saturation. The ReLU function proved to be very effective in solving the problem, partly because it can be calculated very effectively and quickly. This is crucial for neural networks with large numbers of parameters.

However, the ReLU function isn't quite perfect either, as the phenomenon of *dying ReLU* can occur. A neuron that enters this state always delivers the output 0 during learning and therefore makes no contribution to learning, regardless of the input.

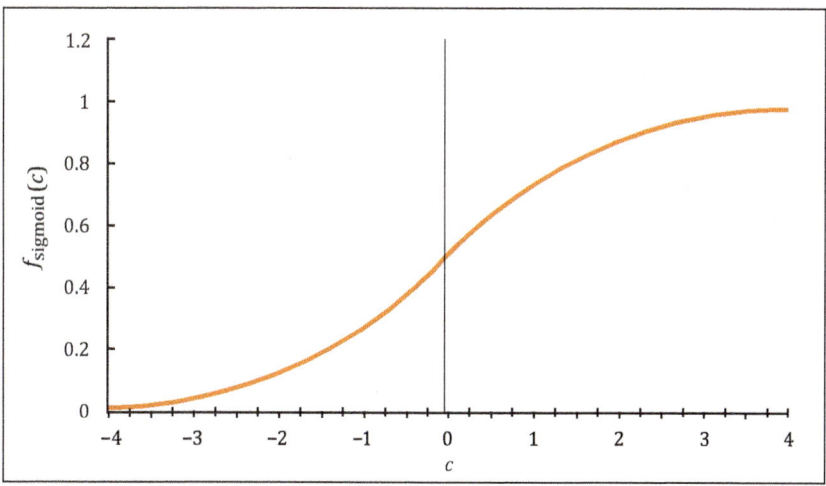

Figure 7.15 Sigmoid Function

However, there are variants of the ReLU function, such as the *leaky ReLU* or the *exponential linear unit (ELU) function*, which reduce or solve this problem. The leaky ReLU function (see Figure 7.16) is defined as follows:

ReLU: $f_{act}(x) = f_{lReLU}(x) = \max(\alpha x, x)$, wo $\alpha = 0.01$ (very small)

The hyperparameter α should be set very small. This ensures that the output signal is never 0, even if it's very small (and negative). But this gives a learning algorithm the chance to train and change this neuron again accordingly.

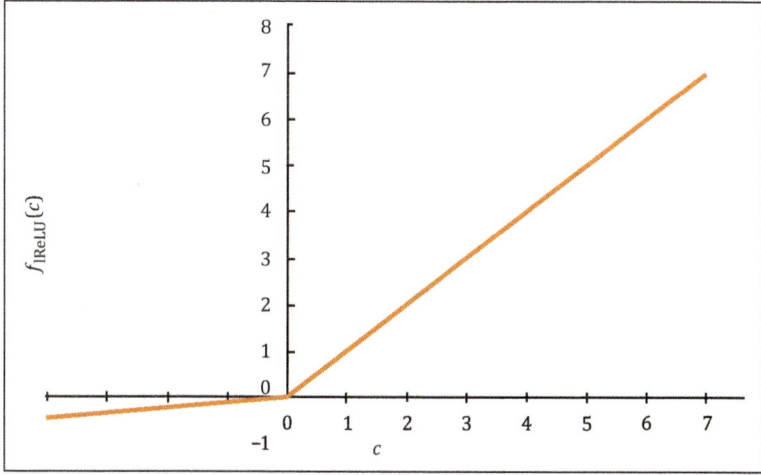

Figure 7.16 The Leaky-ReLU Activation Function

The ELU function is shown in Figure 7.17 and is defined in the following way:

ELU: $f_{act}(x) = f_{ELU}(x) = \begin{cases} \alpha(e^x - 1), & \text{if } x < 0 \\ x, & \text{if } x \geq 0 \end{cases}$

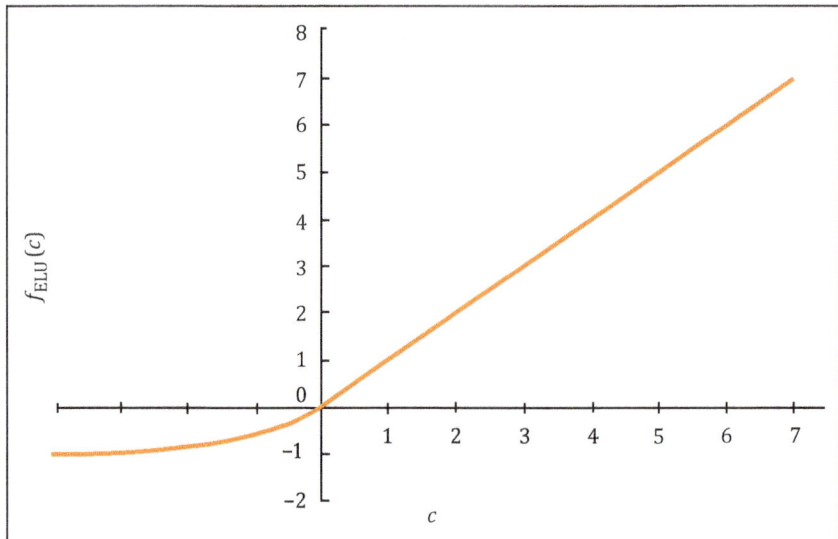

Figure 7.17 The ELU Activation Function

The ELU function does have the disadvantage that it's more complex to calculate. However, it converges faster, which means that training or learning takes less time compared to the ReLU activation function.

7.2   Transformer Neural Networks

Transformer neural networks are a relatively young but revolutionary architecture in the field of AI and ML. Originally presented in the paper "Attention Is All You Need" by Ashish Vaswani et al. in 2017, transformer networks have quickly gained popularity and are now a central component of many advanced systems or *large language models* (LLMs), especially those for natural language processing (NLP).

In contrast to traditional recurrent neural networks (RNNs), which process sequential data step-by-step and usually only from left to right, transformers work with a mechanism referred to as *self-attention*. This mechanism enables the model to consider and weigh different parts of an input sequence (i.e., a sentence or an entire text) simultaneously, regardless of how far apart these parts are. This makes transformer networks particularly effective when long sequences and complex dependencies are processed.

A major advantage of the transformer architecture is that it can be parallelized. While RNNs are difficult to parallelize due to their sequential nature and therefore often train more slowly, the structure of the transformer allows calculations to be performed independently and simultaneously. This leads to a considerable acceleration of the training process, especially with large datasets.

The transformer architecture essentially consists of an encoder-decoder system based on multiple layers of self-attention and feed-forward networks. Each layer in this system helps to filter out and combine the relevant parts of the input sequence to enable the most accurate prediction or translation possible.

In the following sections, we'll take a detailed look at the basic concepts and functionality of transformer neural networks. In particular, we'll focus on embedding, position coding, and the self-attention mechanism, which are crucial for the model's ability to capture the context and order of information in a sequence without errors.

7.2.1 The Network Structure

Let's take a look at the structure and functioning of a transformer neural network using a specific NLP task: we want to generate a text based on a short piece of text, which is called a *prompt*.

> **Prompt and Prompt Engineering**
>
> A *prompt* is an input or request given to an AI model, for example, a language model, to generate a specific response or output. This can be a single word, a sentence, or even a complex command that provides the model with the context and the direction in which the response should go.
>
> *Prompt engineering* refers to the art and science of designing and optimizing prompts in such a way that an AI model delivers the desired results. Careful consideration is given to how a prompt must be formulated to maximize the quality, relevance, and precision of the response generated by the model. This can be done by experimenting with different formulations, contextual information, and special instructions.
>
> Let's look at a quick example:
>
> - Prompt: "Explain the difference between weather and climate."
> - Prompt engineering: "Describe in simple terms the difference between weather and climate, and give examples of each."
>
> Prompt engineering is sometimes regarded as the profession of the future, but in the meantime, LLMs are already very good at improving prompts on their own.

Transformer neural networks are actually nothing more than next-word estimators that add the best matching word (or *token*, to be more precise) based on an input text. This process is repeated iteratively until an end-of-sentence token gets output by the model (see Figure 7.18).

For this purpose, the transformer neural network must specify a list of words (or tokens) and their probability in each step, as shown in Figure 7.18. The sum of the probabilities usually adds up to 100%. This sounds simple at first, but arriving at this list requires a complex but ingenious network structure and a very large amount of text (i.e., training data).

Based on the groundbreaking work mentioned earlier by Vaswani and his colleagues (see Section 7.6), we know a transformer neural network consists of an encoder and a decoder block.

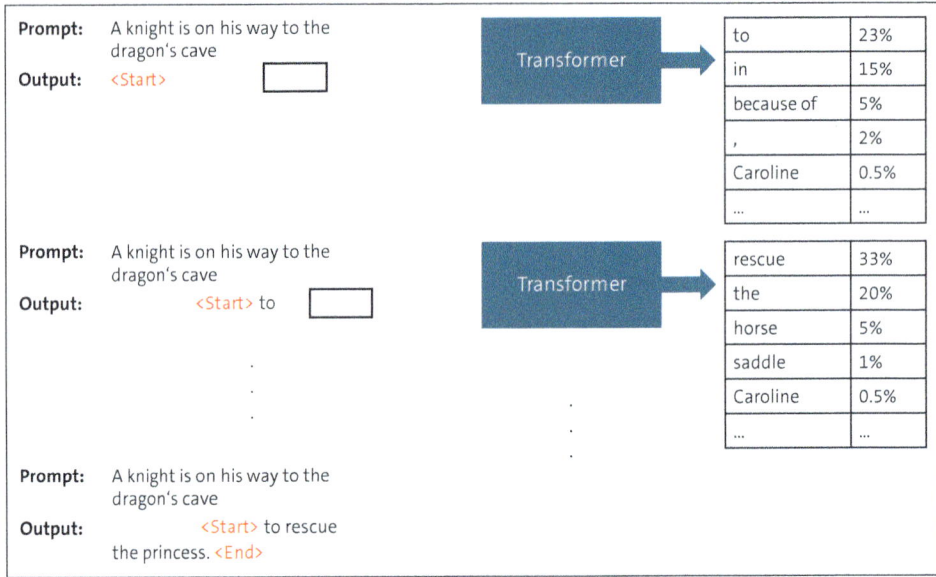

Figure 7.18 Transformer as Next-Word Estimator

Figure 7.19 provides a somewhat simplified overview of the structure of a transformer neural network. There are a lot of derivatives with minor or major modifications, but the basic structure with an encoder and a decoder block is present in all of them. To feed text into the network, we must first convert it into numbers through embedding.

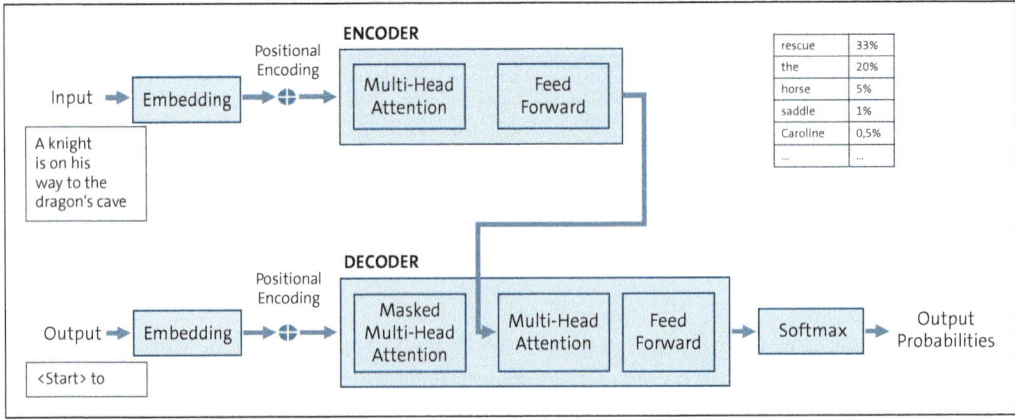

Figure 7.19 Architecture of a Transformer Neural Network

7.2.2 Embeddings

First, the input (i.e., sentence, paragraph, or entire book) must be split up. A natural division is the individual words and punctuation marks, and we can retain this as a model of thought. However, we're actually talking about tokens, which can be entire words, but often also parts of words or punctuation marks. There are also two special tokens: the <Start> and <End> tokens, which mark the start and end of the output. Each word (or token) is then represented by a vector.

Word embedding in Figure 7.20 is a simplified representation and is an anticipation of Chapter 11 in which we deal with the preprocessing of text data. You can imagine a word as a point in a high-dimensional space, where ideally words with a similar meaning are in close proximity to each other. When we talk about a multidimensional space here, it's not a five-dimensional space, as suggested in Figure 7.20, but can comprise several thousand dimensions in LLMs. (The embedding space for GPT-3 has more than 12,000 dimensions, for example.)

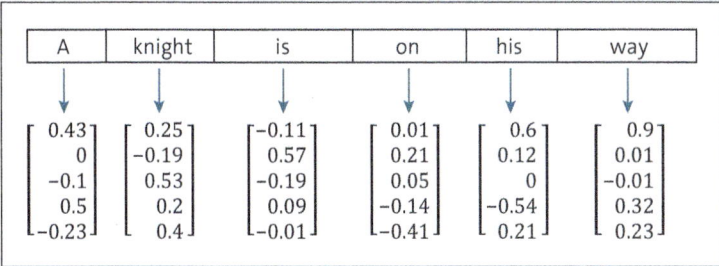

Figure 7.20 Simple Word Embedding

How is the embedding (i.e., vector values) determined? This happens in the course of training this neural network with a huge dataset. These word embeddings are nothing new in themselves and have been used in NLP for a long time. However, the meaning of a word (or token) depends not only on the word itself but also on the context in a text. Transformer networks manage to incorporate this context, adapt this original embedding vector depending on the context, and place it in the correct position in the embedding space (or meaning space).

7.2.3 Positional Encoding

Until the development of transformer neural networks, NLP tasks were solved with RNNs. The text was presented to this form of neural network word by word or token by token. The position of a word in a text is automatically given by the sequential feeding of the RNN with words or their embedding vectors and is therefore not an issue. Transformer neural networks, on the other hand, receive the input text in parallel, that is, at the same time. This makes it necessary to encode the position; otherwise, the difference in sentence meanings can't be recognized (see Figure 7.21).

7 Examples of Deep Neural Networks

> The fox chases the wolf
>
> The wolf chases the fox

Figure 7.21 Two Sentences with Different Meanings But the Same Words (Tokens)

A simple solution is an *absolute positional encoding* with digits, as shown in Figure 7.22, by numbering the words of the input. However, several problems can arise in this context. First, words with a higher numbering also have a higher meaning. Second, the inputs to a transformer neural network can be of different lengths and also contribute to unintentional shifts in meaning due to the absolute numbering. If the word combination "the fox" occurs again later, as in our example, then the higher numbering would change the meaning, even though it's the same fox in the same context.

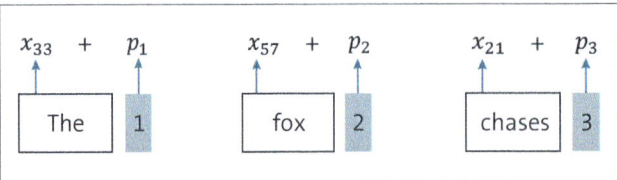

Figure 7.22 Absolute Positional Encoding

We want to use a variant of positional encoding that has the following properties:

- **Unambiguous**
  The position corresponds to the value.
- **Deterministic**
  Our variant is based on an underlying rule.
- **Distance invariant**
  This makes it possible to estimate the distance between the positions of any two tokens.
- **Independent of sequence length**
  The input can have a different number of tokens.

In his work on transformer neural networks, Ashish Vaswani proposed a *frequency-based positional encoding* that fulfills these properties (see Section 7.6). With frequency-based positional encoding, a vector is returned depending on the position number. This position vector P has the same length as the vectors of the word "embedding" (in our example, their are 5 vectors; with GPT-3, there are more than 12,000). We refer to the vector length as d_{emb}.

The frequency-based encoding is then based on sine and cosine functions in the following way: There is a run variable k, which takes the value from 1 to the vector length, d_{emb},

Another run variable, p, stands for the position number of a token in the text and can have a maximum value of the sequence length. We also define a constant L, which is usually set to the value 10,000. This constant represents a scaling factor that keeps the values of our position vector within reasonable limits. Now, we can determine the rules that define the values of the position vector with the length, d_{emb}. The term "frequency-based positional encoding" comes from the fact that sine and cosine functions are used for these rules.

What Do Sine and Cosine Have to Do with Frequencies?

Sine and cosine are trigonometric functions that are used to describe harmonic oscillations. An oscillation can be represented by a wave function that crosses the x-axis in a diagram. The more often this crossing occurs per interval, the higher the frequency (see Figure 7.23).

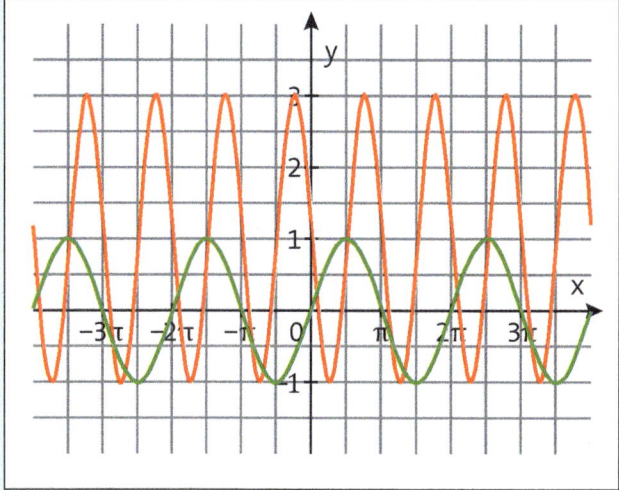

Figure 7.23 Two Oscillations of Different Heights, Which Can Be Represented by Sine and Cosine

The red function in this figure therefore has a higher frequency than the green function; both functions can be represented by sine and cosine.

A sine function is used for the even elements of the position vector and a cosine function for the odd elements. The reason for this is that for the value 0, all sine functions also return the value 0, and you could potentially get a zero vector for a position vector, which you can easily prevent by alternating sine and cosine functions.

So, if our run variable k is even, then we define the value at position k of the position vector as follows:

$$k \text{ even}: P^{k,p} = \sin\left(\frac{p}{L^{\frac{k}{d_{emb}}}}\right)$$

For the odd k, we define it as follows:

$$k \text{ odd}: P^{k,p} = \cos\left(\frac{p}{L^{\frac{k}{d_{emb}}}}\right)$$

Formulas are sometimes not so easy to understand, so we visualized the result in Figure 7.24.

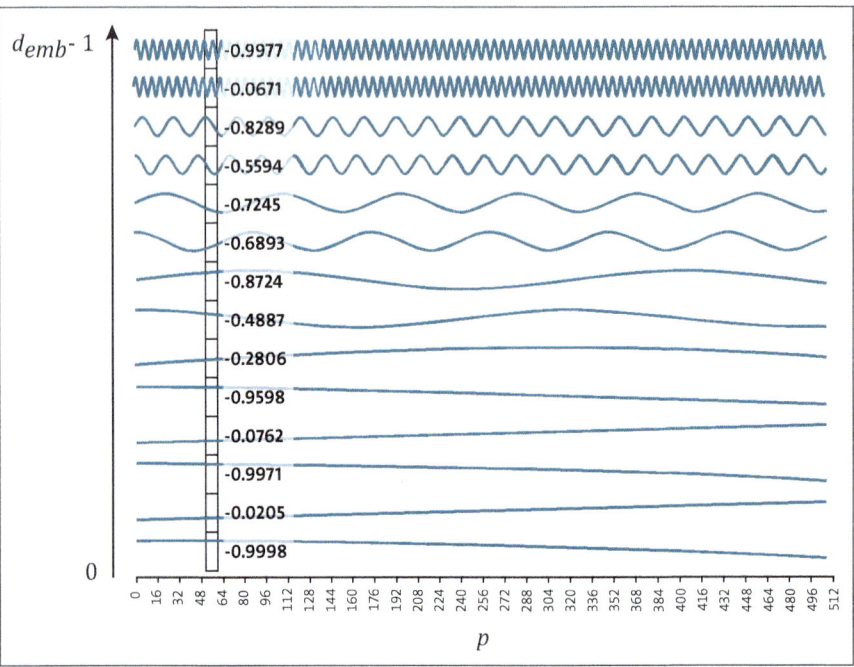

Figure 7.24 Visualized Frequency-Based Position Embedding

The word embedding vector and the position embedding vector are added together and serve as input for the encoder.

7.2.4 Encoder

The main task of the encoder is to convert the input sequence into mathematical vectors. It must take the position (via *positional encoding*) and the context information (via *multi-head attention*) into account accordingly to provide a rich, contextualized representation for further processing.

Self-Attention

One of the basic problems of text processing is the dependence of the meaning of a word on the context, that is, on the entire sequence of an input text. Recurrent networks were already able to cover this to a certain extent. However, the context dependency was only taken into account in one direction in the sentence. The second problem with recurrent networks is the fact that it's difficult to encode the context of longer text sequences. The key to the attention mechanism is that you have access to the entire (very large) sequence at the same time because a sequence is presented in its entirety to the transformer neural network and not word by word, as is the case with recurrent networks.

The task of *self-attention* is to increase the information accuracy of a sequence by taking the context into account. You essentially start with input embedding, that is, at a point in the multidimensional comprehension space, and move this point to the appropriate location in the *comprehension space* depending on the context. To achieve this, the relationship with all words (tokens) in the sequence is considered for each word (or token).

In Figure 7.25, you can see symbolically what these connections should look like and what influence a word change can have on them. This illustration shows the relationship of the word "it" to all other words. This representation could also be implemented as a square matrix in which the strength of the connection is entered. (In the figure, the darker the connecting line is colored, the higher the value.) The strength of the connection indicates how much attention "it" should pay to the other words in the sentence.

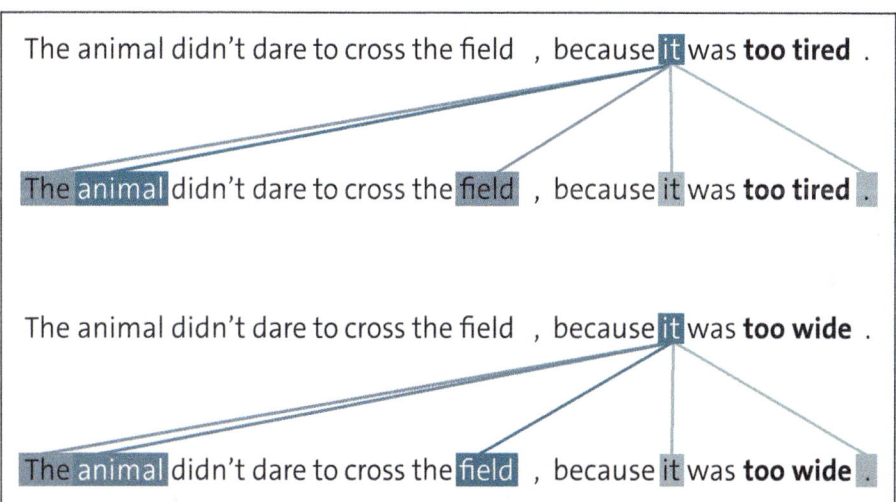

Figure 7.25 Illustration of a Self-Attention Layer in the Network

How can you calculate these connections? For each word, a *query vector*, a *key vector*, and a *value vector* are generated during training, which allow the relevance to be calculated.

> **Query, Key, and Value Vectors**
>
> For each word (token), a query, key, and value vector is calculated. Let's take a look at a sentence: "The cat chases the mouse."

- **Query**
  What is a particular word looking for or asking for? For the word level, the query vector could represent the following: "What other words am I related to?"
- **Key**
  How can a word be compared with other words? The key vector is a description of each word. The key vector of the word "mouse" could contain the following information: "I am an object that is being chased."
- **Value**
  What information can a word contribute as a result? If the model knows that "cat" and "mouse" are important, then the value vector will contain the relevant information that the model should use.

At the end of the self-attention step, each word has a new representation (embedding) that takes into account how strongly it's related to the other words in the sentence and is therefore ideally located in the appropriate place in the comprehension space.

If you've read this section carefully, you'll have noticed that in Figure 7.19, we don't talk about *self-attention*, but about *multi-head attention*. Instead of just one self-attention calculation (a single "head"), multiple self-attention calculations, that is, query, key, and value vectors, are executed in parallel. Each head can focus on different aspects of the word relationships (e.g., one head looks at the grammatical relationship, another at the semantic proximity of the words, etc.). The results of the different heads are combined to provide a more comprehensive representation of the input sequence.

7.2.5 Decoder

The decoder in a transformer neural network has the task of generating an output for the representations of the input sequence generated by the encoder. In a model for translations, the representations of an English text from the encoder could be used to create a German sentence through the decoder.

A decoder usually works autoregressively; that is, it generates the output (e.g., a word or token) step-by-step, with each new word being based on the previously generated words. This process is usually started with start information or a start token <Start>; the next word or token is generated from the text created so far, and this process is continued until an end symbol <End> is reached.

Masked Self-Attention

Of course, only the new tokens that have already been created may be used to generate the output and not tokens that may be created in the future. *Masked self-attention* is used to prevent future information from being processed. For example, if the model predicts the third word, it may only access the first two words, whereas the future words are masked. In terms of math, this is usually achieved by assigning very low or negative values to the strength of the connection to future words.

Encoder-Decoder Attention

With *encoder-decoder attention*, the decoder uses the representation of the encoder to learn which parts of the input sequence (generated by the encoder) are important for generating the next token. In this step, the relevance of the input sequence (e.g., the original text in German) is taken into account to correctly determine the next token (e.g., for a translation into English).

Output

We already know that transformer neural networks are next-word estimators. So, you might think that a word gets output in every step. But this isn't really the case; instead, the entire list of possible tokens and the corresponding probability are output. The tokens with the highest probabilities are then also the most likely candidates for the next position in the output sequence. For this step, a softmax function is used to ensure that the sum of the probabilities is equal to 1. In some transformer neural networks, you can set a hyperparameter called Temp(erature), which can assume a value between 0 and 1. If it's set to 0, the token with the highest probability is always used; otherwise, a word from the most probable tokens is selected at random. This randomness can be used to increase the "creativity" of a transformer neural network.

7.2.6 Training Transformer Neural Networks

Training a transformer neural network is hardly any different from training a CNN or any other neural network. However, transformer neural networks are usually trained with much larger datasets and require correspondingly more computing power.

The training dataset of transformer neural networks consists of input-output pairs that depend on the specific task:

- **Translation:**
  Enter: "The wolf chases the fox."
  Output: "Der Wolf jagt den Fuchs."
- **Text completion:**
  Input: "The wolf chases"
  Output: "the fox"
- **Text classification:**
  Input: "This movie was a disaster. I would never watch it again!"
  Output: "negative"

In Chapter 11, you'll learn how texts are divided into tokens. This list of tokens then forms the basis for further processing of this data.

7.3 The Optimization Method

As we've explained in Chapter 6, the optimization method of choice is a gradient descent method. This is a very unwieldy name for a very natural behavior, comparable to hiking in the mountains with the goal of getting down to the valley very quickly—you always choose the steepest descent.

With the amount of weights to be trained for convolutional or transformer neural networks, training can take a lot of time. This is why we're always on the lookout for faster gradient-based methods, which can often speed up the training process considerably.

We already know that the weights in the network are constantly undergoing small changes during training in the search for the optimum. In mathematical terms, for our weights W and the associated *cost function* $C(W)$ (also referred to as the *loss function*), the weights are changed by a small amount so that we follow the gradient direction, that is, the steepest direction, with the aim of minimizing the cost function. The gradient is therefore a direction vector that is derived from the first derivative over all weights:

$$\nabla_W C(W) = \begin{pmatrix} \frac{\partial}{\partial W_0} C(W) \\ \frac{\partial}{\partial W_1} C(W) \\ \vdots \\ \frac{\partial}{\partial W_n} C(W) \end{pmatrix}$$

This means that a change vector $\vec{h}$ is subtracted from the weight vector. This change vector then looks as follows in the standard gradient method:

$$\vec{h} = \eta \nabla_W C(W)$$

Here, η is our already known learning rate. We therefore obtain the new weight vectors by subtracting the change vector:

$$W^{new} = W^{old} - \vec{h}$$

The following two examples of optimization variants modify this change vector to speed up the training convergence.

7.3.1 Momentum Optimization

The term *momentum* stands for the physical quantity "impulse," which determines the mechanical state of motion of an object. Momentum optimization aims to take into account the momentum already picked up on the descent to reach the optimum faster, like a ball rolling downhill faster and faster. Our change vector $\vec{h}_{new}$ also takes into account the previous change vector $\vec{h}_{old}$:

$$\vec{h}_{new} = \eta_{momentum} \vec{h}_{old} + \eta \nabla_W C(W)$$

Although we have the disadvantage of another hyperparameter $\eta_{momentum}$ (it lies between 0 and 1), we also gain in this context. Imagine that the minimum is located in a very flat valley. With the standard method, we would only reach the optimum very slowly, whereas with momentum optimization, we take the momentum with us and reach our goal more quickly.

7.3.2 ADAM Optimization

Adaptive Moment Estimation (ADAM) takes into account past gradients and, in addition, their squared gradients calculated on a coordinate-by-coordinate basis. However, in contrast to the momentum method, only the average of the previous gradients and the squared gradients is used to change the weights:

$$W_{new} = W_{old} - \eta_{ADAM}(h_{old}, h_{old}^2)$$

What is special about this algorithm is that the learning rate is implicitly reduced or adapted, and the steeper the error terrain, the faster this occurs. This allows you to move more directly toward the optimum in the error space. Another advantage is that there is virtually no need to fine-tune the learning rate η.

The details of this method can be found in a paper by Diederik Kigma and Jimmy Lei Ba (see Section 7.6).

7.4 Preventing Overfitting

Due to the large number of parameters, another problem can occur, namely that the network gets trained exactly on the training dataset. If the trained network is then applied to new unknown datasets, it will fail. This is also referred to as *overfitting*, which means that the network has no generalization capability.

At this point we want to describe two very popular methods that address this problem.

7.4.1 Early Stopping

A popular and simple method to prevent overfitting is *early stopping*. The aim of learning is to minimize the network error by training with the training data. The error curve is similar to the blue curve shown in Figure 7.26; it improves after each training cycle and approaches zero.

The error curve of the validation dataset (shown in red in Figure 7.26) is similar, but the error increases again from a certain point. At that point, you enter the area of overfitting, where the network adapts too much to the training dataset and loses its ability to generalize.

7 Examples of Deep Neural Networks

> **Training and Validation Data**
>
> Every ML process needs a training dataset and a validation dataset. The validation dataset must not be used to train the neural network, but serves the sole purpose of determining the quality of the previously trained network, that is, to validate it.

Early stopping means ending the training at this point, although the training error could be reduced even further.

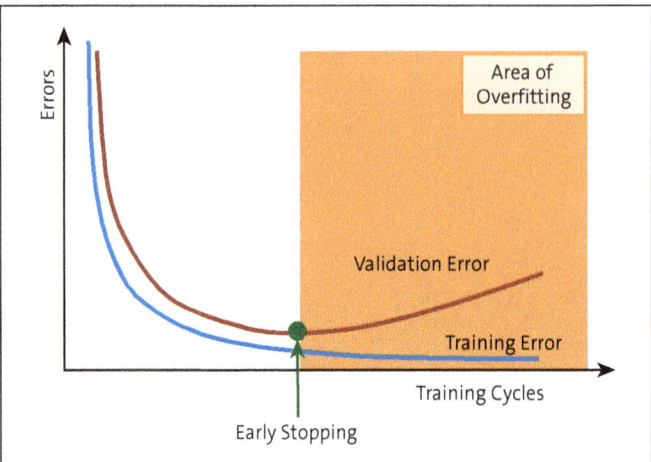

Figure 7.26 Error Curve for Early Stopping

7.4.2 Dropout

The *dropout* is a very interesting idea proposed by Geoffrey E. Hinton and Nitish Srivastava (see Section 7.6), which they have proven to be particularly successful for deep neural networks. With this method, a certain number of neurons are randomly left out of each training cycle. This means that they are simply not taken into account for the weight update (see Figure 7.27). The method is implemented by a new hyperparameter, the dropout rate p, which indicates the probability that a neuron will or won't be included in the training cycle. However, this probability is of course only assigned to the input neurons and the neurons of the hidden layers—but not to the output layer.

With the dropout method, it takes more training cycles for the network to converge, but the computing effort per training cycle is lower.

It's not easy to explain why dropout works so well. One attempt to explain this is that the neurons themselves become more independent. This means a neuron in one layer can't behave similarly to a neighboring neuron, as it's not guaranteed that both neurons are always present in the training cycle. In addition, it may also be due to the fact that you actually train a different network in each training cycle. However, these networks aren't completely independent of each other, as they share weights.

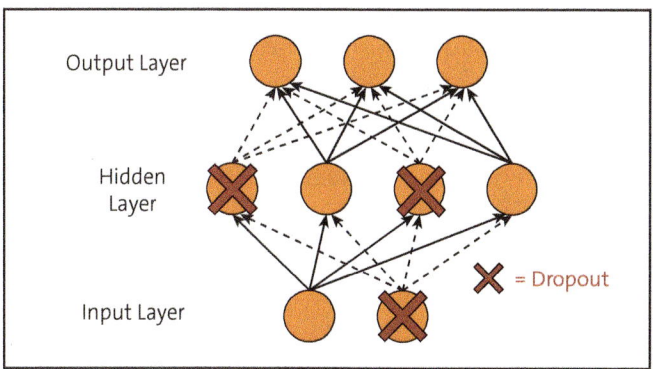

Figure 7.27 Network with Dropout Neurons (Dropout Rate p = 50%)

7.5 Summary

In this chapter, you've learned about two very powerful network types from the family of deep neural networks: CNNs and transformer neural networks. CNNs belong to the class of discriminative AI, whose task is to differentiate and categorize, and which are primarily used for image and video data. Transformer neural networks, on the other hand, are representatives of generative AI, as they generate new data in the form of text to fulfill NLP tasks.

Both network types can contain millions of trainable parameters, which is an indicator of their performance.

Now that we've dealt with the different types of networks in this chapter, without being distracted by the programming, we'll turn to the hands-on application of our newfound knowledge in the following chapter.

7.6 Further Reading

- "ADAM: A Method for Stochastic Optimization" by Diederik Kigma and Jimmy Lei Ba (2015, *https://arxiv.org/pdf/1412.6980v8.pdf*)
- "Attention Is All You Need" by Ashish Vaswani et al. (2017, *https://arxiv.org/abs/1706.03762*)
- "Delving Deep into Rectifiers: Surpassing Human-Level Performance on ImageNet Classification" by Kaiming He et al. (2015, *https://arxiv.org/pdf/1502.01852v1.pdf*)
- "Dropout: A Simple Way to Prevent Neural Networks from Overfitting" by Nitish Srivastava (2014, *http://jmlr.org/papers/volume15/srivastava14a/srivastava14a.pdf*)
- "Gradient-Based Learning Applied to Document Recognition" by Yann LeCun et al. (1998, *https://hal.science/hal-03926082/document*)

- "Improving Neural Networks by Preventing Co-Adaptation of Feature Detectors" by Geoffrey E. Hinton et al. (2012, *https://arxiv.org/pdf/1207.0580.pdf*)
- "Receptive Fields of Single Neurones in the Cat's Striate Cortex" by David H. Hubel and Torsten N. Wiesel (1959, *www.ncbi.nlm.nih.gov/pmc/articles/PMC1363130*)
- "Transformers (How LLMs Work) Explained Visually," (3Blue1Brown, *https://youtu.be/wjZofJX0v4M?si=GvUiEwEyYJ4YITTK*
- "Understanding the Difficulty of Training Deep Feedforward Neural Networks" by Xavier Glorot and Yoshua Bengio (2010, *http://proceedings.mlr.press/v9/glorot10a/glorot10a.pdf*)

Chapter 8
Programming Deep Neural Networks Using TensorFlow 2

Train, analyze, and win.

In this chapter, we'll focus on programming convolutional neural networks (CNNs) and transformer neural networks using the TensorFlow 2 Python library. (See Appendix C for more details on TensorFlow 2 and the integrated Keras library.) Based on the knowledge you acquired in Chapter 7, we'll now venture to implement our own CNN for handwriting recognition in Section 8.1. Then, we use pretrained networks for image classification (Section 8.2) and for natural language processing (NLP) tasks (Section 8.3).

8.1 Convolutional Networks for Handwriting Recognition

For this task, we use the *Modified National Institute of Standards and Technology (MNIST) dataset*, which provides a dataset of handwritten digits. This dataset is often used to compare handwriting recognition methods, partly because the data no longer needs to be preprocessed. Yann LeCun was the first to use CNNs for handwriting recognition based on the MNIST dataset in 1998. This was the starting point for the wave of success of this special network architecture.

8.1.1 The MNIST Dataset

The MNIST dataset consists of 60,000 images (for training) and an additional 10,000 images (for testing) with handwritten digits, as shown in Figure 8.1. Each digit image consists of 28×28 grayscale pixels in the value range from 0 to 255.

> **Using the Code Listings**
>
> The code listings used in this book mostly build on each other and would therefore not work on their own. We've summarized the related code listings in the download area for this book in Jupyter Notebooks. The code listings in the individual cells of the notebook must be executed one after the other.

Figure 8.1 Excerpt from the MNIST Dataset

Before we start loading the dataset, we need to load the necessary libraries as usual. We use the Keras library integrated in TensorFlow for this purpose, as shown in Listing 8.1.

```python
from tensorflow.keras.datasets import mnist
from tensorflow.keras.utils import to_categorical

(train_images, train_labels), (test_images, test_labels) = mnist.load_data()

print("Shape training data: {}".format(train_images.shape))
print("Dimension of image No. 5: {}".format(train_images[5].shape))
print("Label for image No. 5 {}".format(train_labels[5])
# Output:
Shape training data: (60000, 28, 28)
Dimension of image No. 5: (28, 28)
Label for image No. 5: 2
```

Listing 8.1 Loading the MNIST Dataset from the tensorflow.keras Library

> **Keras**
>
> The Keras library developed by François Chollet deserves a special mention. Keras is a Python library that provides a standardized interface for various neural network libraries, such as TensorFlow, Microsoft Cognitive Toolkit, or Theano. The aim of Keras is to make the use of these libraries as user-friendly as possible for beginners. Although Keras remains an independent library, it was integrated from TensorFlow version 1.4 onward and has been further adapted and extended to TensorFlow from version 2.0 onward.

As you would expect, an image has 28×28 pixels, and the corresponding label is an integer, which is shown in Listing 8.1 using the example of image No. 5. But that's not quite enough for us because we need to reshape our dataset.

First, we need to add a dimension to our image dataset. This additional dimension is required for the number of image channels in the image. In the case of the MNIST data, these are grayscale images, that is, images that have only one color channel. Color images need three channels for red, green, and blue. In addition, we transform the pixel values from the 0–255 range into the 0–1 range. It's advantageous to standardize the input data to a certain standard range because neural networks can handle this better. This applies in particular to training data whose features originate from different value ranges or physical units.

But we also have to change the label or the class. Because this is a multiclass task with exactly 10 classes, we want to determine the membership of each class as an intermediate step. The highest class ranking then ultimately determines the class. However, this means that our initial result doesn't consist of just 1 neuron, but of 10 neurons—one for each class. We achieve this by using the to_categorical() function, which, for example, converts the class with the number 3 into a 10-digit binary vector with a 1 at the fourth digit. The position of the vector is as follows:

(0,0,0,1,0,0,0,0,0,0)

This type of converting a number into a vector is also referred to as *one-hot encoding*. You can find out more about this in Chapter 11, Section 11.3. In Listing 8.2, we've now carried out the preprocessing.

```python
train_images = train_images.reshape((60000, 28, 28, 1))
train_images = train_images.astype('float32')
train_images /= 255

test_images = test_images.reshape((10000, 28, 28, 1))
test_images = test_images.astype('float32')
test_images /= 255

train_labels = to_categorical(train_labels)
test_labels = to_categorical(test_labels)

NrTrainimages = train_images.shape[0]
NrTestimages = test_images.shape[0]
```

Listing 8.2 Loading and Preprocessing the MNIST Dataset

Before we continue, let's display some information about our MNIST dataset, as shown in Listing 8.3.

```python
print("Training dataset:{}".format(train_images.shape))
print("Test dataset:{}".format(test_images.shape))

print("We have {} training images and {} test images.".format(NrTrainimages,
NrTestimages))
```

```
# Output:
Training dataset:(60000, 28, 28, 1)
Test dataset:(10000, 28, 28, 1)
We have 60000 training images and 10000 test images.
```

Listing 8.3 Format of the Modified Training and Test Dataset

Let's now display a randomly selected image so that we can be sure we're using the correct data. In addition, we want to display the corresponding class as a vector in the title. We use the `matplotlib` library to display the images and `random` to generate a random number, as shown in Listing 8.4.

```
%matplotlib inline
import matplotlib.pyplot as plt
import random

# Random number between 0 and 60000,
# since our training dataset comprises 60000 images

randindex = random.randint(0,60000)
plttitle = "Training image No. {} \n Class: {}".format(randindex,train_labels[randindex])
plt.imshow(train_images[randindex].reshape(28,28), cmap='gray')
plt.title(plttitle)
plt.axis('off')
plt.show()
```

Listing 8.4 Display of a Randomly Selected MNIST Image

In Listing 8.4, we use `matplotlib` to display the image. The commands in this library are almost self-explanatory due to their name. As `matplotlib` is used to display mathematical objects and functions, an x-axis and a y-axis always appear in displays. This is important for functions, but not necessarily for images. We therefore simply switch off the display by using `plt.axis('off')`. This listing creates a result like the one shown in Figure 8.2.

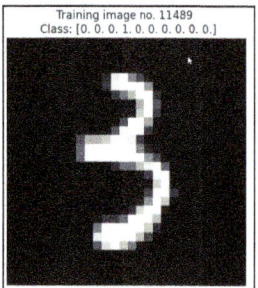

Figure 8.2 Randomly Selected Training Image of the MNIST Dataset

8.1.2 A Simple Convolutional Neural Network

Based on our dataset, we'll now build a CNN. Because we also want to know what the calculation graph looks like and how the loss and accuracy develop, we need to output this information accordingly to display it visually using *TensorBoard*. You can see the structure of this network in Figure 8.3.

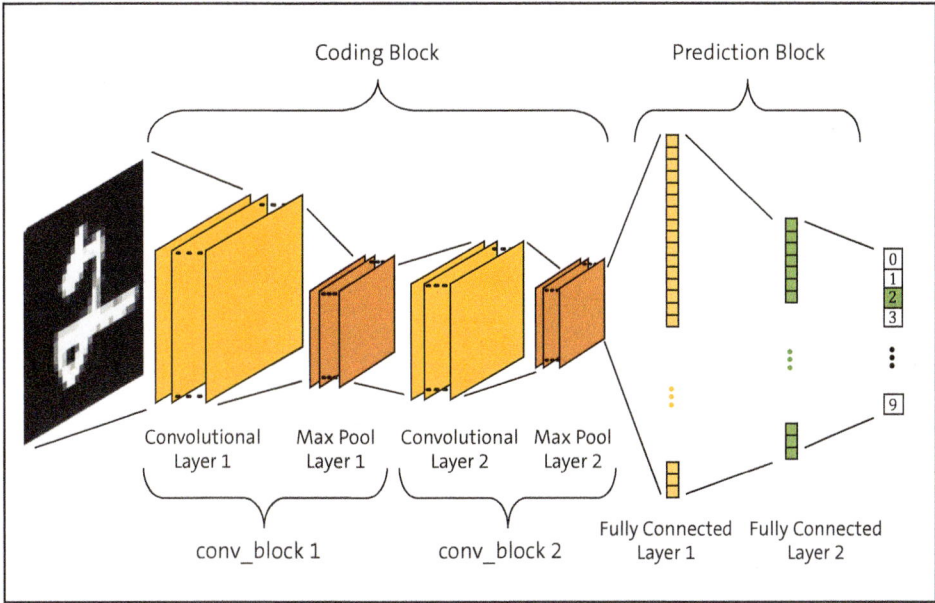

Figure 8.3 The Structure of Our CNN for the MNIST Dataset

The coding block of our CNN consists of two convolutional blocks, whereby a block always consists of a convolutional layer and a max pooling layer. The prediction block consists of two fully connected layers.

For the first convolutional block, we want to create 32 feature maps with a 5×5 filter, followed by a max pooling layer with a 2×2 filter.

The second convolutional block consists of 64 feature maps and a 5×5 filter, followed by a max pooling layer with a 2×2 filter. It's quite common for the later blocks to generate more feature maps because in the first block, simple elements in the image are recognized, such as horizontal or vertical lines and/or edges, while in the subsequent blocks, more complex combinations of these simple elements are added.

After the second convolutional block, we add a dropout.

We start by importing all the necessary libraries, as shown in Listing 8.5.

```
import os
import numpy as np
```

```python
from tensorflow.keras.callbacks import TensorBoard
from tensorflow.keras.models import Sequential
from tensorflow.keras.layers import Dense, Dropout, Flatten
from tensorflow.keras.layers import Conv2D, MaxPooling2D
from tensorflow.keras import backend as K
```

Listing 8.5 Importing the Libraries

You can already recognize all the functions for producing our CNN from the import: `Conv2D()` for our convolutional layer, `MaxPooling2D()` for the pooling layer in the coding block, and `Dense()` for the fully connected layer in the prediction block.

The Model

The next step is to specify a model consisting of the following elements:

- Network architecture with hyperparameters
- Loss function (or *loss*)
- Optimizers
- Model evaluation

After that, we can start training and evaluating our model.

The network architecture is created as shown in Listing 8.6.

```python
# Defining the format of the input data
mnist_inputshape = train_images.shape[1:4]

# The network architecture
model = Sequential()

# Coding block
model.add(Conv2D(32, kernel_size=(5,5),
       activation = 'relu',
       input_shape=mnist_inputshape))
model.add(MaxPooling2D(pool_size=(2,2)))
# Conv_Block 2
model.add(Conv2D(64, kernel_size=(5,5),activation= 'relu'))
model.add(MaxPooling2D(pool_size=(2,2)))
model.add(Dropout(0.5))

# Prediction block
model.add(Flatten())
model.add(Dense(128, activation='relu', name='features'))
model.add(Dropout(0.5))
```

```
model.add(Dense(64, activation='relu'))
model.add(Dense(10,activation='softmax'))
model.summary()
```

Listing 8.6 Code Section for the Network Architecture

Using `model.summary()`, we can display the network architecture very clearly, with some interesting additional information (see Figure 8.4). The few lines from Listing 8.6 have enabled us to create a CNN with almost 200,000 parameters!

```
Layer (type)                 Output Shape              Param #
=================================================================
conv2d_8 (Conv2D)            (None, 24, 24, 32)        832
_____
max_pooling2d_8 (MaxPooling2 (None, 12, 12, 32)        0
_____
conv2d_9 (Conv2D)            (None, 8, 8, 64)          51264
_____
max_pooling2d_9 (MaxPooling2 (None, 4, 4, 64)          0
_____
dropout_8 (Dropout)          (None, 4, 4, 64)          0
_____
flatten_4 (Flatten)          (None, 1024)              0
_____
features (Dense)             (None, 128)               131200
_____
dropout_9 (Dropout)          (None, 128)               0
_____
dense_7 (Dense)              (None, 64)                8256
_____
dense_8 (Dense)              (None, 10)                650
=================================================================
Total params: 192,202
Trainable params: 192,202
Non-trainable params: 0
_____
```

Figure 8.4 Output of the model.summary() Function

You heard about padding and stride in Chapter 7, Section 7.1.2, but we haven't yet defined anything of this kind in Listing 8.6. TensorFlow uses the usual default values here; we're happy with this, but you could also set parameters for padding and stride yourself.

In the next step, we define the loss function and the optimization method and compile, that is, create, our model (see Listing 8.7).

```
model.compile(loss='categorical_crossentropy',
              optimizer='Adam',
              metrics=['accuracy'])
```

Listing 8.7 Defining the Loss Function and Optimization

We use *categorical cross entropy* as the loss function and the Adam optimizer, which you already learned about in Chapter 7, Section 7.3.2, as the optimization method. The categorical cross entropy loss function L_i for a data point i gets defined as follows:

$$\sum_j y_{i,j} \log(\hat{y}_{i,j})$$

Here, j denotes the class, y the target label (i.e., 0 or 1), and $\hat{y}$ the prediction (i.e., a value between 0 and 1).

The Training

Now that we've defined our model, it's time to run and train the model. We also need to specify certain settings, namely the number of training epochs and the batch size.

A *training epoch* consists of the presentation of a training dataset and the adaptation of the model. The number of training epochs tells us how often we present the training data to the network and how often the model gets adjusted. We could present exactly one single dataset per training epoch, which is called *stochastic gradient descent*, and then update our model. A real disadvantage, however, is the significantly higher computing effort. This is still possible for our example, but it poses a problem for larger networks and input data. If we present the entire training dataset for a training epoch and update the model only afterwards, this method is referred to as a *batch gradient descent*.

Although the batch gradient descent method requires fewer model updates, it's not yet ideal because for very large amounts of training data, the model update and thus the training process itself can become very slow. The method that has established itself in this context is referred to as the *mini-batch gradient descent*. We only present the network with a certain amount of training data and modify the network model after that; that is, we change its weights. Although we have to introduce an additional hyperparameter, the *batch size*, the advantages clearly outweigh the disadvantages. TensorFlow is prepared for this and provides a corresponding parameter in the model.fit() function (referred to as batch_size in Listing 8.8).

In Listing 8.8, we define the hyperparameters, prepare TensorBoard, and carry out the training.

```
# TensorBoard - preparation
LOGDIR = "logs"
my_tensorboard = TensorBoard(log_dir = LOGDIR,
      histogram_freq=0,
      write_graph=True,
      write_images=True)
# Hyperparameters
my_batch_size = 128
my_num_classes = 10
```

8.1 Convolutional Networks for Handwriting Recognition

```
my_epochs = 12

history = model.fit(train_images, train_labels,
     batch_size=my_batch_size,
     callbacks=[my_tensorboard],
     epochs=my_epochs,
     verbose=1,
     validation_data=(test_images, test_labels))

# Output:
Train on 60000 samples, validate on 10000 samples
Epoch 1/12
60000/60000 [==============================] - 114s 2ms/step - loss: 0.3730 -
acc: 0.8801 - val_loss: 0.0554 - val_acc: 0.9831
...
```

Listing 8.8 Training the Model Using model.fit()

Listing 8.8 already shows the results for the loss function (`loss`) and accuracy (`acc`) for the training dataset and for the test dataset.

> **Task**
>
> Change the model in such a way that a 3×3 filter is used for the first convolutional block and a 3×3 filter for the second convolutional block. The dropout layer in the prediction block is supposed to be omitted.
>
> Compare loss and accuracy!

8.1.3 The Results

Figure 8.5 shows a section of the network structure as a TensorBoard graph.

> **TensorBoard**
>
> To see the results yourself using TensorBoard, you can start it in a terminal (preferably from Anaconda Navigator in the corresponding environment) using the following command:
>
> `tensorboard -logdir "logs"`
>
> Warning! In the terminal, you must first change to the directory in which the Jupyter Notebook is located. Instead of `"logs"`, you must enter the directory you selected in Listing 8.8 for the `LOG_DIR` variable.
>
> Then, open a web browser, and start the TensorBoard web application via the URL *http://localhost:6006/* (or, if this doesn't work, use *http://computername:6006*).

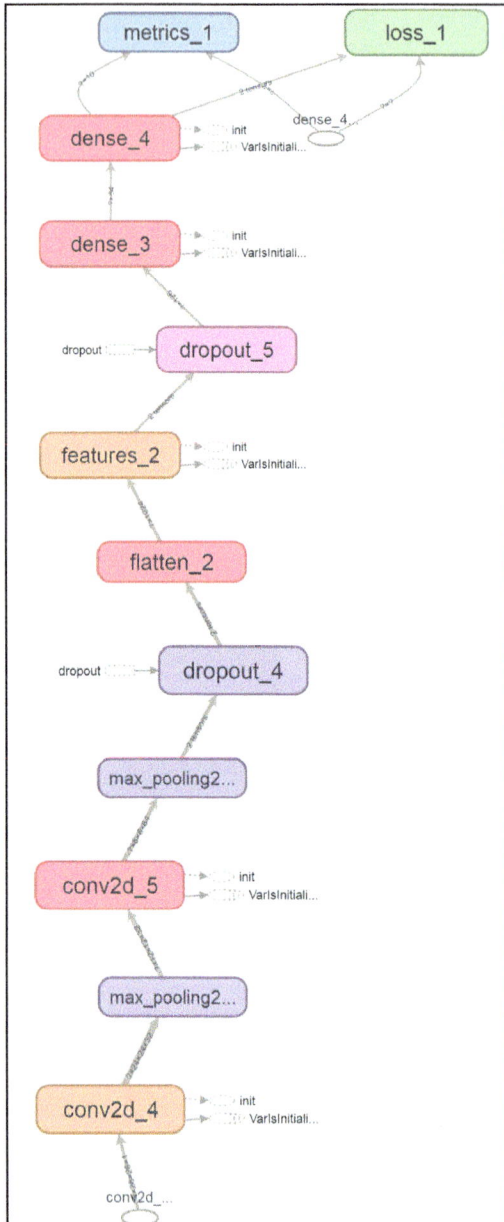

Figure 8.5 The CNN in the TensorBoard (Excerpt)

We can use the model.evaluate() function to display the accuracy and the loss function based on the test dataset, as shown in Listing 8.9.

```
score = model.evaluate(test_images, test_labels)
print('Test loss:', score[0])
print('Test accuracy:', score[1])
```

```
# Output:
10000/10000 [==============================] - 6s 650us/step
Test loss: 0.0206829725912
Test accuracy: 0.994
```

Listing 8.9 Loss and Accuracy of Our Trained CNN

We've achieved a very good accuracy of 99.4% after just a few epochs.

Now, let's look at the progression of loss and accuracy during training. In Figure 8.6 and Figure 8.7, we show the accuracy (**acc**) or the loss (**loss**) of the training data or test data across the training epochs.

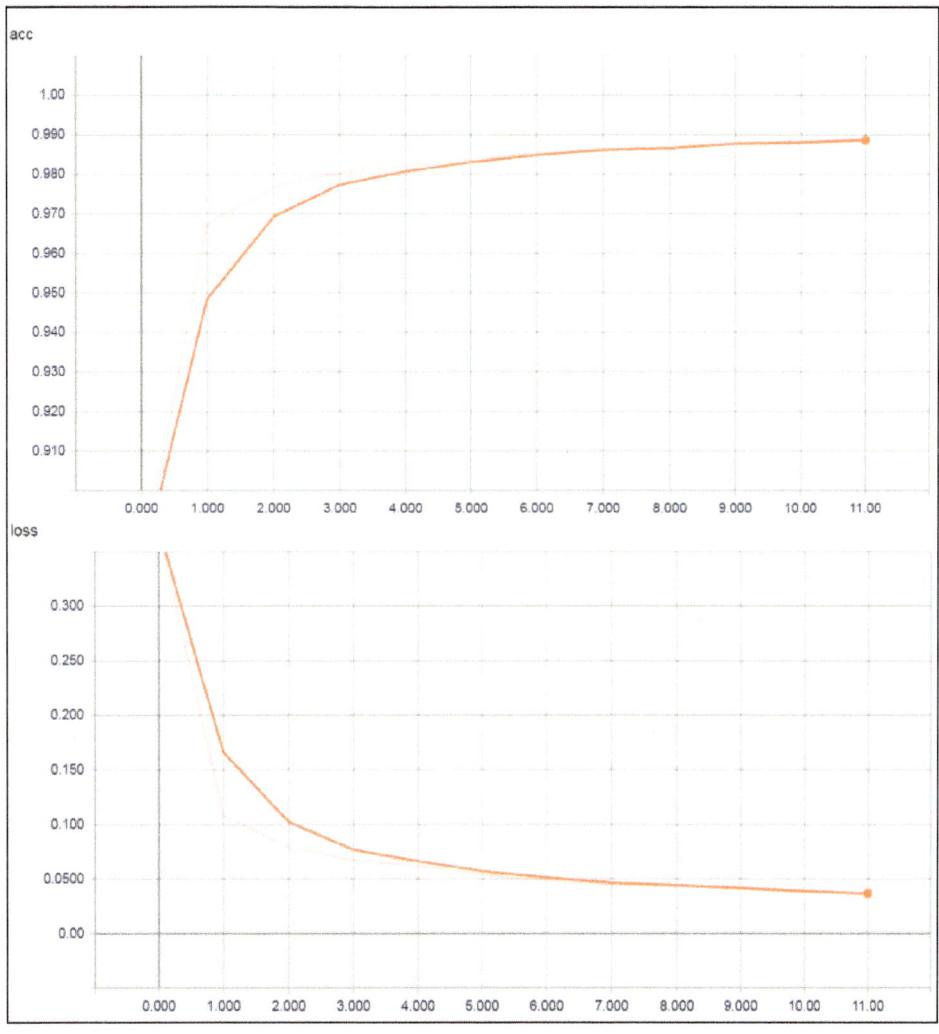

Figure 8.6 Progression of Accuracy and Loss of the Training Data

In Figure 8.6 and Figure 8.7, you can see a thick orange curve and a curve with a slightly paler color. TensorBoard provides a smoothing function, as the curves often don't have a nice shape. The thick orange curve shows the smoothed curve, while the pale curve shows the actual curve.

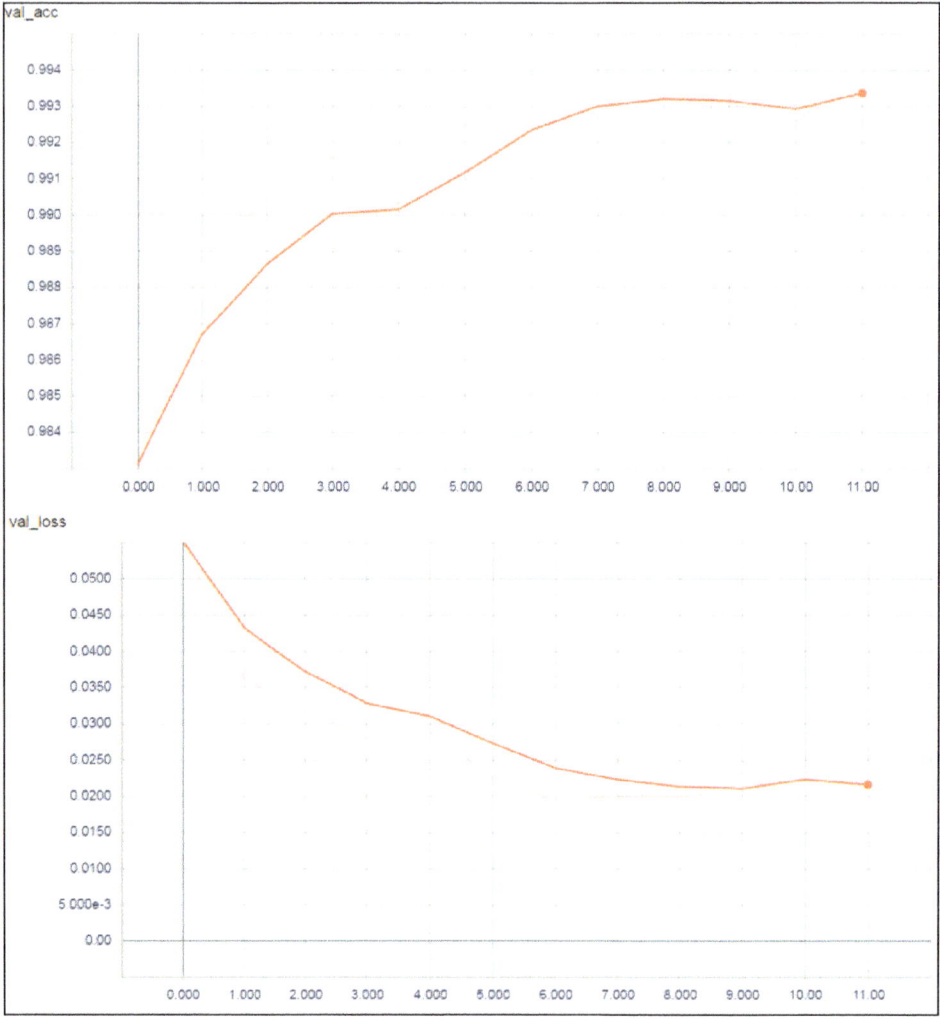

Figure 8.7 Progression of Accuracy and Loss of the Test Data

Of course, this isn't the only result. We now also have a pretrained network that we can apply to any data.

TensorFlow allows us to save the network architecture with the trained weights. For example, we could pass on the network, incorporate it into a website for number recognition, or save it for later comparison with other models, as shown in Listing 8.10.

```python
from tensorflow.keras import models

# Save the model in Keras HDF5 format
model.save('SimpleCNN_MNIST.h5')
```

Listing 8.10 Saving the Model and Weights (h5 Format)

Listing 8.10 makes sure that the model, including the trained weights, gets saved in *HDF5* format in the *SimpleCNN_MNIST.h5* file. The file contains the network architecture, the weights, the training configuration (i.e., the loss function and the optimization method), and the status of the optimization method.

TensorFlow 2 also provides the option to save the trained model in TensorFlow's own format (SavedModel, .pb format, as shown in Listing 8.11), which will be discussed more in Appendix C. The HDF5 format comes from the original Keras library, while the SavedModel format is a pure TensorFlow development.

```python
from tensorflow import models
model_directory = 'models'

# Save the model in TensorFlow-SavedModel format
model.save(model_directory, save_format = 'tf')
```

Listing 8.11 Saving the Model and Weights (pb Format)

> **Hierarchical Data Format 5**
>
> The HDF5 format is used in particular in scientific applications for storing large amounts of data.

Reusing a model is very simple. It doesn't matter which format you choose, as shown in Listing 8.12.

```python
from tensorflow.keras.models import load_model
model_directory = 'models'

# Loading the HDF5 model (h5)
# h5 is the typical file extension for this format.
# Everything is saved in one file
new_model_h5 = load_model('SimpleCNN_MNIST.h5')

# Loading the SavedModel model (pb)
# pb is the typical file extension for this format.
# The assets and variables directories are also created
new_model_pb = load_model(model_directory)
```

```
prediction_h5 = new_model_h5.predict(test_images)
prediction_pb = new_model_pb.predict(test_images)
```

Listing 8.12 Loading and Using the Pretrained Model

In Listing 8.12, the model was reloaded from different formats, and in Listing 8.13, we check whether we get the same result in both cases using our test images. Using that listing (which is similar to Listing 8.4), we also randomly check whether our trained network is actually classifying correctly.

```
import matplotlib.pyplot as plt
import random

randindex = random.randint(0,10000)
# The argmax() function returns the index of the highest value
# of the result vector
h5_class = prediction_h5[randindex].argmax()
pb_class = prediction_pb[randindex].argmax()
plttitle = "Test image No. {} \n (h5) Class: {} \n (pb) Class:
{}".format(randindex,h5_class, pb_class)
plt.imshow(test_images[randindex].reshape(28,28), cmap='gray')
plt.title(plttitle)
plt.axis('off')
plt.show()
```

Listing 8.13 Comparison of the Results and Testing of the Trained Network

Listing 8.13 should provide the output shown in Figure 8.8. As we use a random number generator, the test image and therefore your result will probably be different.

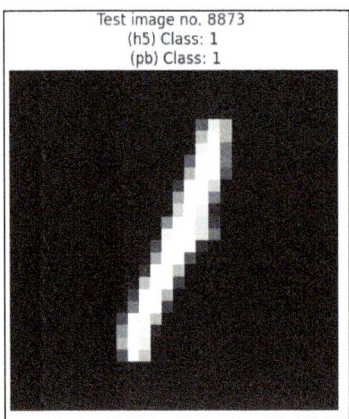

Figure 8.8 Classification Result of a Test Image

With regard to the storage format, there are certain differences in what is stored. However, we don't need to go into these differences in detail here. You can find out more about the reuse of trained networks in Section 8.3.

8.2 Transfer Learning with Convolutional Neural Networks

The MNIST example shows that this concept works very well for the supervised classification of images. A simple CNN is sufficient for this type of image. Training, the most computationally intensive element, can be accomplished in an acceptable amount of time even with simple computers.

But what if we had to distinguish between 100 or 1,000 image classes? We would need considerably more computing power to train the neural network, but above all, we would have a huge amount of training data, that is, image class pairs. Let's link this to a specific task (see the Task box).

Task

Using the images shown in Figure 8.9, implement an image classifier that can recognize the following six classes: mountain, badger, steam locomotive, German shepherd, strawberry, and forklift truck.

Figure 8.9 Sample Images for Our Task

This is a task that posed a major challenge around two decades ago and whose solution was reserved for image processing experts.

In this section, we'll demonstrate how this task can be solved using *transfer learning* with just a few lines of Python code.

The first arduous step in solving this task is to collect sufficient training data. Fortunately, this can be avoided by using pretrained CNNs (e.g., our MNIST model created in Section 8.1), which of course must contain the classes from Figure 8.9. But how did these pretrained networks come about, and why can we use them?

In 2010, Stanford University provided a dataset and announced a competition known as the *ImageNet Large Scale Visual Recognition Challenge* (ILSVRC). The dataset is often referred to as the *ImageNet dataset* and consists of 150,000 images with 1,000 classes. The three best teams were invited to present their solutions during a prestigious scientific event at the *IEEE Conference on Computer Vision and Pattern Recognition (CVPR)*. In 2012, a CNN (*AlexNet*) outperformed the other competitors for the first time with a previously unattained error rate of 15.3% (the runner-up was at 26.2%).

Figure 8.10 shows the development of error rates since 2010, with a reduction from 28% to less than 3% within six years. The human error rate is 5% to 10%.

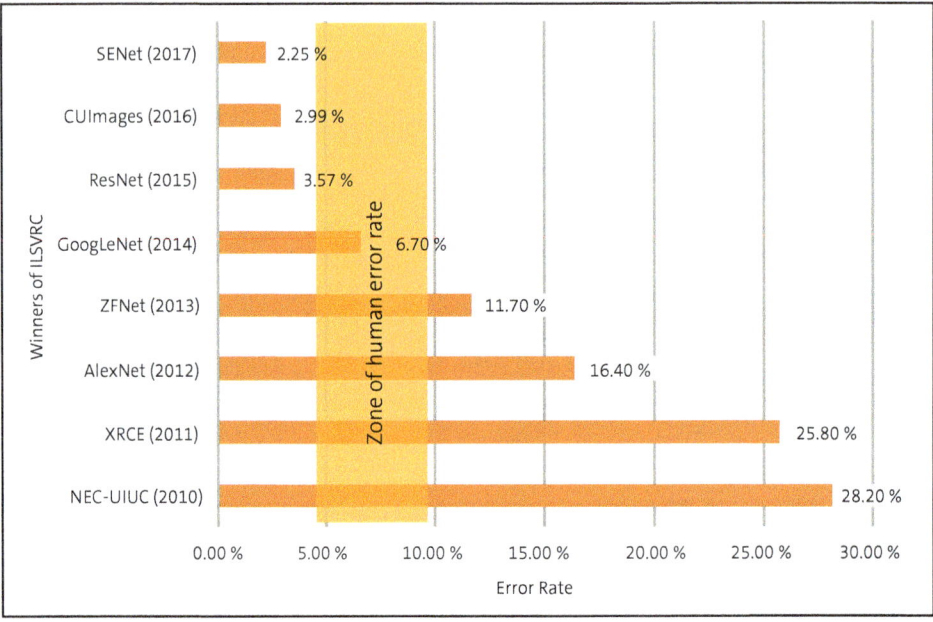

Figure 8.10 Winners of the ILSVRC Since 2010 with an Indication of the Error Rate

This result for a dataset with 1,000 classes is really impressive.

8.2.1 The Pretrained Network

For our example, we use a network structure called *Inception-v3* with trained weights. This network was trained with the previously mentioned ImageNet dataset.

Figure 8.11 shows the schematic structure of the Inception-v3 network and gives an idea of how challenging it is to train such a network.

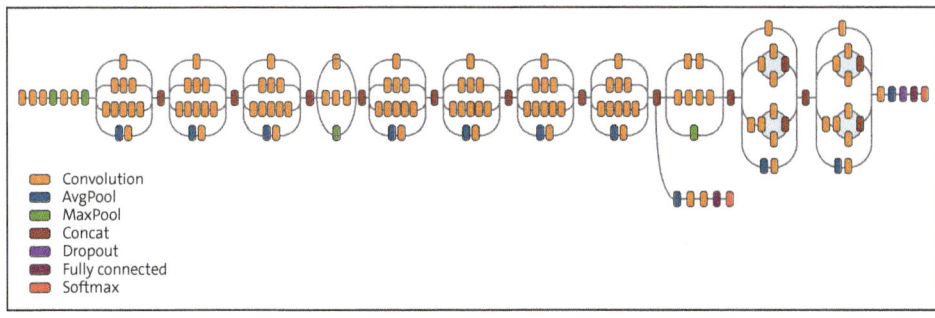

Figure 8.11 The Network Structure of Inception-v3

Why Is It Called Inception?

Rumor has it that the name actually originated in connection with the *Inception* movie. Figure 8.12 shows a meme that is said to have significantly influenced the naming.

Figure 8.12 Memes Based on the Inception Movie (Source: https://meme-generator.net)

The inspiration comes from the following idea: When developing the network architecture, you always have to decide for each layer whether a 3×3 filter or a 5×5 filter or average pooling or max pooling will follow. But why decide? Why not take everything and let the model decide? That's exactly what happens with an Inception-v3 module: you apply everything and then combine the results, which is called *filter concatenation*, in the way shown in Figure 8.13.

Tip
Details of this network can be found at *https://arxiv.org/abs/1409.4842*.

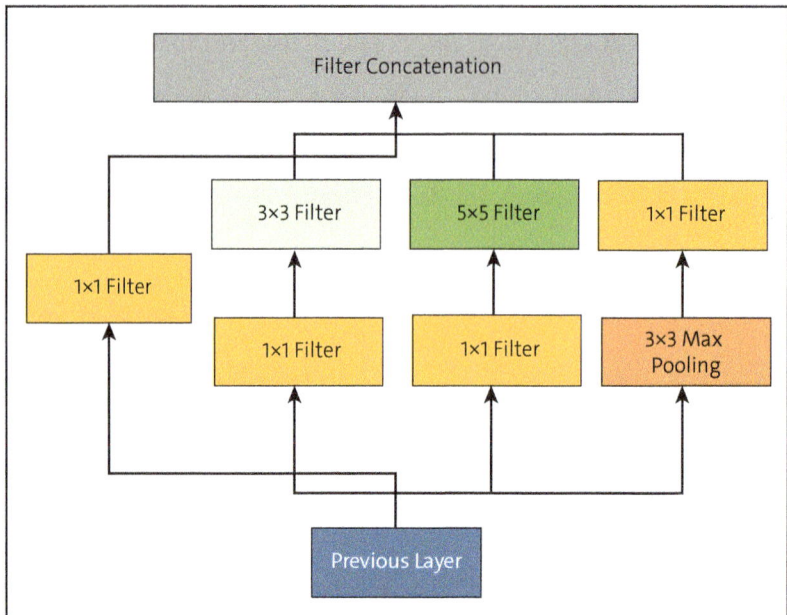

Figure 8.13 One of the Inception Modules of Inception-v3

8.2.2 Data Preparation

As usual, we start by importing the required libraries and packages, as shown in Listing 8.14.

```
import tensorflow.keras as tf

from tensorflow.keras.applications.inception_v3 import InceptionV3
from tensorflow.keras.applications.inception_v3 import preprocess_input, decode_predictions
from tensorflow.keras.preprocessing import image
```

Listing 8.14 Import of Libraries for Neural Networks

In Listing 8.14, you can already see that we use the proven `tensorflow` and `keras` libraries. With the import of `InceptionV3`, we've committed ourselves to a specific model of a CNN.

All pretrained CNNs require images of a certain size; unfortunately, the images don't do us the favor of being available exactly in the way the CNN needs them. In this regard, TensorFlow is very helpful because it provides functions such as `preprocess_input` and `image`, which help us to process the images.

However, we need additional libraries for loading, displaying, and transforming (see Listing 8.15).

```python
%matplotlib inline
import matplotlib.pyplot as plt
import numpy as np
import os
```

Listing 8.15 Important Little Helpers for Image Processing

We use the `matplotlib` library to display the image or the result and `numpy` to transform the image into a NumPy matrix, which is required as input by `tensorflow.keras`.

8.2.3 The Pretrained Network

Now we load the Inception-v3 model with the weights created by training with the images from the ImageNet dataset, as shown in Listing 8.16.

```python
model = InceptionV3(weights='imagenet')
model.summary()
```

Listing 8.16 Loading and Displaying the Inception-v3 Model

Figure 8.14 shows the beginning and end of the model structure of Inception-v3. The model consists of almost 24 million parameters.

```
Model: "inception_v3"
_____
Layer (type)                    Output Shape            Param #    Connected to
=================================================================================
input_1 (InputLayer)            [(None, 299, 299, 3)    0
_____
conv2d (Conv2D)                 (None, 149, 149, 32)    864        input_1[0][0]
_____
batch_normalization (BatchNorma (None, 149, 149, 32)    96         conv2d[0][0]
_____
activation (Activation)         (None, 149, 149, 32)    0          batch_normalization[0][0]
_____
conv2d_1 (Conv2D)               (None, 147, 147, 32)    9216       activation[0][0]
_____
batch_normalization_1 (BatchNor (None, 147, 147, 32)    96         conv2d_1[0][0]
_____
activation_1 (Activation)       (None, 147, 147, 32)    0          batch_normalization_1[0][0]
_____
conv2d_2 (Conv2D)               (None, 147, 147, 64)    18432      activation_1[0][0]
          .                              .                                  .
          .                              .                                  .
          .                              .                                  .
_____
concatenate_1 (Concatenate)     (None, 8, 8, 768)       0          activation_91[0][0]
                                                                   activation_92[0][0]
_____
activation_93 (Activation)      (None, 8, 8, 192)       0          batch_normalization_93[0][0]
_____
mixed10 (Concatenate)           (None, 8, 8, 2048)      0          activation_85[0][0]
                                                                   mixed9_1[0][0]
                                                                   concatenate_1[0][0]
                                                                   activation_93[0][0]
_____
avg_pool (GlobalAveragePooling2 (None, 2048)            0          mixed10[0][0]
_____
predictions (Dense)             (None, 1000)            2049000    avg_pool[0][0]
=================================================================================
Total params: 23,851,784
Trainable params: 23,817,352
Non-trainable params: 34,432
```

Figure 8.14 Section of the Inception-v3 Network Structure

The last layer, predictions (Dense), outputs a vector of length 1000, which is to be expected because the ImageNet dataset contains 1,000 classes.

We've now loaded our model. All that is missing now is an image to classify. In our example, we use an image named *germanshepherd.jpg* (see Figure 8.15 and Listing 8.17), but you can also use your own image. Ideally, you should make sure that the image contains an object that corresponds to one of the 1,000 classes of the ImageNet dataset (*https://gist.github.com/yrevar/942d3a0ac09ec9e5eb3a*). Of course, the 6 classes for our task are also taken from the 1,000 classes of the ImageNet dataset.

Figure 8.15 Image of a German Shepherd Dog (by Budinho, CC-BY-SA 3.0, https://commons.wikimedia.org/wiki/File:Grauer_Deutscher_Schäferhund_Standbild.jpg)

```
path = 'input_images/'
file = 'germanshepherd.jpg'
img_file = os.path.join(path,file)

img = image.load_img(img_file, target_size=(299, 299))

x = image.img_to_array(img)
print(x.shape)
print(x)
# Output:
(299, 299, 3)
[[[219. 224. 228.]
  [212. 217. 221.]
  [213. 218. 222.]
  ...
```

Listing 8.17 Loading the Image to Be Classified

It should be noted here that the path in Listing 8.17 has been set relative to the start directory of the Jupyter Notebook. You can also set an absolute path, such as D:\path_to_the_images\input_images. It's advisable to use the os.path.join function, as it combines the path and filename according to the operating system used. You can save yourself a lot of trouble with the correct use of slashes or backslashes because the os function takes over this task.

The imgage.load_img() function performs two tasks, namely loading the image data from the file and transforming it into the image size 299×299 pixels. This is necessary because the Inception-v3 model expects images of exactly this size. Because we're dealing with color images—that is, with three channels for red, green, and blue—the dimension is (299, 299, 3).

The entire image matrix is output using the print(x) function. If you compare the outputs from Listing 8.17 and Listing 8.18, you'll see that the pixel values have changed. The pixel values are usually in the numerical range from 0 to 255. However, Inception-v3 expects images with pixel values in the value range between 0 and +1. We achieve this by using the preprocess_input() function provided by tensorflow.keras.

In addition, our image is expanded by one dimension. This wouldn't be necessary for the image classification itself. However, the Inception-v3 model needs this additional dimension for the training phase, namely for the batch processing of images. Because the same network is clearly used for training and classification, we must also use images of the same dimension.

```
x = np.expand_dims(x, axis=0)
x = preprocess_input(x)
print(x.shape)
print(x)
# Output:
(1, 299, 299, 3)
[[[[ 0.7176471   0.75686276  0.7882353 ]
   [ 0.6627451   0.7019608   0.73333335]
   [ 0.67058825  0.70980394  0.7411765 ]
   ...
```

Listing 8.18 Further Image Preprocessing and Output of the Shape and Pixel Values

8.2.4 The Results

The image is now ready and can be classified. Listing 8.19 classifies our image, which means the task has been solved! As expected, the model.predict(x) function returns a vector of length of 1000. A value between 0 and 1 is output for each coordinate, representing the class to which it belongs. The higher the value, the more likely the image belongs to the class that corresponds to this position in the vector. Fortunately, the decode_predictions() function is also available in tensorflow.keras, which returns the class name, returns the class membership, and sorts the result according to the class membership probability. By default, this function outputs the first five results in readable form—the first three are sufficient for us.

```python
prediction = model.predict(x)

decoded = decode_predictions(prediction, top=3)[0]
print(decoded)
# Output:
[('n02106662', 'German_shepherd', 0.92125314), ('n02105162', 'malinois',
0.02156584), ('n02105412', 'kelpie', 0.0059969528)]
```

Listing 8.19 Classification of the Image

We've almost reached our goal, but we still want to display the image with the first three classes in Jupyter Notebook, as shown in Listing 8.20.

```python
plt.figure(figsize=(8, 8))
plt.axis('off')

pos_x, pos_y = 1,10
for id, lab, prob in decoded:
    plt.text(pos_x, pos_y, 'class: {} - {:.2f}%'.format(lab,prob*100),
            fontsize=12, color='black',
            bbox=dict(boxstyle="round", pad=0.1, fc='white'))
    pos_y += 10

plt.imshow(img)
```

Listing 8.20 The Classified Image with Class Legend

In Listing 8.20, we use matplotlib to display the image. We can use the matplotlib.text() function (contained as plt.text() in the listing) to integrate texts into images (see Figure 8.16). We use this function to display the class names and probabilities.

The image achieves a classification probability of 92.13% for the German shepherd class, which isn't a bad result. In second place follows malinois (Belgian Shepherd Dog), and in third place kelpie (Australian Kelpie), albeit with very low values, that is, classes that aren't so dissimilar to the German Shepherd Dog.

In Figure 8.16, you can see the input image for the Inception-v3 network with the dimension of 299×299 pixels. The distortion makes very little difference to the classification, which is also a quality feature for our classifier.

Figure 8.16 Classification Result

Task

In this example, we've used Inception-v3 for image classification. Now try to perform the same classification, but with the *ResNet50* model, and rebuild the code accordingly. The area that requires particular attention is the import of the corresponding tensorflow.keras library (tensorflow.keras.applications.resnet50). When you define the model (refer to Listing 8.16), the weights should also be used by training with the ImageNet dataset. The size of the input images should also be noted—it's (224,224,3).

Try out your code with the other images in the download area or with your own images!

8.3 Transfer Learning with Transformer Neural Networks

Some typical tasks in the field of *natural language processing* (NLP) are described in the following list:

- **Text classification**
  Assignment of a text to one or more predefined categories, such as spam detection or sentiment analysis.
- **Machine translation**
  Automatic translation of texts between different languages.
- **Text summary**
  Automatic creation of a shorter version of a text that contains the essential content.

- **Question answering**
  Answering questions based on a trained knowledge base.
- **Text generation**
  Automatic creation of new texts based on *prompts*.

Using the *transformer* library from *Hugging Face*, we now demonstrate how we can use pretrained transformer networks for different tasks.

> **Hugging Face**
>
> *Hugging Face* is a company and an open source platform that originally specialized in NLP technologies. Today, Hugging Face is best known for its extensive library of pretrained models for all kinds of AI tasks, as well as tools for developing and implementing AI solutions. The main goal is to make the use of modern machine learning models more accessible to researchers and developers.
>
> On this platform, you can find the following (and much more):
>
> - A collection of datasets for training neural networks
> - A central repository with thousands of pretrained models for various tasks
> - A platform for trying out applications (here, referred to as *spaces*) with the help of user-friendly interfaces

8.3.1 The Transformer Library

Before you start, you must install the necessary libraries. You should already be familiar with the installation in the Anaconda environment and in a Google Colab. In Listing 8.21, we specify the installation command in a Jupyter Notebook or Google Colab cell.

```
!pip install transformers datasets evaluate accelerate
```

Listing 8.21 Installing the Transformer Libraries

To access one of the models for a specific NLP task, we create a `pipeline()` instance and define the type of task. In Listing 8.22, we show an example that performs a text classification—more precisely, a sentiment analysis. The aim is to classify a text in terms of its positivity or negativity.

```
from transformers import pipeline
sentclassifier = pipeline("text-classification")
result= sentclassifier("I am very happy with this book on Neural Nets!")
print(result)
# Output:
([{'label': 'POSITIVE', 'score': 0.9998326301574707}])
```

Listing 8.22 Text Classification with Pretrained Default Model

8.3 Transfer Learning with Transformer Neural Networks

We've only specified the task here; the `pipeline()` function then uses a default transformer network, which is a *BERT transformer* in this example: distilbert/distilbert-base-uncased-finetuned-sst-2-english. However, this model only works for the English language.

You'll also find models for other languages on Hugging Face, which you'll then have to specify in more detail, as shown in Listing 8.23.

```python
from transformers import pipeline
sentclassifier = pipeline("text-classification", model=" nlptown/bert-base-multilingual-uncased-sentiment",device="cuda")
result= sentclassifier("I am very happy with this book on neural networks!")
print(result)
# Output:
[{'label': '5 stars', 'score': 0.8118821382522583}]
```

Listing 8.23 Text Classification with a Specified Model

You can see that the output is slightly different, but the text is still rated positively. In addition, we've specified the `device="cuda"` parameter in the definition of `sentclassifier` so that the function also uses the GPU for sentiment analysis.

Of course, it's also possible to transfer multiple sentences at the same time, as shown in Listing 8.24.

```python
from transformers import pipeline
sentclassifier = pipeline("text-classification", model="nlptown/bert-base-multilingual-uncased-sentiment",device="cuda")
results = sentclassifier(["I am very happy with this book on Neural Nets!", "I don't like the topic at all!"])
print(results)
# Output:
[{'label': '5 stars', 'score': 0.7789900898933411}, {'label': '1 star', 'score': 0.673673152923584}]
```

Listing 8.24 Text Classification with Multiple Inputs

> **Task**
>
> Experiment with your own sentences, and randomly check the quality of the text classification models. Search for other models on the Hugging Face website and test them in Google Colab.

8.3.2 Tokenizers and Models

Let's now take a closer look at the internal processes. As you've learned in Chapter 7, the first step is to transform text into tokens or token IDs. In the previous examples, this happens automatically, but now we want to see exactly what happens here in Listing 8.25.

```python
from transformers import pipeline
from transformers import AutoTokenizer, AutoModelForSequenceClassification
model_name = "nlptown/bert-base-multilingual-uncased-sentiment"
model = AutoModelForSequenceClassification.from_pretrained(model_name)
tokenizer = AutoTokenizer.from_pretrained(model_name)
sentclassifier = pipeline("sentiment-analysis", model=model, tokenizer=tokenizer)
```

Listing 8.25 Definition of a Tokenizer and a Pretrained Model

We use the same model as shown earlier in Listing 8.24 and take a look at what a token list looks like. The `AutoTokenizer()` function provides us with the tokens as it runs through the model. It's important to note that all models have their own tokenizers and that these are usually different, as shown in Listing 8.26.

```python
sequence = "I still like this book about neural networks!"
# Transformation to tokens
tokens = tokenizer.tokenize(sequence)
print("Tokens: ", tokens)
# Transformation in IDs
ids = tokenizer.convert_tokens_to_ids(tokens)
print("IDs: ", ids)
# Back-transformation of IDs into a sentence
decode_ids = tokenizer.decode(ids)
print("IDs for sentence: ", decode_ids)
# Output:
Tokens: ['I', 'still', 'like', 'this', 'book', 'about', 'neural', 'networks', '!']
IDs: [151, 12440, 11531, 10372, 11768, 10935, 86165, 28310, 106]
IDs for sentence: i still like this book about neural networks!
```

Listing 8.26 Representation of Tokens and IDs

Here, you can see that not every word is in fact a token, but that some words are subdivided. Each token has its own ID, which is ultimately used. It's also interesting that both the tokens and the back-transformed sentence are in lowercase. Why is that? Well, if we look at the name of the model used, bert-base-multilingual-uncased-sentiment, we read uncased, which means *case-insensitive*.

8.3.3 The Model Hub from Hugging Face

So how do we find the right model for our NLP tasks? To get started, we want to find a model that summarizes a text, so go to *https://huggingface.co/models* to get an overview of all the models Hugging Face makes available to us (see Figure 8.17).

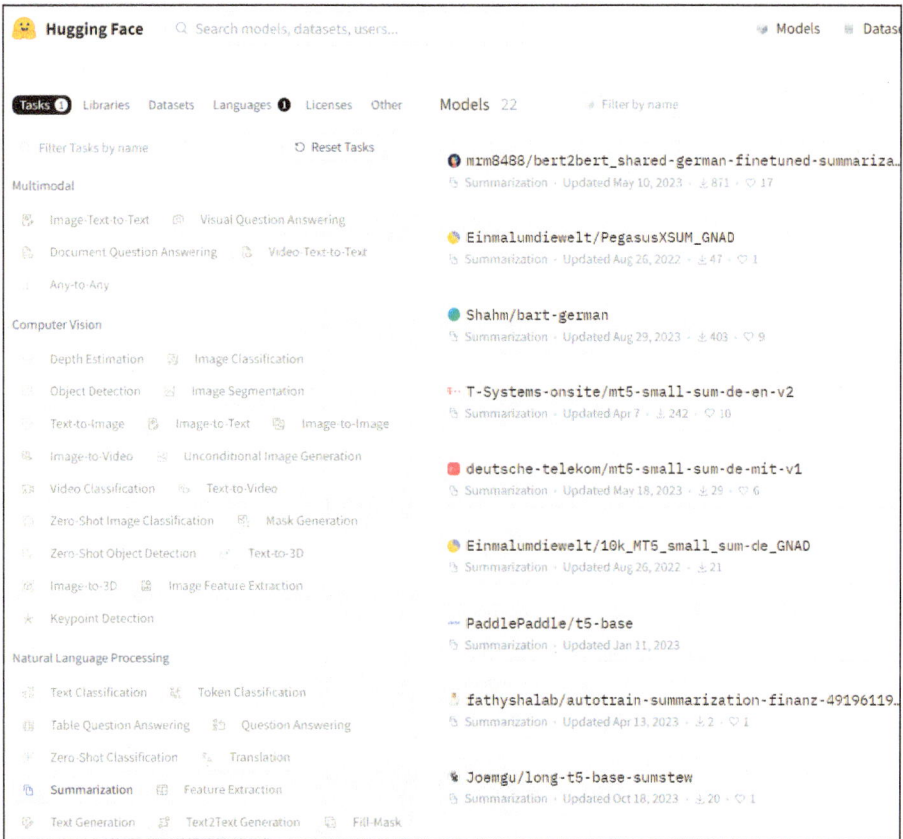

Figure 8.17 Model Selection on Hugging Face

Then, we select the **Summarization** task under **Tasks** (in Figure 8.17 bottom left) and the **English** language under **Languages**. Let's now summarize the text of this chapter.

We can select any model from the model selection. Models that have already been downloaded several times are ideal. This often indicates acceptable quality. We therefore select the facebook/bart-large-cnn model, which is based on the facebook/bart-base model and has been customized with a special dataset (Listing 8.27).

```
from transformers import pipeline


summarizer = pipeline("summarization",model="facebook/bart-large-cnn", device=
"cuda")
```

```
text = "Having acquired the theoretical knowledge of Convolutional Neural
Networks and Transformer Neural Networks in the previous chapter, we will now
implement the tools using TensorFlow and the integrated Keras library. We first
created our own network model to generate a classifier for the MNIST dataset.
We then used (very complex and powerful) pre-trained Deep Neural Nets to apply
them to tasks. On the one hand, we used a Convolutional Neural Net called
Inception-v3, which is already very good at extracting essential features from
an image. Secondly, we used the Hugging Face library to solve typical NLP
tasks. This approach, also known as transfer learning, saves us the time-
consuming training of our own complex network."

summary = summarizer(text, min_length=10, max_length=100)
print(summary)
# Output:
Device set to use cuda
[{'summary_text': 'We first created our own network model to generate a
classifier for the MNIST dataset. We then used (very complex and powerful) pre-
trained Deep Neural Nets to apply them to tasks. This approach, also known as
transfer learning, saves us the time-consuming training of our own complex
network.'}]
```

Listing 8.27 Text Summary with a Selected Model

Remember that the output of transformer neural networks doesn't always have to be the same. So, you should expect to receive a different summary of the text.

8.4 Summary

After acquiring the theoretical knowledge of CNNs and transformer neural networks in Chapter 7, you've become familiar with the tools for practical implementation using TensorFlow, the integrated Keras library, and the transformer library. You first created your own network model to generate a classifier for the MNIST dataset. Then, you used (the very complex and powerful) pretrained deep neural networks to apply them to different tasks. On the one hand, you've used a CNN called Inception-v3, which is already very good at extracting key features from an image. Secondly, you've used the Hugging Face transformer library to solve typical NLP tasks. This approach, also known as transfer learning, saves you the time-consuming training of your own complex network.

8.5 Further Reading

- "Gradient-Based Learning Applied to Document Recognition" by Yann LeCun (1998, *http://vision.stanford.edu/cs598_spring07/papers/Lecun98.pdf*)

- Hugging Face (*https://huggingface.co*)
- Inception-v3: "Going Deeper with Convolutions" by Christian Szegedy et al. (2014, *https://arxiv.org/abs/1409.4842*) and "1000 Synsets for Task 2 (Same as in ILSVRC2012) (2014, *http://image-net.org/challenges/LSVRC/2014/browse-synsets*)
- Keras (*https://keras.io*)

PART II
Deep Dive

In the second part, we stick with the practical side, but we'll also focus on the theoretical foundations. To help you can think outside the box of feed-forward networks and learn how to use other network types for your projects, this part of the book covers more about the biology and biological mechanisms. We describe the historical development of artificial neural networks (ANNs), describe the machine learning (ML) process, present different learning methods, and provide a compilation of application areas and practical examples. So, let's move on to the next round!

Chapter 9
From Brain to Network

Using biology as a template for mathematical models means thinking about thinking. Is this a contradiction? Can a brain fathom itself?

Our focus so far has been on feed-forward networks, as they are the most popular and favorite variant of artificial neuron network (ANN) models. Of course, the models that actually exist go far beyond this. The biological template provides many points we can take into account in the models. Of course, we can already give some examples of such models, although we'll explain the biological basis for them later:

- *Spiking neural networks* (SNN) are networks that don't use the *signal amplitude* (i.e., the output value) to encode the information, but the *signal frequency*, that is, how often a neuron fires in a certain period of time. These networks are useful for time-dependent tasks such as speech processing and sensor data analysis.

- *Self-organizing map* (SOM) simulates *topological representations* in the brain, which are used to localize bodily functions in the *cortex*. Every sensory organ and every sequence of movements is represented in the brain in defined areas (representation), whereby neighboring areas of the body also retain their proximity in the brain. This aspect is referred to as *topology-preserving*.

- *Neocognitron* is based on the structure and function of the visual cortex. The Neocognitron consists of several layers that are arranged hierarchically. Each layer contains cells that react to certain characteristics. These features become increasingly complex from the lower to the upper layers. The concept of the Neocognitron has contributed significantly to the development of convolutional neural networks (CNNs).

- *Recurrent neural networks* (RNNs) and *long short-term memory* (LSTM) model the time-bound dependencies and memory capabilities of the human brain, especially in the processing of sequential data. They are used in language processing, time series analysis, and machine translation.

Details on some of these models and how they work are discussed in Chapter 10.

9.1 Your Brain in Action

Let's start with a short example. You're reading a book that you're actually holding in your hands. Now you've reached the end of the page, and the next line is almost unreachable on the next page, which unfortunately can only be reached by turning the page. Incredibly, your hand reaches for the book page, takes the page between your index finger and thumb, and turns it over. You've successfully turned the page and continue reading without consciously paying attention to this moment of maximum neural interaction. If we look at your accomplished feat in detail, then the various neurons are busy supporting you with such skills. First, the *central nervous system* activates the hand and arm muscles with the help of *motor neurons*. *Sensory* information, such as the position of your hand or fingers and finger pressure, is constantly sent to your brain by sensory neurons, where it's controlled by neurons and sent back to the muscles via motor neurons. There you go! You have successfully, but probably unconsciously, turned the page.

In this chapter, we want to take a look under the skullcap and draw inspiration for ANNs from biology. Our motivation is to give you a basic understanding of the biological processes and to motivate you to come up with your own creative, technical solutions. As a branch of science, *neurology*, the science of neurons, is the food source for our activities as "machine learners." Artificial neurons and networks are undoubtedly motivated by their biological counterparts. Through abstraction and simplification, that is, by *modeling* properties and functionalities, we're able to extract profitable behaviors for our needs. In this section, we bring biology to our workbench and extract the important elements for the creation of ANNs. Of course, our human brain and some of its components serve as our primary model, even if we as humans haven't yet understood all of its modes of action.

9.2 The Nervous System

Our nervous system comprises all the nerve cells in our body. It serves to process *stimuli* of our organism and leads to *reactions* to them.

A distinction is made between the *central nervous system* (CNS) and the *peripheral nervous system* (PNS) (see Figure 9.1).

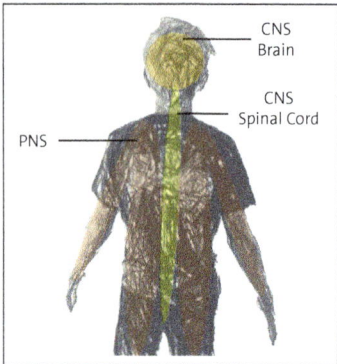

Figure 9.1 The Nervous System and Its Subsystems

We're interested in the CNS, which comprises the *nerve tracts* in the brain and spinal cord. The CNS is well protected in the skull and in the *spinal canal* of the spinal column.

9.3 The Brain

The brain (*encephalon* is ancient Greek, meaning "in" + "head") is a spongy mass weighing approximately 3.1 lbs, in which the incredible number of around 80-100 billion nerve cells with an average of 1,000 synapses per nerve cell is located. This makes around 100 trillion connections that link neighboring cells, but also cells that are far apart (*connectome*). Of course, these cells aren't organized as clusters but in different regions, which are roughly divided into four areas (see Figure 9.2):

- Cerebrum
- Diencephalon
- Cerebellum
- Brain stem

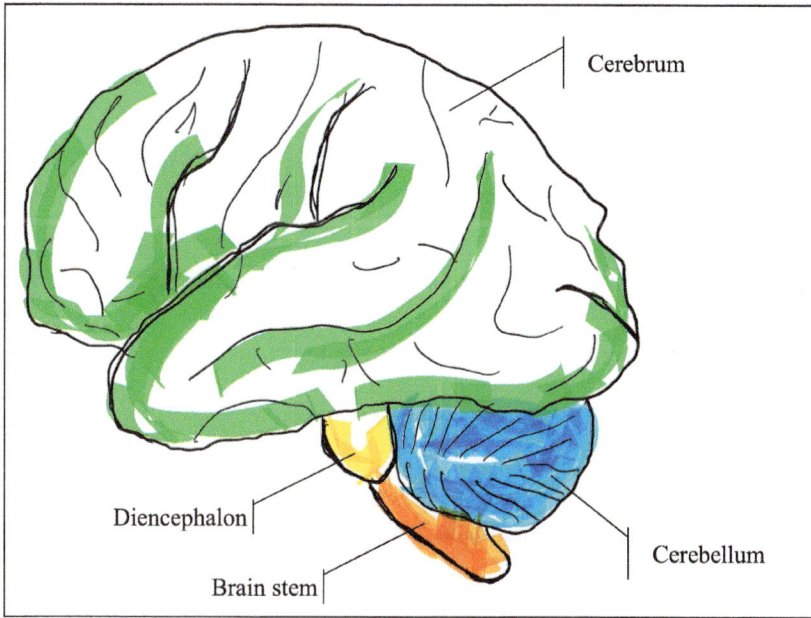

Figure 9.2 The Brain

9.3.1 The Parts

The cerebrum is divided into two *hemispheres* by a large furrow. There's a fat highway between them called the *corpus callosum*.

The surface is heavily folded and is reminiscent of a walnut. At 0.25 square meters (around 2.5 square feet), its surface area is considerable, as is the number of 16 billion nerve cells located there—almost a fifth of the estimated total number of neurons in the brain.

On the *cerebral cortex* (*cortex cerebri*), *cortical areas* are localized that are responsible for processing specialized information. They are called *primary areas*, and a well-known example of this is the *visual cortex*. It's extremely important for us connectionists because the most diverse types of ANNs have tried their hand at its architecture and functionality. Fukushima's Neocognitron and CNNs are particularly noteworthy in this respect.

> **Note**
>
> We also want to take this opportunity to mention David Marr and his major work *Vision: A Computational Investigation into the Human Representation and Processing of Visual Information* (see Section 9.8). Marr is considered one of the founders of neuroinformatics and developed a model of vision as information processing in the brain.

The *diencephalon* is another area of the brain that houses, among other things, the *thalamus*, which transmits sensory and motor signals from and to the cerebrum. All signals from the sensory organs are collected and distributed here.

Two hemispheres can also be identified on the *cerebellum*, which resembles a cauliflower in appearance. It's responsible for balance, movement, and coordination.

The *brain stem*, the oldest part of the brain, connects and processes incoming sensory impressions and outgoing motor information. It's responsible for elementary and reflex control mechanisms, for example, for the life-sustaining functions of breathing and heart rate control. The brain stem establishes the connection to the *spinal cord*.

9.3.2 A Section

Finally, here's a view that shows you what the interconnectedness can look like in the brain. The section you see in Figure 9.3 is taken from the cerebral cortex, which is about 4 mm (approx. 0.16 inches) thick in humans. The neurons are divided into six layers, whereby the stratification is based on the different cell types that occur in each layer. Incidentally, the largest cells in the cerebral cortex are the *pyramidal cells*, whose name is derived from their shape.

In the following sections, we'll take a closer look at the neuron and describe its individual components. The names and designations for the elements of the neurons already create the analogy to ANNs.

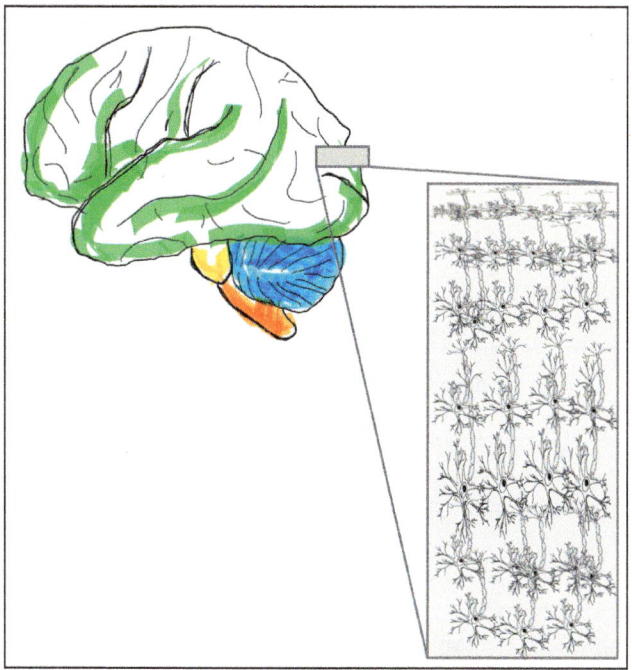

Figure 9.3 A Schematic Representation of a Cross-Section through the Cortex

9.4 Neurons and Glial Cells

The mission of a neuron could be described like this: "specialize in excitation conduction and excitation transmission." Of course, the neuron isn't alone, but performs various tasks with 80-100 billion other neuron colleagues.

Building on Chapter 1, Section 1.6.1, we want to discuss additional aspects of the cells, so we'll quickly rush through the basics. Each nerve cell consists of a *cell body* (*soma*) and extensions (see Figure 9.4). The extensions that collect signals (excitations) like feelers are called *dendrites*. The outgoing signal-conducting extension is referred to as an *axon* or *neurite*. It can be up to 1 meter (3.3 feet) long and is of course connected to other cells.

The connection points between the neurons are referred to as *synapses* and the signals are transmitted in them (see Figure 9.5). There can be an average of 1,000 of these synapses per neuron.

Electrical *action potentials* (nerve impulses) are sent along the axons in coded patterns. Each pulse has a voltage of approximately 0.1 volts, lasts 1-2 thousandths of a second, and has a maximum speed of 480 km/h. When the electrical impulse reaches a synapse, it's recoded into chemical signals. The information flows from the sending to the receiving neuron. The impulse triggers the release of *messenger substances*, known as *neurotransmitters,,* which are also called *signaling molecules*.

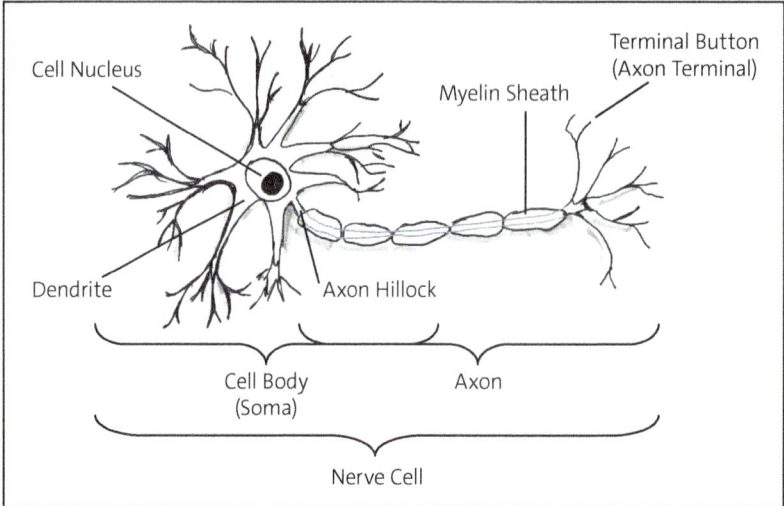

Figure 9.4 The Neuron

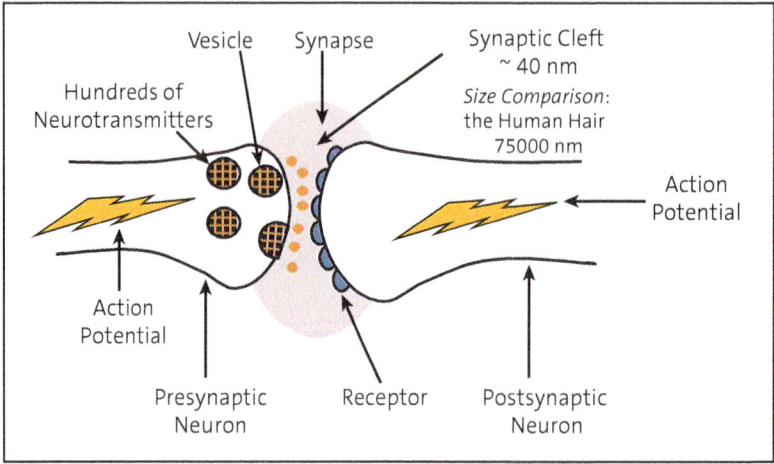

Figure 9.5 Presynaptic and Postsynaptic Interaction

The neurotransmitters are stored in small synaptic *vesicles* enclosed by a membrane. Each nerve ending in the CNS has several hundred of them. Synapses that are attached to the motor end plates of the muscles can have significantly more. After the arrival of the action potential, a molecular process takes place which ensures that the membrane of the vesicle fuses with the membrane of the synapse, causing the neurotransmitters stored in the vesicle to be released into the synaptic cleft.

The transmitter molecules are received on the postsynaptic side by *receptors* that are specialized for certain neurotransmitters. The neurotransmitters have an *excitatory* or *inhibitory* effect. Depending on which neurotransmitters are received, the receptors regulate the electrical properties of the postsynaptic membrane, and the resulting change in membrane resistance leads to a change in voltage, which is processed by the *postsynaptic* neuron (see Figure 9.5).

The correct functioning and transmission depend on a suitable mix of chemicals surrounding the synapse. The *glial cells* are responsible for this. They take great care of the neurons; they help to connect the neurons in the developing brain, nourish the cells, insulate the axons, dispose of dead cells, recycle used neurotransmitters, and protect the brain from infections. Glial cells form an insulating layer around the axons, the *myelin sheath*, and thus allow faster excitation conduction.

Take, for example, the optical nerve, which comprises around 1 million axons and transmits signals at high speed. If no myelin were used for insulation, thicker axons would have to be used to ensure the high transmission rate, an estimated total of 0.75 m (2.5 ft), and the brain would have a diameter of 20 m (65 ft).

9.5 A Transfer in Detail

Have you ever heard of standby? The standby mode of your laptop is comparable to the *resting potential* of a cell. Your laptop is ready, but not active, and the same applies to the cell. Officially, the resting potential is understood to be the negatively charged state of an unexcited nerve cell.

So that you can understand the official version, we first need to clarify a few terms, such as *membrane potential*, which is also referred to as *membrane voltage* (see Figure 9.6).

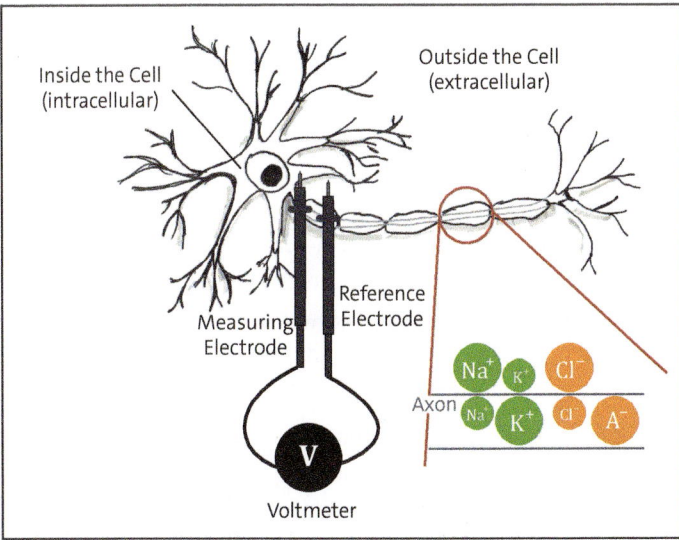

Figure 9.6 Measurement of the Membrane Potential Using Electrodes and a Voltmeter

The membrane potential is the charge difference between the inside of the membrane and the outside of the cell. It fluctuates between -110 mV and +30 mV, depending on the excitation state of the cell. The difference in charge is caused by the positively charged outside and the negatively charged inside of the cell. The concentration of the *ions* is decisive for the type of charge.

> **Ion**
>
> An ion is an electrically charged atom or molecule. Atoms and molecules have the same number of electrons as protons in the neutral state. However, if an atom or molecule has one or more electrons less or more than in the neutral state, it has an electric charge and is referred to as an ion. Ions with a lack of electrons are positively charged; ions with an excess of electrons are negatively charged.

Now that you know what the membrane potential is, we can return to the resting potential. What causes the resting potential? There are many positively charged *potassium ions* (K^+) inside the nerve cell and a few outside the cell. For this reason, there is a *difference in concentration*. Because there is a fundamental urge to balance the concentration and there are channels through which the potassium ions (K^+) can escape, they migrate outward and take their positive charge with them. This gives the outside a positive charge. The negative ions are trapped inside the cell because they are too large to disappear to the outside. This makes it more negative inside the cell, and the more potassium ions (K^+) migrate to the outside, the greater the membrane voltage becomes.

However, the more negative the inside becomes, the stronger the positive potassium ions (K^+) are attracted, and this leads to a halt in the outward migration of potassium. This state is referred to as *electrochemical equilibrium*, and the voltage measured at this point is referred to as the *resting potential*.

Now that you know what the *resting potential* is, we can consider what happens when signals from other neurons are sent to a neuron and what effect this has on the neuron's activity. Just like your TV, which you bring out of standby mode by switching it on, a cell is activated by an action potential (see Figure 9.7).

A nonfiring neuron is in a resting state; that is, it does nothing. The resting potential prevails at the outer skin of the cell, the membrane. The excitations from incoming cells are collected at the dendrites, among other places, and the voltage at the membrane of the axon changes, which propagates to the synapses. This action potential lasts approximately 2 ms in humans.

At the *axon hillock*, it becomes clear whether the *threshold potential* has been exceeded, and an action potential is subsequently triggered, following the *all-or-none law*. This means that it's either sent or not—very binary. If the action potential gets triggered, the neurotransmitters will be released, and an action potential will subsequently get triggered in the receiving cell. The meaning of the signals is determined by the frequency and the number of incoming action potentials; in this case, *coding* is also relevant.

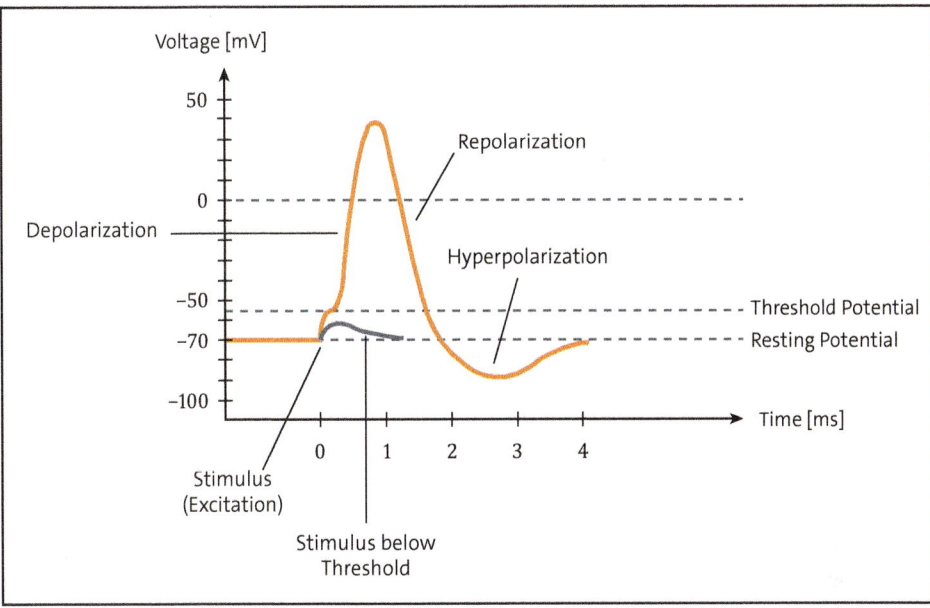

Figure 9.7 Action Potential

> **Coding**
>
> How can we distinguish between quiet, loud, and very loud; that is, how can intensities be coded by neurons? Basically, intensities can be represented by *amplitude modulation* or *frequency modulation*. Because the neuron doesn't allow arbitrarily high voltages (amplitudes), only frequency modulation, that is, the number of action potentials per second, remains for coding purposes.
>
> If, for example, the membrane potential at the axon hillock is only just above the threshold value, few action potentials are formed per second (e.g., 50). If, on the other hand, the membrane potential is far above the threshold value, many action potentials are formed (e.g., 170).

We've thus described the biological basics from the reception to the transmission of the signal. One area that we want to share with you next is the representation of neurons and networks.

9.6 Representation of Cells and Networks

The nerve cells of the nervous system are highly interconnected and therefore highly complex. Each cell is already very complex on its own. Different approaches can be used to make these complexities comprehensible. One of these is to create an image of reality. Santiago Felipe Ramón y Cajal excelled in the creation of drawings of the nervous system, and his work was honored with the Nobel Prize for Physiology and Medicine in 1906. His co-Nobel Prize winner was Camillo Golgi, who developed the *Golgi staining*, which can be used to contrast the nerve tissue and thus distinguish it from the pictorial background. This makes it possible to visualize a nerve cell in isolation. Figure 9.8 shows one of Ramón y Cajal's works. It already shows the basic structures of networking and the structure of cells.

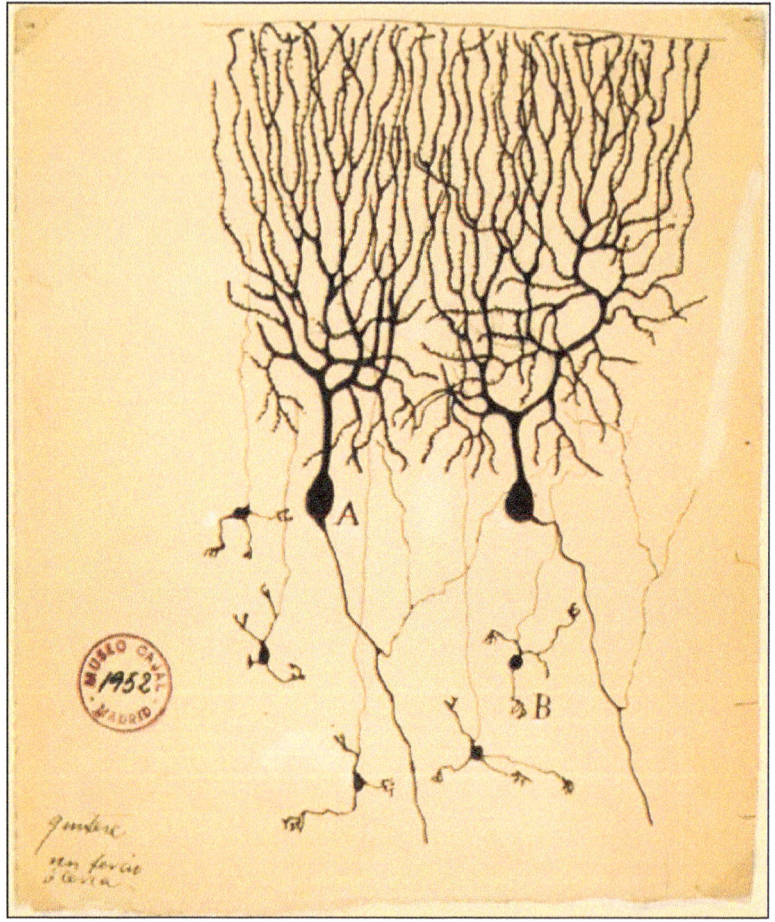

Figure 9.8 Drawing of Two Purkinje cells (A) and Five Granule Cells (B) from the Cerebellum of a Pigeon by Santiago Felipe Ramón y Cajal, 1899 (PurkinjeCell.jpg, https://commons.wikimedia.org/w/index.php?title=File:PurkinjeCell.jpg&oldid=313838888)

More modern *neuroimaging* techniques, which make it possible to "watch" the brain as it thinks, provide no less spectacular insights. A wide variety of technical approaches are in use and provide such impressive images. The example in Figure 9.9 uses a *microscope technique*.

This concludes the in-depth look at biology and gives you the opportunity to play with all the new ideas and suggestions that you can use for your own ANN models.

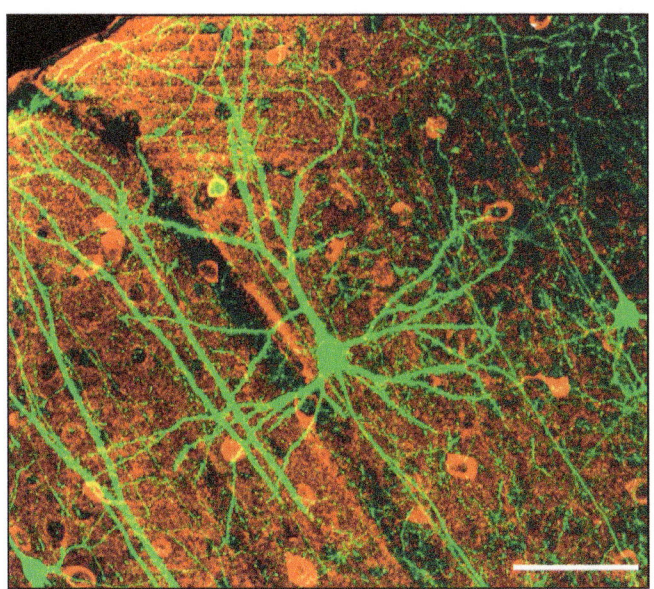

Figure 9.9 Microscopic Image of the Cerebral Cortex of a Mouse (by Wei-Chung Allen Lee, Hayden Huang, Guoping Feng, Joshua R. Sanes, Emery N. Brown, Peter T. So, Elly Nediv. License: CC-BY-2.5)

9.7 Summary

All of these biological, neurophysiological, chemical and physical principles are possible motivations for more abstract and, of course, highly simplified models that we can use and also depict technically. ANNs use the principles of biological networks but also supplement them with principles that are detached from biology, such as the models we presented at the beginning.

The most important principle to remember is that neurons collect input from different sources, integrate these signals, and pass on a newly generated signal to subsequent neurons. This is the simple principle that forms the basis of the ANNs we use and that we've learned from nature.

9.8 Further Reading

- The Brain (*www.mpg.de/brain*)
- *https://en.wikipedia.org/wiki/Neuron*
- "An Introduction to Bio-Inspired Artificial Neural Network Architectures" by B. Fasel (2003, *www.actaneurologica.be/pdfs/2003-1/01-fasel.pdf*)
- *Vision: A Computational Investigation into the Human Representation and Processing of Visual Information* by David Marr (2010, MIT Press)

Chapter 10
The Evolution of Artificial Neural Networks

Evolution! In excerpts.

In this chapter, we delve into the history of neural networks. You'll learn about different network approaches, areas of application, tools, and influential publications in this context. Our aim is to show you the wide range and variety of artificial neural network (ANN) alternatives that should stimulate your imagination and encourage independent experimentation. In doing so, we follow history and always summarize 10 years in a block of history. Let's start with the first blocks, the 1940s to 1980s (see Figure 10.1).

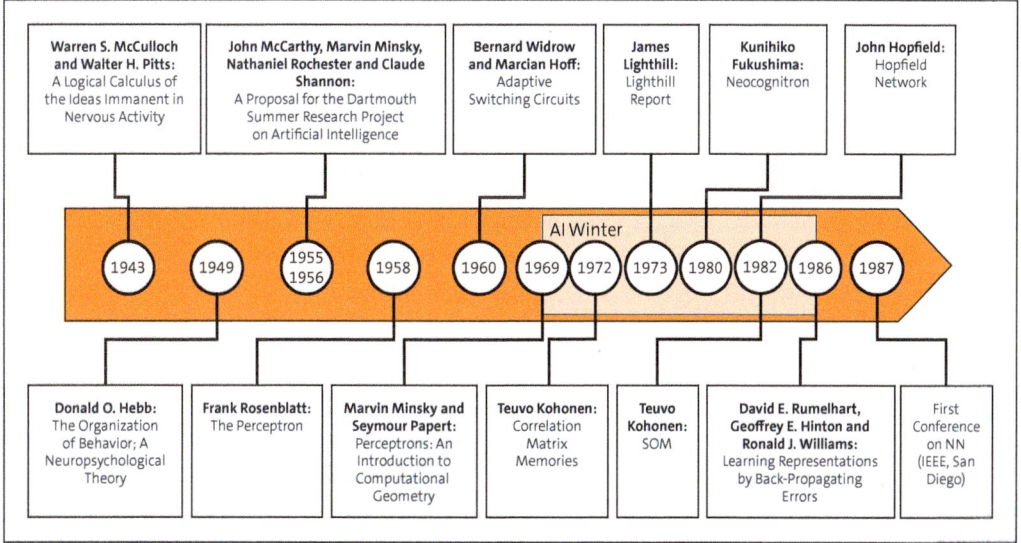

Figure 10.1 ANN History, Part 1

Before we get started, note that we compiled the selection of publications subjectively and it's not an exhaustive compilation. We base our selection on historical relevance plus some statistics on citation frequencies.

10 The Evolution of Artificial Neural Networks

10.1 The 1940s

Without Santiago Felipe Ramón y Cajal, who proved the existence of synapses as early as 1911, the following development would hardly have been possible. Not only was he an excellent artist and produced detailed drawings (e.g., of the rodent hippocampus), he also provided the basis for the first abstractions of neural networks.

10.1.1 1943 McCulloch-Pitts Neurons

In 1943, the neurophysiologist Warren McCulloch and the logician Walter Pitts presented a simple neuron model known as the *McCulloch-Pitts* (*MCP cell*) or *MCP neuron*.

They used electrical circuits to show how the brain works from their point of view. They referred to the basic components they used as *threshold logic units*, as they converted analog input into digital input. This corresponds entirely to the all-or-nothing principle of a neuron. McCulloch and Pitts showed that neurons with binary signals ("binary threshold activation function") are analogous to statements in mathematical logic.

Figure 10.2 shows an excerpt from the original drawings of the networks. An MCP doesn't sends an output until two *excitatory* inputs and no *inhibitory*) input reach the neuron. Excitatory inputs are shown with a dot, and inhibitory inputs are shown with a loop.

Figure 10.2 Graphical Representation (Extract) of McCulloch-Pitts Networks from Their "A Logical Calculus of the Ideas Immanent in Nervous Activity" Paper in 1943

The signals in McCulloch-Pitts networks are exclusively binary. This means, among other things, that each neuron can only have 0 or 1 at the output. Inhibitory signals can be processed in the same way as in biological networks, but the threshold value is a real number. The output is calculated like this (see Figure 10.3):

- At the input, there are n excitatory signals (input $\vec{x}$) and m inhibitory signals (input $\vec{y}$), which can have the value 0 or 1.
- If $m \geq 1$ and one of the inhibitory signals = 1, then the output is 0.
- Otherwise, the sum of the excitatory signals, that is, $x_1 + x_2 + \cdots + x_n$, is calculated and compared with the threshold value θ. If the sum is greater than or equal to the threshold value, the output is 1, otherwise 0.

If you're more of the formal type, then you'll want a compact presentation, which we're happy to provide, enriched with color:

$$net_x = \sum_{i=1}^{n} x_i$$

$$net_y = \sum_{i=1}^{m} y_i$$

$$f_{act}(net_x, net_y, \theta) = \hat{y}_j = \begin{cases} 1, & \text{if } net_x \geq \theta \text{ and } net_y = 0 \\ 0, & \text{else} \end{cases}$$

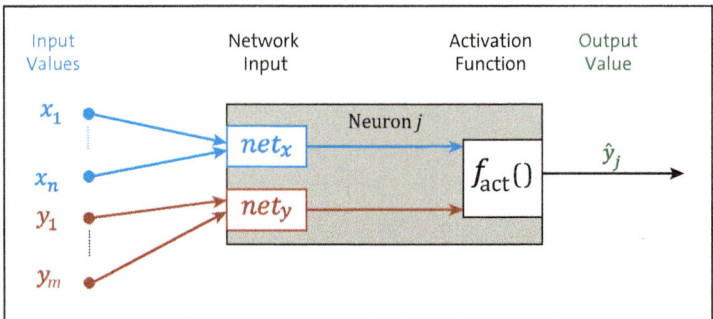

Figure 10.3 Illustration of a McCulloch-Pitts Neuron

10.1.2 1949: Donald Hebb

In 1949, Donald Hebb published the *Hebb learning rule* in his book *The Organization of Behavior*. It describes the following effect: the more often a neuron A is active at the same time as neuron B, the more preferentially the two neurons will react to each other.

For this reason, Hebb is regarded as the discoverer of *synaptic plasticity*, which is the neurophysiological basis of learning and whose principle is still used in today's learning strategies. According to Hebb, "what fires together, wires together."

Hebb's learning rule is used in ANNs and is historically the oldest of all learning rules. You've already gotten familiar with the *perceptron learning rule*, the *delta rule* (Widrow/Hoff), and *error backpropagation* as other popular learning rules.

10.2 The 1950s

In the 1950s, computer technology developed further, making it possible to simulate neural networks.

10.2.1 1951: Marvin Minsky and Dean Edmonds – SNARC

In 1951, Marvin Minsky and Dean Edmonds, motivated by the work of McCulloch and Pitts, used vacuum tubes and motors to build the first hardware-based neural network

(*SNARC – Stochastic Neural Analog Reinforcement Calculator*), which simulated a rat finding its way through a maze. They used 40 randomly connected Hebbian synapses, which have a memory that stores the probability value for the arrival of a signal via an input and the departure of a signal via the output. The learning or rewarding of the network is carried out manually by a human operator.

10.2.2 1955/1956: Artificial Intelligence

We've already mentioned and acknowledged the year 1956 and John McCarthy, Marvin Minsky, Claude Shannon, and Nathaniel Rochester in Chapter 1, but we still want to mention these scientists again in the chronology and give them the room they deserve.

In the context of AI, it's worth mentioning another approach: *cybernetics*. It was founded by Norbert Wiener, among others, in the 1940s and deals with the *control* and *regulation* of machines, living organisms, and organizations. The approach therefore involves, for example, machines that can learn and control themselves. It's interesting to note at this point that Warren McCulloch and Walter Pitts, among others, marked the beginning of cybernetics with their work in 1943.

10.2.3 1958: Rosenblatt's Perceptron

Frank Rosenblatt, who developed the *perceptron* at the Cornell Aeronautical Laboratory, is often regarded as a pioneer in the field of neural networks. The perceptron consists of 400 photosensitive input cells. An *association layer* is used to connect the input with the output, which consists of eight units.

You've already gotten very familiar with Rosenblatt's perceptron and implemented it in Python. For this reason, there isn't much to say here, except that it's worthwhile to study the *convergence theorem* from 1958 intensively because it shows that all possible solutions to linearly separable problems can be learned using perceptron learning.

10.2.4 1960: Bernard Widrow and Marcian Hoff – Adaline and Madaline

In 1960, Bernard Widrow and Marcian Hoff published their models for neural networks at Stanford University under the names of *Adaline* (Adaptive Linear Neuron) and *Madaline* (Multiple Adaptive Linear Neuron).

The motivation behind Adaline (see Figure 10.4) was to predict the next bit in a sequence of bits that were transmitted over a phone line. Madaline was used as an adaptive filter to eliminate echoes from a phone line. You've already implemented Adaline yourself and learned about a particularly interesting feature: even if two classes aren't perfectly separable, Adaline still finds the best and most stable solution.

You've come to know the delta rule as *gradient descent* or *least mean square* (LMS); it was first used by Widrow and Hoff in 1960 to train Adaline.

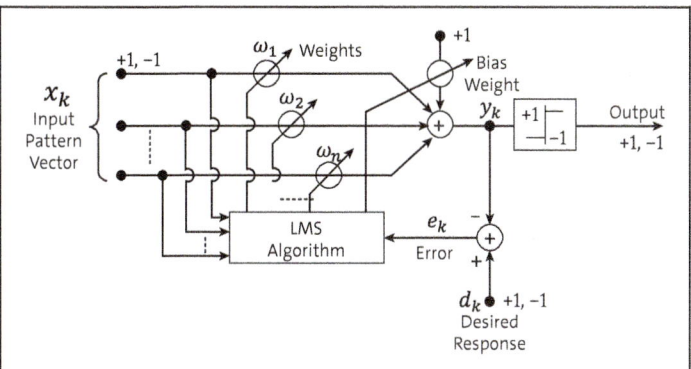

Figure 10.4 Adaline Circuit Diagram by Widrow and Hoff (Source: Bernard Widrow and Marcian Hoff, "Adaptive Switching Circuits." In: Technical Report 1553-1, Stanford Electron. Labs., Stanford, CA, June 30, 1960)

10.3 The 1960s

As promising as the 1960s began for ANN, the highs were brought to an abrupt end by a single publication.

10.3.1 1969: Marvin Minsky and Seymour Papert

Marvin Minsky and Seymour Papert wrote the book *Perceptrons: An Introduction to Computational Geometry*, in which they proved mathematically that the concept of the perceptron can't represent the exclusive OR (XOR); you know this aspect as *linear separability*. Unfortunately, the publication caused a decline in funding in the ANN area for 15 years (the "AI winter").

However, later neural network algorithms, such as the perceptron-based backpropagation network, remedied this shortcoming. Ironically, Minsky himself turned to neural networks in the 1980s, and his research was based on Rosenblatt's findings.

10.4 The 1970s

In the 1970s, important approaches emerged that gave ANNs a new boost, especially the training of multilayer networks.

10.4.1 1972: Kohonen – Associative Memory

Teuvo Kohonen presented "correlation matrix memories" in 1972, a model of a special *associative memory* with linear activation functions and continuous values for weights, activations, and outputs.

10.4.2 1973: Lighthill Report

The *Lighthill Report* (James Lighthill: "Artificial Intelligence: A General Survey")—a report that reflected the status quo of research in the field of AI from James Lighthill's point of view—presented the *combinatorial explosion* as one of the most important points of criticism of AI, thereby denying its applicability to real-world tasks. In response, the UK government decided to stop funding research into AI, except at the universities of Edinburgh, Sussex, and Essex.

10.4.3 1974: Backpropagation

Paul Werbos mentioned the applicability of the backpropagation learning method to ANNs in his 1974 PhD thesis at Harvard University. He published the thesis in 1982 in the formalization currently in use, as we discussed it in Chapter 6.

10.5 The 1980s

This decade was characterized by a resurgence of ANNs and the emergence of a large number of different models.

10.5.1 1980: Fukushima's Neocognitron

An absolute must in the ancestral line of neural network types is the *neocognitron* by Kunihiko Fukushima, which is based on his *cognitron*. With his approach, he was the motivator, inspiration, and companion for convolutional neural networks (CNNs). The neocognitron, in turn, was inspired by the models developed by Hubel and Wiesel in 1959. They found two types of cells in the visual cortex, the *simple cells* and *complex cells*.

In the neocognitron (see Figure 10.5), the S-cells and C-cells are created from the simple and complex cells. The *S-cells* have the task of recognizing local features (e.g., edges), whereas the *C-cells* are responsible for handling deformations of the features, for example, shifts in position or scaling. The S and C layers integrate the features at increasingly higher levels, making complex pattern recognition possible (see Figure 10.6).

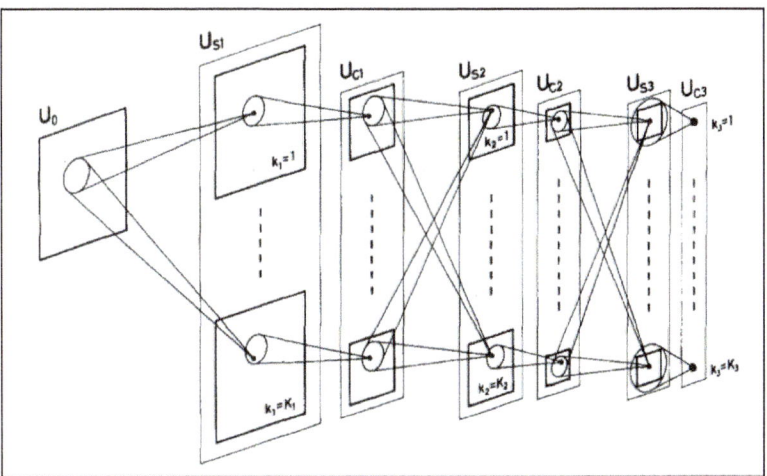

Figure 10.5 Diagram of the Interconnections in Fukushima's Neocognitron (Source: Kunihiko Fukushima, Neocognitron: "A Self-Organizing Neural Network Model for a Mechanism of Pattern Recognition Unaffected by Shift in Position," www.rctn.org/bruno/public/papers/Fukushima1980.pdf)

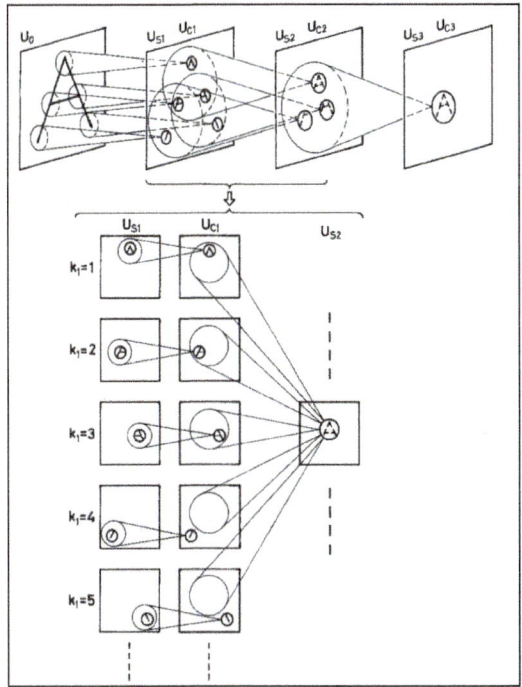

Figure 10.6 Neocognitron Example for the Extraction of Features (Source: Kunihiko Fukushima, Neocognitron: "A Self-Organizing Neural Network Model for a Mechanism of Pattern Recognition Unaffected by Shift in Position," www.rctn.org/bruno/public/papers/Fukushima1980.pdf)

The Neocognitron is capable of position-invariant and scaling-invariant recognition of handwritten characters from a layered sequence of simple and complex cells, such as those found in the biological visual system of cats.

The input connections to the S-cells can be trained to respond to certain features in their *receptive field*. The input connections to the C-cells can't be changed, and the C-cells receive inputs from S-cells that react to the same feature, but at a shifted position. The C-cell reacts as soon as one of the S-cells provides an output; this allows the C-cells to respond to features in a position-invariant manner (see Figure 10.7).

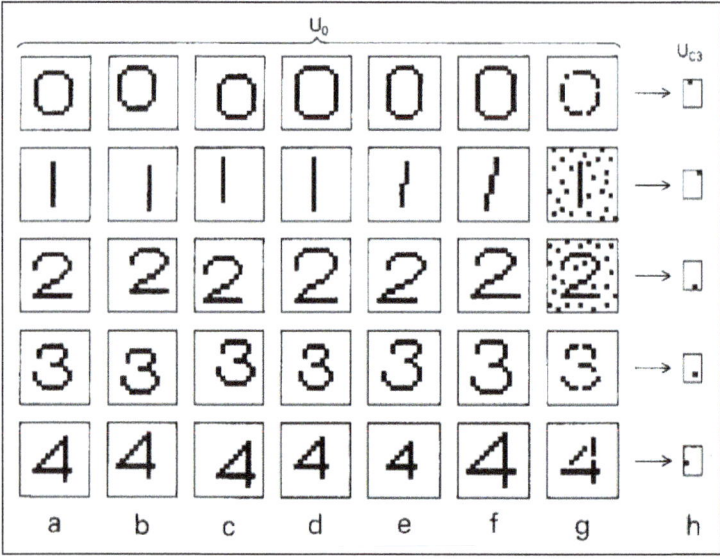

Figure 10.7 Neocognitron Example of Different Variations and the Response in the Highest Layer (Source: Kunihiko Fukushima, "Neocognitron: A Self-Organizing Neural Network Model for a Mechanism of Pattern Recognition Unaffected by Shift in Position," www.rctn.org/bruno/public/papers/Fukushima1980.pdf)

10.5.2 1982: John Hopfield

The AI winter slowly turned into spring when the Japanese Ministry of International Trade and Industry (MITI) announced at the US-Japan Conference on Cooperative/Competitive Neural Networks in 1982 that it would undertake funding for the development of the fifth generation of computer hardware with AI in mind, and John Hopfield presented his binary *Hopfield Network*.

History of Computer Hardware
- First generation: Circuits, tubes and cables (machine language).
- Second generation: Transistors and diodes (Assembler).

- Third generation: Solid-state technology, for example, integrated circuits (higher programming languages).
- Fourth generation: Code generation and microprocessors (domain-specific languages).
- Fifth generation: (From 1982 onward) high number of CPUs and AI.

A Hopfield network is a *feedback network* or *recurrent neural network* (RNN), that is, a network that feeds the calculated output back into the network.

Feedback Networks

RNNs are ANNs specifically designed to process sequential data. RNNs are designed to capture the dependencies between elements in sequences. RNNs have *feedback loops* that allow them to store information about previous inputs and incorporate this information into the processing of current inputs. These feedback loops create a kind of memory that helps them recognize sequence dependencies, which makes them particularly useful for tasks such as time series prediction, language processing, and machine translation.

The Hopfield network was motivated by the associative abilities of the human brain, which can still make sense of unclear representations (see Figure 10.8).

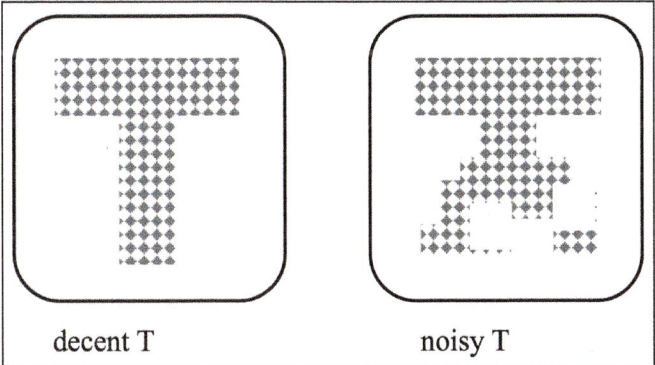

Figure 10.8 Associative Memory

The Hopfield network (see Figure 10.9) has only one layer, which serves as the input and output layer. Each of the neurons, more precisely McCulloch-Pitts neurons, is connected to every other neuron except itself. The outputs of the neurons can take the values +1 or -1, and the weights of the connections between two neurons are identical regardless of the direction, that is, $w_{ij} = w_{ji}$.

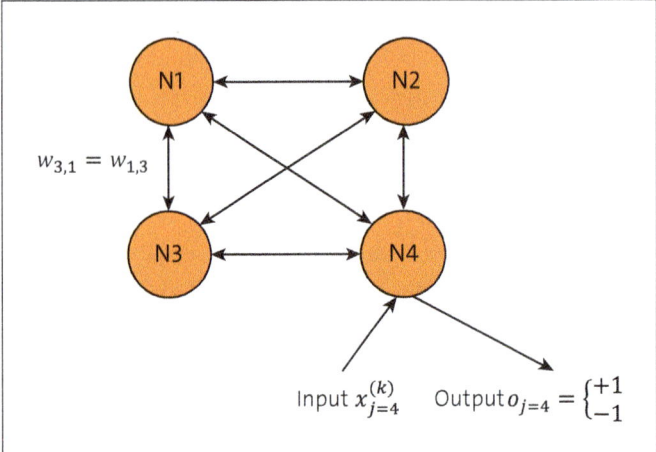

Figure 10.9 Example of a Binary Hopfield Network

The Hopfield network learns by adjusting the weights for all training examples, for example, by iterating through all training examples and then adjusting the weights for all neurons. The calculation of the weights is defined as follows:

$$w_{ij} = \sum_{k=1}^{N} x_i^{(k)} \cdot x_j^{(k)}$$

This equation has the following:

- w_{ij} represents the weight of the connection from neuron i to neuron j.
- N is the number of inputs that are represented as a vector as usual, for example, the k-th input $\vec{x}^{(k)} = \left(x_1^{(k)}, x_2^{(k)}, \ldots\right)$.

The calculation of the outputs can either be performed asynchronously by randomly selecting one neuron per iteration, or synchronously, by calculating the output of each neuron until no more changes occur or a maximum number of iterations has been reached.

The output of a neuron is in turn calculated in a tried and tested way. The following therefore applies to a Hopfield network with n neurons:

$$net_j = \sum_{i=1}^{n} w_{ij} \cdot o_i$$

$$f_{\text{step}}(net_j) = o_j = \begin{cases} +1, & \text{if } net_j \geq \theta_j \\ -1, & \text{if } net_j < \theta_j \end{cases}$$

If the network has now learned and is presented with an input that is unknown or a noisy example of a learning example, then the network finds the output that best matches the learned examples. There are limits to the Hopfield network with regard to the examples that can be learned, which can be calculated using the following formula:

$N < 0.14 \cdot n$

A *recall error* is tolerated in a maximum of 5% of cases. For example, a network with $n = 10$ neurons can learn a maximum of $1.4 > N$, that is, 1 example.

A Small Example

In Figure 10.10, we've prepared two inputs the network is supposed to learn. The gray boxes are given the value +1 and the white boxes the value -1. The values are determined row by row and linked to form a vector. The T is represented by the following vector:

$$\vec{x}^{(T)} = (+1, +1, +1, -1, +1, -1, -1, +1, -1)$$

The U is represented by the following vector:

$$\vec{x}^{(U)} = (+1, -1, +1, +1, -1, +1, +1, +1, +1)$$

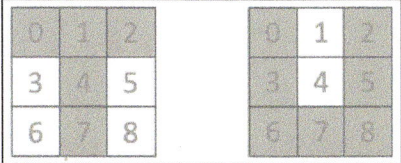

Figure 10.10 A T on the Left and a U on the Right with the Indexes for the Vector

The next step is to determine the weights, which are again represented in a matrix; the first index for the weight w_{ij}, that is, i, corresponds to the row in the matrix, and the second index, that is, j, corresponds to the column. Furthermore, the neurons have no connections to themselves, which means that $w_{ii} = 0$. Furthermore, the weights are symmetrical: $w_{ij} = w_{ji}$. This means we only need to calculate the values above the diagonal of the matrix.

We'll now calculate the values per vector and then simply add the matrixes. Let's start with the first vector $\vec{x}^{(T)}$ and calculate $w_{11} = x_1^{(T)} \cdot x_1^{(T)} = (+1) \cdot (+1) = +1$ and $w_{41} = x_4^{(T)} \cdot x_1^{(T)} = (-1) \cdot (+1) = -1$. If we take this further, we get the following matrix:

$$W_T = \begin{pmatrix} 0 & 1 & 1 & -1 & 1 & -1 & -1 & 1 & -1 \\ 1 & 0 & 1 & -1 & 1 & -1 & -1 & 1 & -1 \\ 1 & 1 & 0 & -1 & 1 & -1 & -1 & 1 & -1 \\ -1 & -1 & -1 & 0 & -1 & 1 & 1 & -1 & 1 \\ 1 & 1 & 1 & -1 & 0 & -1 & -1 & 1 & -1 \\ -1 & -1 & -1 & 1 & -1 & 0 & 1 & -1 & 1 \\ -1 & -1 & -1 & 1 & -1 & 1 & 0 & -1 & 1 \\ 1 & 1 & 1 & -1 & 1 & -1 & -1 & 0 & -1 \\ -1 & -1 & -1 & 1 & -1 & 1 & 1 & -1 & 0 \end{pmatrix}$$

Here's a little hint: if you're too lazy to calculate, like us, then the following small Python program will do the matrix calculation for you. We assume that you've already created a Jupyter Notebook for this chapter and will now create one code cell after the other and program them with us, as shown in Listing 10.1.

```python
import numpy as np
# T
xT = np.matrix([+1,+1,+1,-1,+1,-1,-1,+1,-1])
WT = np.matmul(xT.T,xT)
print(WT)
# Output:
[[ 1  1  1 -1  1 -1 -1  1 -1]
 [ 1  1  1 -1  1 -1 -1  1 -1]
 [ 1  1  1 -1  1 -1 -1  1 -1]
 [-1 -1 -1  1 -1  1  1 -1  1]
 [ 1  1  1 -1  1 -1 -1  1 -1]
 [-1 -1 -1  1 -1  1  1 -1  1]
 [-1 -1 -1  1 -1  1  1 -1  1]
 [ 1  1  1 -1  1 -1 -1  1 -1]
 [-1 -1 -1  1 -1  1  1 -1  1]]
```

Listing 10.1 Multiplying the Transposed T-Vector by Itself Produces a Matrix

Next, we calculate the second matrix—for the vector $\vec{x}^{(U)}$—also using the same program (see Listing 10.2).

```python
import numpy as np
# U
xU = np.matrix([+1,-1,+1,+1,-1,+1,+1,+1,+1])
WU = np.matmul(xU.T,xU)
print(WU)
# Output:
[[ 1 -1  1  1 -1  1  1  1  1]
 [-1  1 -1 -1  1 -1 -1 -1 -1]
 [ 1 -1  1  1 -1  1  1  1  1]
 [ 1 -1  1  1 -1  1  1  1  1]
 [-1  1 -1 -1  1 -1 -1 -1 -1]
 [ 1 -1  1  1 -1  1  1  1  1]
 [ 1 -1  1  1 -1  1  1  1  1]
 [ 1 -1  1  1 -1  1  1  1  1]
 [ 1 -1  1  1 -1  1  1  1  1]]
```

Listing 10.2 The Matrix for the U-Vector

Finally, the two matrixes W = WT + WU are added together, as shown in Listing 10.3: the diagonal contains 0 values because there is no self-reference to the neuron to itself, thus we may simply ignore the diagonal values from a pragmatic point of view. However, if that bothers you, feel free to explicitly set the diagonal to 0 in your program.

```python
import numpy as np
# T + U
```

```
W = WT + WU
print(W)
# Output:
[[ 2  0  2  0  0  0  0  2  0]
 [ 0  2  0 -2  2 -2 -2  0 -2]
 [ 2  0  2  0  0  0  0  2  0]
 [ 0 -2  0  2 -2  2  2  0  2]
 [ 0  2  0 -2  2 -2 -2  0 -2]
 [ 0 -2  0  2 -2  2  2  0  2]
 [ 0 -2  0  2 -2  2  2  0  2]
 [ 2  0  2  0  0  0  0  2  0]
 [ 0 -2  0  2 -2  2  2  0  2]]
```

Listing 10.3 The Sum of the Two Weight Matrixes W = WU + WT

What follows now is the step of recognizing a created pattern because we want to use the Hopfield network as an associative memory:

1. To do this, we append the first vector $\vec{x}^{(0)}$ to the network and hope it gets recognized. The output must be calculated for each neuron, which is nothing other than the following multiplication (you'll remember the many matrix-vector multiplications we carried out during backpropagation):

    $\vec{x}_{t+1} = W \cdot \vec{x}_t$

    In this equation, $\vec{x}_t$ is the vector created at time t. We start at time 0, so $\vec{x}^{(0)}$ is the vector that is created at the start.

2. The activation function must still be applied to each component of the vector.

Of course, we've also put together a program for the evaluation. This allows you to experiment and "noise" the images, that is, the vectors, and explore the limits of when the network recognizes an input as T or U.

In addition, note that we've encapsulated the output of the letter in the `drawLetter` function, and we've implemented the step function as a lambda (see Listing 10.4).

```
# Matrix and vector operations
import numpy as np
# Draw
import matplotlib.pyplot as plt
%matplotlib inline

# Row dimension of the matrix
rowDim = 3
# One matrix in one row
lineDim = rowDim * rowDim
```

```python
# Draw a letter
def drawLetter(letter):
    global rowDim
    fig = plt.figure(111)
    ax1 = fig.add_subplot(111)
    # Output of the matrix in black and white, cm = Colormap,
    # interpolation: discrete colors
    ax1.imshow(letter.reshape(rowDim,rowDim),cmap=plt.cm.binary, interpolation='nearest')
    plt.show()

# Lambda for step function
f_step = lambda w: 1 if w >= 0 else -1
# Vectorize lambda
vectorized_step = np.vectorize(f_step)

# Learn
# Identity
Wi = np.identity(lineDim)
# T
xT = np.matrix([+1,+1,+1,-1,+1,-1,-1,+1,-1])
WT = np.matmul(xT.T,xT) - Wi
print('T - Pattern')
drawLetter(xT)
print('T - Matrix minus unit matrix')
print(WT)

# U
xU = np.matrix([+1,-1,+1,+1,-1,+1,+1,+1,+1])
WU = np.matmul(xU.T,xU) - Wi
print('U - Pattern')
drawLetter(xU)
print('U - Matrix minus unit matrix')
print(WU)

# Sum of the pattern matrixes
W = WT + WU
print('T + U - Matrix')
print(W)

### Evaluation
x0 = np.matrix([-1,-1,-1,+1,-1,+1,+1,+1,+1])
print('Input vector')
drawLetter(x0)
```

```
vt = x0.T

# Search for attractor
for t in range(2):
    vt1 = vectorized_step(W * vt)
    print('Step {}'.format(t))
    Wt1 = vt1.reshape(rowDim,rowDim)
    drawLetter(vt1)
    vt = vt1
```

Listing 10.4 Learning and Evaluation with Hopfield Networks

The output looks as follows:

T - Pattern

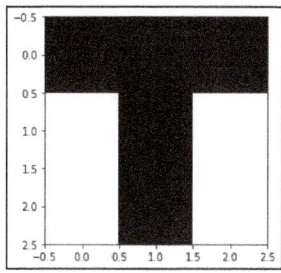

```
T - Matrix minus unit matrix
[[ 0.  1.  1. -1.  1. -1. -1.  1. -1.]
 [ 1.  0.  1. -1.  1. -1. -1.  1. -1.]
 [ 1.  1.  0. -1.  1. -1. -1.  1. -1.]
 [-1. -1. -1.  0. -1.  1.  1. -1.  1.]
 [ 1.  1.  1. -1.  0. -1. -1.  1. -1.]
 [-1. -1. -1.  1. -1.  0.  1. -1.  1.]
 [-1. -1. -1.  1. -1.  1.  0. -1.  1.]
 [ 1.  1.  1. -1.  1. -1. -1.  0. -1.]
 [-1. -1. -1.  1. -1.  1.  1. -1.  0.]]
```
U - Pattern

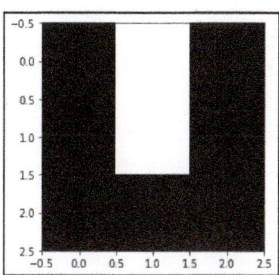

```
U - Matrix minus unit matrix
[[ 0. -1.  1.  1. -1.  1.  1.  1.  1.]
 [-1.  0. -1. -1.  1. -1. -1. -1. -1.]
 [ 1. -1.  0.  1. -1.  1.  1.  1.  1.]
 [ 1. -1.  1.  0. -1.  1.  1.  1.  1.]
 [-1.  1. -1. -1.  0. -1. -1. -1. -1.]
 [ 1. -1.  1.  1. -1.  0.  1.  1.  1.]
 [ 1. -1.  1.  1. -1.  1.  0.  1.  1.]
 [ 1. -1.  1.  1. -1.  1.  1.  0.  1.]
 [ 1. -1.  1.  1. -1.  1.  1.  1.  0.]]
T + U - Matrix
[[ 0.  0.  2.  0.  0.  0.  0.  2.  0.]
 [ 0.  0.  0. -2.  2. -2. -2.  0. -2.]
 [ 2.  0.  0.  0.  0.  0.  0.  2.  0.]
 [ 0. -2.  0.  0. -2.  2.  2.  0.  2.]
 [ 0.  2.  0. -2.  0. -2. -2.  0. -2.]
 [ 0. -2.  0.  2. -2.  0.  2.  0.  2.]
 [ 0. -2.  0.  2. -2.  2.  0.  0.  2.]
 [ 2.  0.  2.  0.  0.  0.  0.  0.  0.]
 [ 0. -2.  0.  2. -2.  2.  2.  0.  0.]]
```
Input vector

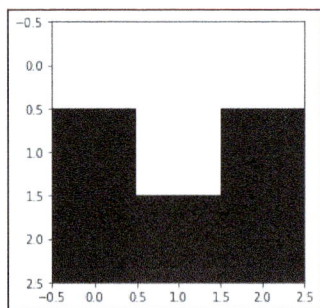

Step 0

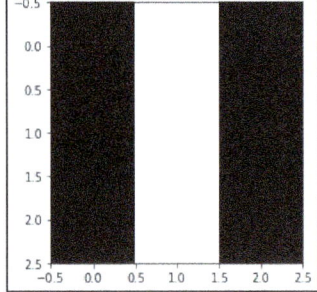

Step 1

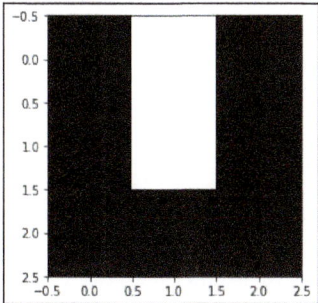

At the start of the program, we learn the two patterns by performing a matrix multiplication of the weights with the individual vectors, and then subtract the *unit matrix* from the result so that there are actually only zeros in the diagonal. The unit matrix has all ones in the diagonal and otherwise only zeros.

We add both matrixes together to preserve the associative memory. This is followed by the calculation of the outputs of the neurons, as with the perceptron. The loop at the end is used to iterate during the evaluation and to repeatedly apply the association matrix to the output of the previous iteration. In our case, this is done five times, but it can also depend on whether there has been a change compared to the last iteration.

10.5.3 1982: Kohonen's SOM

We'll discuss the SOMs by Teuvo Kohonen in more detail in Chapter 12, and also implement them. This approach is motivated by neurobiological studies which show that sensory information is mapped to corresponding areas in the cerebral cortex with the following two properties:

- The incoming information is kept in its appropriate context.
- Neurons that process related information are spatially close to each other so that they can interact with each other through short connections.

This gives rise to the principle of the *topographic map*. The spatial positioning of an output neuron in a topographic map corresponds to a feature that was extracted from the input. This means that a SOM maps a high-dimensional vector onto a one- or two-dimensional map in such a way that it preserves the topology.

10.5.4 1986: Backpropagation

In 1986, Rumelhart, Hinton, and Williams published the backpropagation learning method. You've already implemented this procedure in full earlier in the book so we won't explain it any further here.

10.5.5 1987: NN Conference

The first conference on neural networks was organized by the Institute of Electrical and Electronics Engineers (IEEE) in San Diego. The event drew 200 speakers and 2,000 visitors.

10.5.6 1989: Yann LeCun: Convolutional Neural Networks

Yann LeCun's Tech Report CRG-TR-89-4 from 1989 deals with generalization problems and network design strategies in neural networks. A central aspect of the report is the introduction of convolutional neural networks (CNNs), as discussed in Chapter 7, which were developed specifically for processing image data.

This brings us to part 2 of our ANN story (see Figure 10.11).

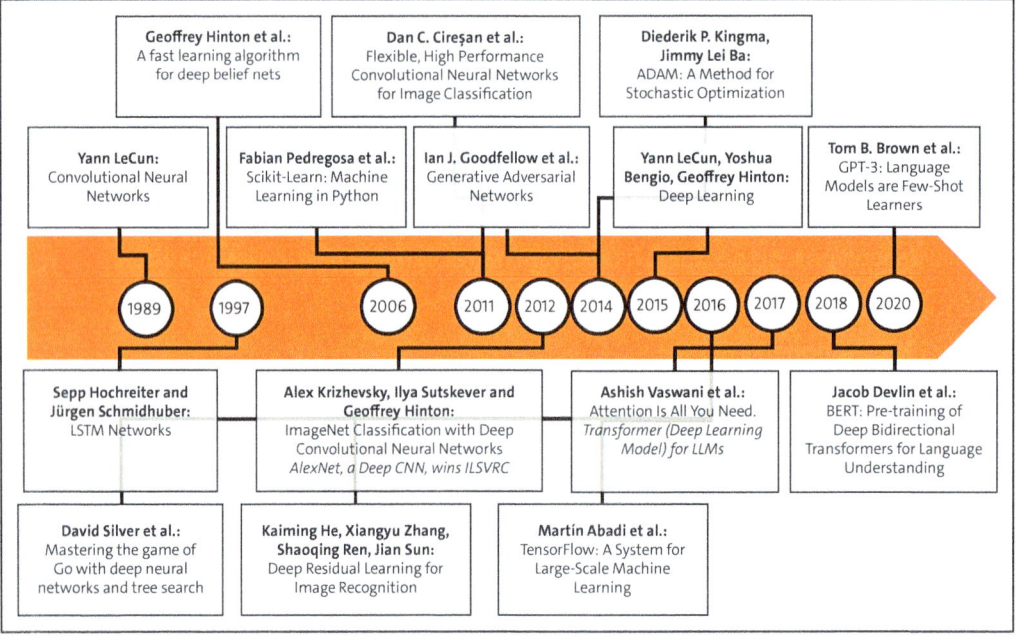

Figure 10.11 ANN History, Part 2

10.6 The 1990s

A highlight in the 1990s was the consideration of time dependencies in the data, such as in the case of spoken language. RNNs, which have been known for some time, were used as the basis for this.

10.6.1 1997: Sepp Hochreiter and Jürgen Schmidhuber – Long Short-Term Memory

In 1997, Sepp Hochreiter and Jürgen Schmidhuber published an article that proposed a new architecture for networks that have a *long short-term memory* (LSTM). There are approaches (as you'll see, for example, with regard to *Q-learning* in Chapter 12) that enable classic feed-forward architectures to build up a memory. However, LSTM delivers much better performance and can solve problems and tasks that can't be solved by classic architectures.

LSTM networks are a specialized form of RNNs designed to capture long-term dependencies in sequence data and overcome the problems of vanishing and exploding gradients. LSTMs use a *cell state*, which retains information almost unchanged over many time steps, and three *gates*:

- **Input gate**
  Determines which new information is included in the cell state.
- **Forget gate**
  Controls which information is removed from the cell state.
- **Output gate**
  Controls which information from the cell state gets passed on as output.

This allows LSTMs to store important information over longer periods of time and discard irrelevant data, making them particularly useful for tasks such as language processing, time series analysis, and image caption generation.

10.7 The 2000s

From the 2000s onward, much more data was available for training ANNs, and computing power increased accordingly. These two aspects have reshuffled the cards for the learning of ANNs in terms of learning speed and quantity.

10.7.1 2006: Geoffrey Hinton et al.

In 2006, Geoffrey Hinton, Simon Osindero, and Yee Whye Teh presented the successful training of a multilayer neural network in their paper, "A Fast Learning Algorithm for Deep Belief Nets," thereby reigniting research in the field of ANNs. Based on this, the time had come for *deep neural networks* and their use in a wide variety of areas, for example, in the game of Go. Soon after that AlphaGo became the talk of the town. The Google DeepMind company developed *AlphaGo*, a program that plays the board game Go and that beat the world's best professional player Lee Sedol from South Korea in March 2016. But that wasn't the end of the story: In December 2017, DeepMind presented the AI called *AlphaZero*. Within just a few hours, it learned to play chess, Go, and

Shogi (the Japanese version of chess) at a higher level than any human being, simply by knowing the rules of the game.

10.8 The 2010s

A further increase in computing power was achieved through the use of GPUs on graphics cards. This opened up the possibility of creating even more complex networks with many more parameters, for example, a higher number of neurons, connections, and layers—even networks that compete against each other and stimulate the other network to learn.

10.8.1 2014: Ian J. Goodfellow et al. – Generative Adversarial Networks

Ian J. Goodfellow and his research group, then at the University of Montreal, are credited with the idea of two complementary or opposing networks that lead to a constantly improving result.

The main concept is to have two networks compete against each other (see Figure 10.12), with one network, called the *generator*, generating images from noise—that is, random numbers—and the other network, called the *discriminator*, predicting whether a given input is a real image from the training examples or an image generated by the generator.

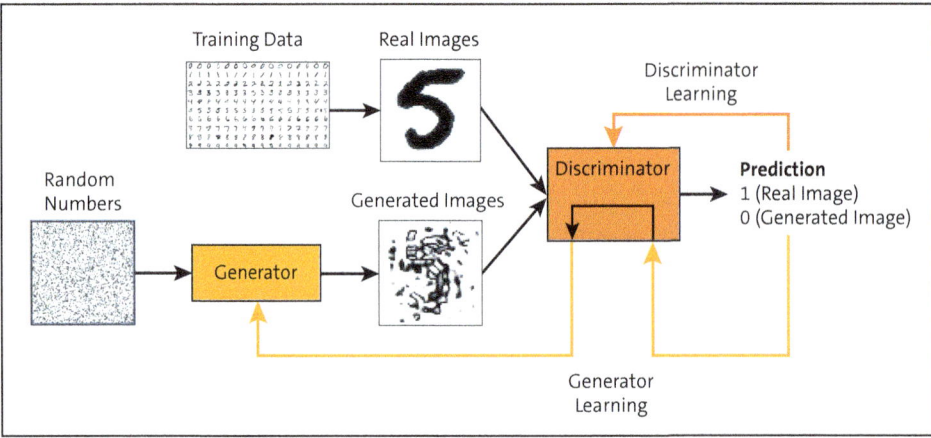

Figure 10.12 Generative Adversarial Networks: Basic Structure

The interaction between generator and discriminator can be summarized as follows:

- The generator receives a vector with random numbers, referred to as a *latent sample*, and generates an image from it (see Figure 10.13). Over time, the generator learns to create images that look similar to real images.

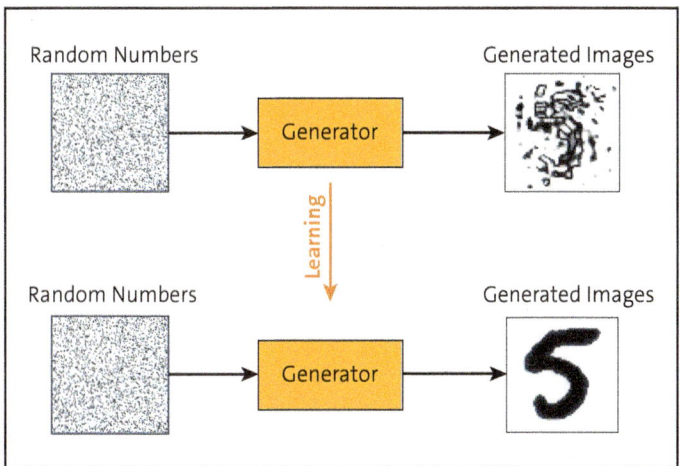

Figure 10.13 The Generator Learns to Create Images

- Then, a generated image gets passed to the discriminator, which is trained to classify images using error backpropagation. Images from a training set of real images (e.g., images of people as positive examples) and the generated images from the generator (as negative examples) are used as training input.

The discriminator provides a prediction, that is, a number between 0 and 1, about the authenticity of the image, where 0 stands for "fake" and 1 for "real" (see Figure 10.14).

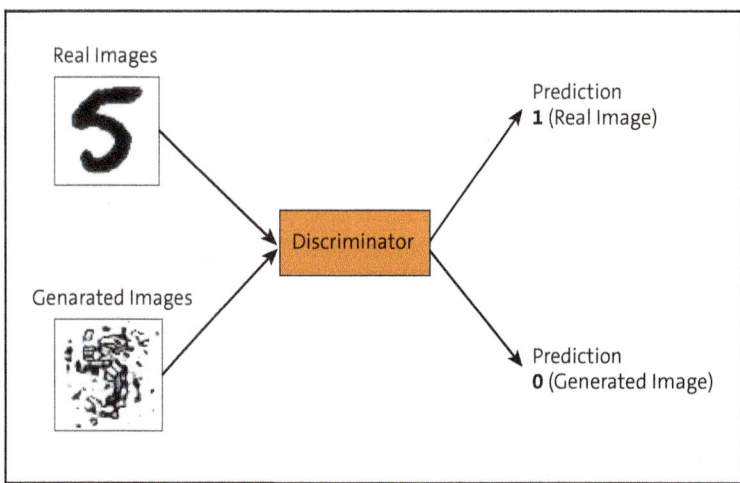

Figure 10.14 The Discriminator Learns to Differentiate between Images

The possible errors in the classification are used for the error backpropagation learning of the discriminator.

- The calculated classification value of the discriminator is used, among other things, to adjust the weights of the generator by means of backpropagation so that better images are produced. The error backpropagation is performed by the discriminator network and then the generator network without adjusting the weights of the discriminator.

- The training of the discriminator and the generator takes place alternately. Whenever one is trained, the other is fixed; that is, the weights aren't changed. In the course of training, the generator should continuously improve, and the discriminator's classification performance should deteriorate. In the best case, that is, when the generator works perfectly, the discriminator can only guess, achieving a prediction probability of 0.5, whether it's a real or a generated image.

This game is played over and over again, and so the generator learns to produce increasingly better examples, and the discriminator learns to discriminate better and better.

10.8.2   2017: Ashish Vaswani et al. – Attention Is All You Need

In their seminal paper, "Attention Is All You Need" (2017), Ashish Vaswani and his colleagues introduced the *transformer model*, which revolutionized the architecture for many modern natural language processing (NLP) tasks. In contrast to traditional RNNs and LSTM networks, the transformer completely dispenses with recurrent structures and instead relies on a *self-attention* mechanism that processes the relationships between all words in a sentence in parallel. In Chapter 7 and Chapter 8, we took an in-depth look at transformer models.

10.9   Summary

This chapter contains a lot of history, and we've also shown you some alternative models for neural networks. We wanted to inspire you and present the diversity of neural networks, which can't be reduced to individual approaches. Furthermore, you've seen that the area is far from exhausted and still allows for many variations. Deep networks, RNNs, and generative approaches are at the cutting edge and are waiting to be implemented and further developed by you. You should start right away with a Hopfield network!

10.10   Further Reading

- **1943**: Warren S. McCulloch und Walter H. Pitts, "A Logical Calculus of the Ideas Immanent in Nervous Activity." In: *Bulletin of Mathematical Biophysics*, Vol. 5, p. 115–133 (1943), *www.cse.chalmers.se/~coquand/AUTOMATA/mcp.pdf*

- **1949:** Donald O. Hebb, *The Organization of Behavior; A Neuropsychological Theory*. John Wiley & Sons, New York, 1949, *https://archive.org/details/in.ernet.dli.2015.168156*
- **1955:** John McCarthy, Marvin Minsky, Nathaniel Rochester, and Claude Shannon, "A Proposal for the Dartmouth Summer Research Project on Artificial Intelligence," *http://jmc.stanford.edu/articles/dartmouth/dartmouth.pdf*
- **1958:** Frank Rosenblatt, "The Perceptron: A Probabilistic Model for Information Storage and Organization in the Brain." *Psychological Review*, Vol. 65, No. 6, 1958, *www.ling.upenn.edu/courses/cogs501/Rosenblatt1958.pdf*
- **1960:** Bernard Widrow and Marcian Hoff, "Adaptive Switching Circuits." 1960 I.R.E. Wescon Convention Record: Sessions at Los Angeles, August 23-26, 1960, Vol. 4, Part 4, p. 96-104 Institute of Radio Engineers, *https://isl.stanford.edu/~widrow/papers/c1960adaptiveswitching.pdf*
- **1972:** Teuvo Kohonen, "Correlation Matrix Memories." In: *IEEE Transactions on Computers*, Vol. c-21, No. 4, April 1972, *https://lucidar.me/fr/neural-networks/files/1972-correlation-matrix-memories.pdf*
- **1973:** James Lighthill, "Lighthill Report," *www.chilton-computing.org.uk/inf/literature/reports/lighthill_report/p001.htm*
- **1980:** Kunihiko Fukushima. "Neocognitron: A Self-Organizing Neural Network Model for a Mechanism of Pattern Recognition Unaffected by Shift in Position," *www.rctn.org/bruno/public/papers/Fukushima1980.pdf*
- **1982:** John Hopfield, "Neural Networks and Physical Systems with Emergent Collective Computational Abilities." In: *Proc. Natl Acad. Sci.* USA Vol. 79, pp. 2554-2558, April 1982 *Biophysics, www.pnas.org/content/pnas/79/8/2554.full.pdf*
- **1982:** Teuvo Kohonen, "Self-Organized Formation of Topologically Correct Feature Maps." In: *Biol. Cybern.* 43, p. 59-69 (1982). *https://tcosmo.github.io/assets/soms/doc/kohonen1982.pdf*
- **1986:** David E. Rumelhart, Geoffrey E. Hinton, and Ronald J. Williams, "Learning Internal Representations by Error Propagation." September 1985, ICS Report 8506. *https://gwern.net/doc/ai/nn/fully-connected/1986-rumelhart.pdf*
- **1989:** Yann LeCun, "Generalization and Network Design Strategies." Y. le Cun Department of Computer Science University of Toronto Technical Report CRG-TR-89-4 June 1989, *https://www.academia.edu/2813343/Generalization_and_network_design_strategies*
- **1997** Sepp Hochreiter and Jürgen Schmidhuber, "Long Short-Term Memory," *Neural Computation* 9(8):1735-1780, 1997, *www.bioinf.jku.at/publications/older/2604.pdf*
- **2006:** Geoffrey Hinton et al., "A Fast Learning Algorithm for Deep Belief Nets." In: *Neural Computation* (2006) 18 (7):1527-1554. *www.mitpressjournals.org/doi/pdf/10.1162/neco.2006.18.7.1527*

- **2014**: Ian J. Goodfellow et al., "Generative Adversarial Nets," *Advances in Neural Information Processing Systems* 27 (2014). *https://arxiv.org/pdf/1406.2661.pdf*
- **2017**: Ashish Vaswani et al., "Attention Is All You Need." 31st Conference on Neural Information Processing Systems (NIPS 2017), Long Beach, CA, USA. *https://arxiv.org/pdf/1706.03762*

Chapter 11
The Machine Learning Process

Learning something and getting more and more practiced at it over time, isn't that also a joy?
—Confucius (551–479 BC)

So far, we've focused on the neural networks themselves. You now know what input and output data is, as well as what training, test, and validation datasets are and what you need them for. Although that's a lot of information, you may be wondering what else you need to know. You may have questions like these: What do I want to do with the neural network? Where do I get the data from? How can I then use the finished neural network in a production system? It quickly becomes clear that our neural networks are part of a process that called the *machine learning (ML) process*.

In this chapter, we'll first focus on one of the most common process descriptions for data analysis tasks: the Cross Industry Standard Process for Data Mining (CRISP-DM) model. We'll then look at some aspects that you also need to consider when you develop and use neural networks. These relate to ethical and legal issues, as well as ecological aspects. In the third part of this chapter, we pick out all the elements from the model that involve data preparation. Data preparation is also referred to as feature engineering.

11.1 The CRISP-DM Model

Here's the answer to the question of what's next: the *Cross Industry Standard Process for Data Mining* (CRISP-DM) model, which was developed as a theoretical model at the end of the 1990s in the course of large data mining projects. Figure 11.1 gives you a first impression of its components.

We won't get bogged down in a detailed description of this model (that would be sufficient material for another book), but we'll present you with the relevant information. We're also not confused by the fact that it's a standard process for data mining. Data mining, ML, and neural networks are closely related, although we don't want to make a statement about the nature of the relationship. There are countless attempts to define and differentiate in the vastness of the internet, but no truly universally recognized interpretation exists. *Data mining* is an area in which attempts are made to recognize patterns in a large amount of data, which then provide the basis for business decisions. ML methods can also be used for this purpose.

11   The Machine Learning Process

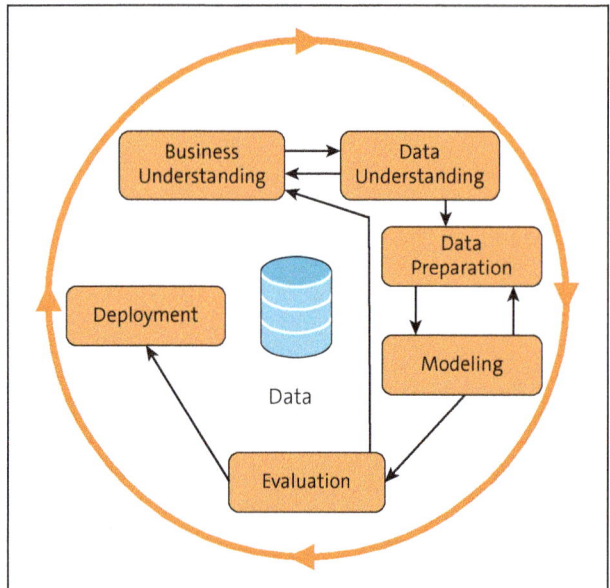

Figure 11.1 The CRISP-DM Model (Source: Wikimedia Commons)

As you can see in Figure 11.1, a data mining project consists of six phases (*business understanding*, *data understanding*, *data preparation*, *modeling*, *evaluation*, and *deployment*), which are arranged in a circle. The circular shape symbolizes the circular structure of data mining projects. It's also interesting to note that there are always backward relationships between the phases, which means that a constant back and forth between the phases is a necessity.

Let's try to garnish these theoretical approaches with an example.

11.1.1 Business Understanding

This phase is used to determine the direction before you start running. This is where both business and data mining goals are defined, as well as success criteria against which the result can be measured. In this initial phase, an inventory of possible data sources is created, and requirements, premises, and conditions are defined. A cost-benefit analysis is just as much a part of this as the clarification of risks and uncertainties. For the sake of completeness, we should mention that some project management is involved as well, as is always the case.

Let's look at the example of an online news publisher who wants to know which features of an article lead to many clicks to increase the number of subscribers or website visits. This could be the writing style, the length of the article, its content, the topic, the format, and so on.

11.1.2 Data Understanding

In this phase, the data doesn't get manipulated or changed, but first collected and then analyzed. A precise description of the data and data quality is an important prerequisite for the subsequent step. Understandably, business understanding and data understanding influence each other greatly because the analysis of the data and its quality can have an impact on the business and data mining goals and lead to a reduction or expansion of the goals.

In our example, the aim is to clarify what data is needed, what data is available, and—above all—in what quality. Quantitative data such as the number of clicks or the length of an article is relatively easy to provide. Qualitative properties such as the writing style aren't available from the outset and must first be defined and determined manually. Properties such as the topic or subject area may already be present, but this also needs to be determined.

11.1.3 Data Preparation

Data preparation and elements of data understanding are referred to as *feature engineering* in ML jargon. This phase is about preparing the data so that it can be used for an ML model (and that includes neural networks, of course). This phase is full of challenges and is also very time-consuming. Typical tasks include selecting, cleaning, constructing, and integrating data to transform it into a dataset to which ML algorithms can be applied. We then refer to this dataset as an *ML-ready dataset*.

To illustrate the difficulties that need to be overcome in this phase, let's look at the *number of clicks* property. At first glance, this seems very simple, but it could be that you first have to laboriously pick out this data from *log files*. In our case, these would be web server log files. You would then have to write a computer program that extracts this information from the log files and writes it to a database for further reuse. You must first determine whether all log files are still available for the selected time period. The log files can change as a result of a change to the website.

> **Log Files**
> Log files contain the automatically maintained log of process actions on a computer system. Log files typically contain a timestamp with the date, the exact time, and the type of action. Web server log files also contain access information, such as users and IP addresses, and also record from which website or with which browser an action was carried out.

You can already see that data collection can be very time-consuming.

11.1.4 Modeling

You already have a good idea of what is needed in this phase. It starts with the selection of models; this can be a conventional ML method, a neural network, or ensemble techniques that use multiple methods simultaneously.

Regardless of the model, the tasks are very similar: A model is created, the model parameters are determined iteratively, and an attempt is made to maximize the model quality. In the case of neural networks, this is the definition of a network architecture, learning rate, loss function, and so on. It's clear that the data preparation and modeling phases can be easily separated in terms of description, but that there is a high level of interaction in operational terms, as the model quality is also very much dependent on the data quality.

For the modeling of our online article example, we have to decide on the input data and the output data. Every change here has an immediate effect on the data preparation and on the model. In our example, we agree that our model should be able to predict the number of clicks for an article and define this as our output data. Input data includes the length of the article, the writing style, the topic, the topicality, the author, and so on.

11.1.5 Evaluation

Here it's important to distinguish the evaluation of a model in terms of accuracy and generalizability from the evaluation as a phase of the CRISP-DM model. The former is part of the modeling process, while the latter concerns the comparison of the model with the business objectives. However, this assessment isn't very simple and is often carried out by means of test applications in production operation and checked in this way. The evaluation also includes the review of results, which provides new, unexpected knowledge as a by-product.

An important output of this phase is also determining whether any changes should be made to the business objectives. This is where the decision is made whether to enter the next phase or whether to start a new iteration by changing the business (and data mining) objectives.

For our online article model, we'll obtain a predictor that can predict the number of clicks or—even better—determine a quality score for an article. This can provide valuable decision support for the online editor. For the evaluation, it's necessary to investigate whether the use of this predictor reveals a change in the number of website visits.

11.1.6 Deployment

This phase deals with the commissioning of the developed tools. It's therefore planned, monitored, and implemented accordingly for day-to-day business. As is usual in project management, the final report is prepared in this phase.

Our online editor receives a tool that expects the article and its properties as input and outputs a quality score. This score provides the editor with a decision-support system for selecting articles. Experience can now be gained in real operation. This may lead to the inclusion of new properties for evaluation, which again triggers our CRISP-DM cycle.

11.2 Ethical and Legal Aspects

The development and application of AI is increasingly affecting people in their personal development. It follows that the CRISP-DM process must also be examined and designed according to ethical and legal aspects. Certain aspects, such as *algorithmic fairness*, affect almost all elements of the CRISP-DM model, and a large part of the clarifications should already be made at the beginning when understanding the business process. The topic of ethics can be approached from a philosophical perspective, but technical tools are of course also needed to implement AI ethics principles. A model that creates an ethical framework for AI from a philosophical perspective consists of the five principles according to Luciano Floridi et al. (see Section 11.5), which is shown in Figure 11.2.

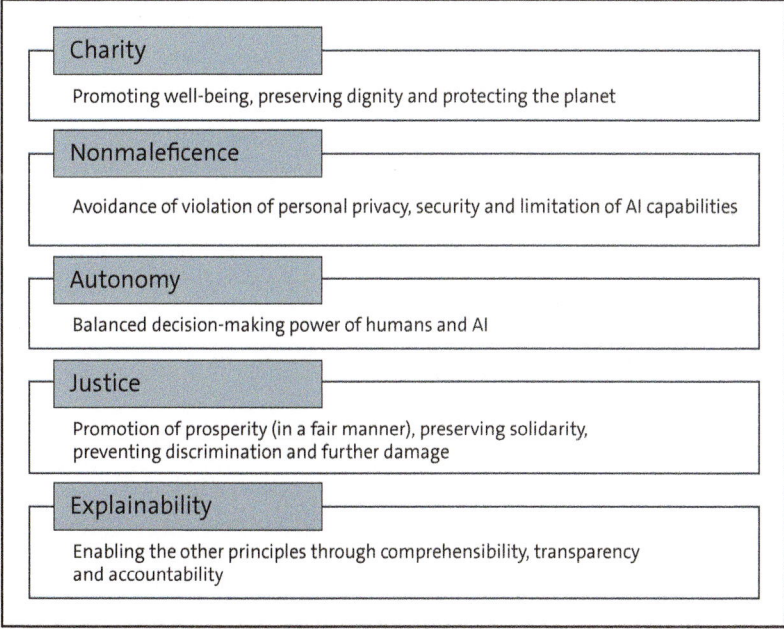

Figure 11.2 The Five Ethical Principles for AI

But why do we talk about law and ethics? Shouldn't the legal framework ensure social ethics? One of the first equations that lawyers learn during their training is the following:

Law ≠ Customs and morals

There is a great deal of overlap between law and ethics and the aim of any legal system is to guarantee socially recognized ethics. However, from a mathematical perspective, ethics can be described as a function of time (history) and place (geography), that is, *Ethics* $[t,(x,y)]$, which doesn't make setting a legal framework any easier. Books have already been written on this subject, and it's something that can be discussed at length. However, talk is cheap—so let's get back to neural networks and AI.

To enable you to use the ethical principles from Figure 11.2 as a guide in your work, we've picked out the following elements for our consideration: *algorithmic fairness*, *explainability* and *interpretability*, *ecological aspects*, and *legal aspects*.

11.2.1 Algorithmic Fairness and Bias

From time to time, we hear in the media about the failures of AI systems—be it AI chatbots that make racist statements through targeted influence, image recognition methods that classify black people as gorillas, recruitment systems that favor men, and even AI tools that favor white people over black people in the American legal system. This list of examples could go on and on.

To ensure algorithmic fairness, you must first determine the causes of these errors. These are based on various types of bias.

> **Bias**
>
> The term *bias* originates from statistics. Statistics are used to attempt to draw conclusions about the entire data or the entire population from a selection (or sample) from an overall dataset (or overall population). An inappropriate sample selection can lead to systematic errors that provide a more or less strongly biased result. The situation is similar with neural networks, which are also trained with a selection of all possible datasets. Errors in the selection or even actual injustices are then learned by the AI system and can even be exacerbated as a result.

Bias can have a variety of causes. We list just a few examples here:

- **Selection bias**
  You only include data in the sample or training dataset that you're comfortable with.

- **Historical bias**
  This isn't actually a bias in the conventional sense, but rather injustices that can also be found in the population as a whole. One example of this is when women earn less than men.

- **Algorithmic bias**
  Systemic errors can also occur if the algorithm (i.e., a neural network we've trained) is less able to cope with certain data. For example, the misclassification of dark-skinned people could be due to the fact that the contours are less recognizable by the image processing algorithm.

But how do you tackle this unfairness? There are three key points on which to focus:

- Modifying the training data
- Modifying the algorithm to be trained
- Modifying the output of the trained algorithm

Figure 11.3 shows an example of how the *income* property can be modified by statistical adjustment so that the (historical) bias can be reduced, and the distribution for men and women is more balanced.

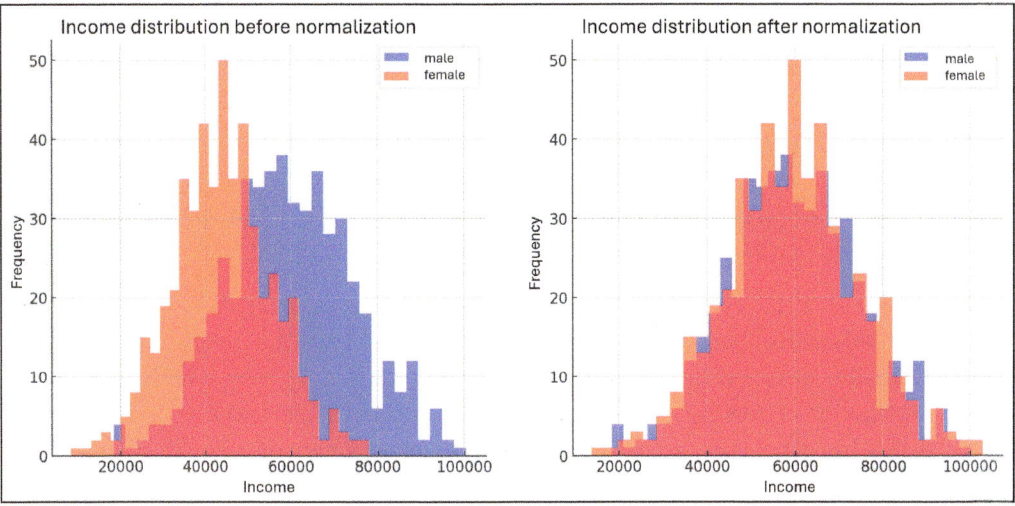

Figure 11.3 (Synthetic) Income Distribution Before and After the Modification of the Training Data

If we've given you the impression that everything has been clarified with the methods of algorithmic fairness, then we're sorry to disappoint you. The following questions still remain:

- **What, after all, is fair?**
  This is where the qualitative field of ethics meets the quantitative field of statistics. Does fairness mean "equal opportunities for all" (equality of opportunity) or "equal results for all" (equality of outcome)? A company must define what it understands by fairness as early as the *business (process) understanding* phase and act accordingly.

- **What about fairness for different groups?**
  If you correct the characteristics of the training dataset for a nonprivileged group, for example, women, it's not guaranteed that the bias will be reduced for other group subdivisions (e.g., by ethnicity, religion, or disability).

- **Does fairness for a group also mean fairness for the individual?**
  Individual fairness means that two similar applicants in an AI system should also be treated similarly. However, most systems aim for *group fairness*, which means that different demographic groups are treated fairly on average. But if the focus on group fairness is too strong, it can happen that similar individuals are treated unequally because they belong to different groups.

Algorithmic fairness therefore also requires nonquantitative elements, and it's helpful if you're aware of the problem. Another important aspect is that you understand how an algorithm arrives at a result. We'll discuss this in the next section.

11.2.2 Explainability and Interpretability

There have been various approaches in the history of AI. One of these is *rule-based AI* in the form of expert systems, domain ontologies, fuzzy rule systems, and others. The aim was to translate the knowledge of experts into formalized, programmable rules and then apply them in an AI system.

These rules can be quite simple, for example: "If the systolic blood pressure is greater than 140 mmHg, then the blood pressure is elevated." A collection of rules can then lead to an overall diagnosis. With this type of AI, you not only have a result but also an explanation of how the AI system arrives at the result thanks to the chain of rules. And that is important because when an AI system gives a patient a cancer diagnosis, the patient also wants an explanation.

The problem is that the most successful AI systems—especially those based on neural networks—are data-based systems. In other words, rules aren't actively formulated, but these systems learn from training data and formulate their own rules. Unfortunately, this often happens in a highly nontransparent manner.

The question is to what extent we can trust such data-based AI systems. Achieving high accuracy metrics during learning or training isn't enough. Not only do we want to know what the result is, but also why the AI system arrives at a result or prediction.

However, there are interpretable AI systems, for example, the *decision tree* algorithm. We'll apply this algorithm to the Iris dataset we already know and visualize the result. You can start as shown in Listing 11.1.

```python
import numpy as np
import pandas as pd
from sklearn.datasets import load_iris
from sklearn.tree import DecisionTreeClassifier, plot_tree
import matplotlib.pyplot as plt

# Load the iris dataset
iris = load_iris()
```

```
X = iris.data
y = iris.target
```

Listing 11.1 Loading the Required Libraries and the Iris Dataset

For this task, we use the scikit-learn library, which provides various algorithms for ML.

```
# Features and classes of th iris data set
feature_name = ['Sepal length', 'Sepal width', 'Petal length', 'Petal width']

class_names = ['Setosa', 'Versicolor', 'Virginica']
```

Listing 11.2 Property Names of the Iris Dataset

Training the decision tree is a two-liner, but we won't go into detail about how the algorithm works here. What is important, however, is that we obtain a decision tree, which, based on subdivisions of the properties, provides a chain of rules that leads us to a prediction (in our case, to one of the three iris species *setosa*, *versicolor*, or *virginica*). Listing 11.3 shows the code for training the decision tree.

```
# Create a decision tree classifier
clf = DecisionTreeClassifier(max_depth=3, random_state=0)
clf.fit(X, y)

# Visualization of the decision tree
plt.figure(figsize=(12, 8))
plot_tree(clf, filled=True, feature_names=feature_names,
class_names=class_names, rounded=True,
        label='none', impurity=False)
plt.title("Decision tree for the iris dataset")
plt.show()
```

Listing 11.3 Training and Visual Representation of the Decision Tree

When defining the `DecisionTreeClassifier` classifier, we set the `max_depth` hyperparameter to 3. This means that the trained decision tree has a maximum of three levels, which of course has an impact on the accuracy. We've chosen this parameter to ensure a simpler presentation that supports our arguments.

How should we understand the decision tree from Figure 11.4? The first line always contains the rule, that is, "Petal width <= 0.8 cm" in box #0. If the petal width property fulfills this rule, that is, if the petal width is less than or equal to 0.8 cm, then we follow the left branch; otherwise, we follow the right branch.

Now let's try to classify an iris date using the decision tree:

- Sepal length = 6.3 cm
- Sepal width = 2.9 cm

- Petal length = 5.6 cm
- Petal width = 1.8 cm

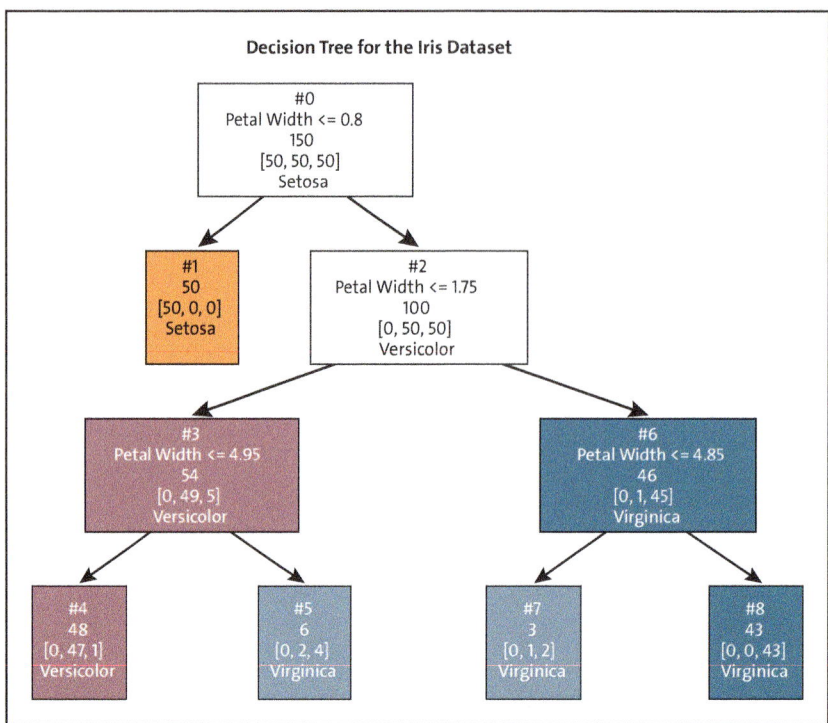

Figure 11.4 The Trained Decision Tree

The petal width is greater than 0.8 cm, so we follow the right branch to box #2. The rule "petal width <= 1.75 cm" isn't fulfilled, so we follow the right branch to box #6. Its rule "petal length <= 4.85 cm" isn't fulfilled, so we continue to follow the right branch to leaf #8, which provides us with this prediction: our dataset would be classified as *Iris virginica*.

We now have a sequence of human-readable and understandable rules that lead us to a classification. However, we also see that the sepal width and length aren't used in this decision tree. We understand this to mean that the petal provides us with more information than the sepal.

Unfortunately, interpretable data-based models are often useful for certain tasks, but aren't nearly as powerful as neural networks, which unfortunately don't provide as nice rules.

There are interpretation methods for these cases, which can be roughly categorized according to the following criteria:

- Model-specific vs. model-agnostic methods
- Local and global methods

Model-specific methods only work for a specific AI model. For neural networks, these are pixel attribution maps, which are particularly popular for image data. The idea behind this is that each pixel of an image to be classified is assigned an attribution value between 0 and 1, which indicates how much a pixel has contributed (i.e., attributed) to a classification.

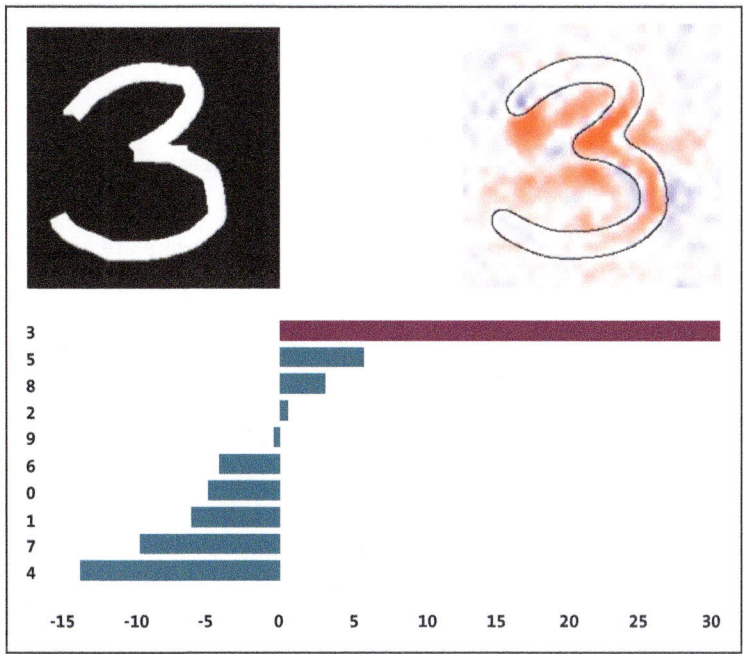

Figure 11.5 An Image of a 3 (Top Left) and the Corresponding Pixel Attribution (Top Right) for Classification (Bottom)

Figure 11.5 shows the attribution map for the classification on the right-hand side: the more the pixels are colored red, the more they contribute to the classification as digit 3. From the image, you can see that the tip in the middle of the number 3 is particularly important for classification. These pixel attribution maps belong both to the model-specific interpretation methods and to the local methods because they explain a classification for a specific dataset.

Model-agnostic methods only look at the input and output data for an AI model. More precisely, they observe how the output data behaves with small changes in the input data. If these changes are viewed across the entire dataset, we speak of a *global model-agnostic method*. If we only do this for a specific dataset, it's a *local model-agnostic method*.

There are a huge number of possible interpretation methods for AI systems, so we can only scratch the surface here. If you want to learn more about this, you should read "Interpretable Machine Learning: A Guide for Making Black Box Models Explainable" by Christian Molnar (see Section 11.5).

11.2.3 Ecological Aspects

The rapid development and increased use of neural networks (especially in the field of generative AI) have led to remarkable progress in various areas of AI in recent years. However, despite their impressive performance, these technologies raise challenging environmental issues that should be considered in their development and application.

A key ecological aspect is the high *energy consumption*, which is particularly high when large neural networks are trained. Training complex models requires enormous computing resources and can lead to a considerable carbon footprint. Studies have shown that training individual large models can result in energy consumption equivalent to the annual consumption of several households. This high energy demand contributes to global warming and poses a challenge for sustainability.

In addition to energy consumption, the *resources required* for hardware are another important factor. The production of specialized hardware components such as GPUs and *tensor processing unit* (TPUs) requires rare and often environmentally harmful materials. In addition, the rapid innovation cycle in the technology industry leads to a shortened lifespan of hardware and thus to electronic waste, which is often not recycled appropriately and causes environmental problems.

Another aspect that is often overlooked is *water consumption*, as water is required both for cooling data centers and in the production of semiconductors and other hardware components. Cooling large data centers where neural networks are trained and executed requires considerable amounts of water. In water-scarce regions, this can lead to additional pressure on local water resources. Furthermore, water in the production of electronic components is often contaminated by processes that are harmful to the environment.

Various strategies and technologies are being developed to meet these challenges. This includes the optimization of algorithms to reduce the computing effort and model compression, which makes it possible to create smaller and more energy-efficient models with comparable performance. The use of energy-efficient hardware, low-water cooling systems, and the relocation of computing processes to regions with clean energy generation and sufficient water resources are further measures to minimize the environmental impact.

In addition, the concept of green AI is gaining in importance, with ecological considerations already being integrated in the development phase of AI systems—in line with the CRIPS-DM model. This also includes the evaluation and transparency of the energy and water consumption of AI models to be able to make informed decisions when using them.

Finally, it's important to weigh up the social benefits of neural networks against their ecological costs. In areas such as medicine, environmental monitoring, and renewable energies, neural networks can help to develop solutions to global challenges and thus have an indirect positive impact on the environment.

Companies must consider and clarify all of these issues in terms of sustainability as early as the business (process) phase.

11.2.4 Legal Aspects

Neural networks and the broader application of AI raise a variety of legal issues that affect both existing laws and new, specific regulations. These legal frameworks are central to ensuring the responsible development and use of neural networks and cover areas such as privacy, copyright, and specific AI laws.

A key topic is *privacy*, which is regulated in Europe in particular by the General Data Protection Regulation (GDPR). The GDPR ensures that personal data may only be processed with the express consent of the data subject. For neural networks, which often require large amounts of data, this means that developers and operators need to carefully consider how data is collected, stored, and processed. *Transparency* is a central principle of the GDPR. This means that data subjects must be clearly and comprehensibly informed about how their data is used, the purposes of the processing, and their rights. This poses a particular challenge, as neural networks often make complex decisions that are difficult to understand, which makes transparent communication difficult.

Anonymization and *pseudonymization* are important tools for meeting the requirements of the GDPR and ensuring the functionality of the networks simultaneously. At the same time, companies must ensure that they meet the transparency requirements of the GDPR by providing clear and understandable privacy statements and giving users the opportunity to make informed decisions about the use of their data.

In the area of *copyright*, the question arises as to who owns the rights to content created by neural networks. When a neural network generates works of art, music, or texts, for example, it's often unclear whether these results should be attributed to the developer of the network, the user, or possibly the originator of the original training data. In many countries, copyright law doesn't provide clear regulations for such works created by machines, which leads to legal uncertainties.

Recently, the legal discussions have also extended to more specific *AI laws*. In Europe, the *Legal AI Act* was proposed, which aims to create a comprehensive set of regulations for the use of AI. The Legal AI Act is designed to ensure that AI systems are used transparently, securely, and to ethical standards. This includes regulations for high-risk AI systems that must meet particularly strict requirements, such as transparency, human oversight, and accountability.

Similar efforts are underway internationally, for example, by the Organization for Economic Cooperation and Development (OECD) and United Nations Educational, Scientific and Cultural Organization (UNESCO), which are developing frameworks for the ethical use of AI. These frameworks aim to set global standards for AI that take into account both the protection of human rights and the promotion of innovation and competitiveness. The USA and China, as leading nations in AI development, also have their own regulatory approaches, but these are often less stringent than European regulations.

Another important topic is *liability for damage* caused by AI systems. Traditional liability rules reach their limits when it comes to assigning responsibility for decisions or actions of autonomous systems. There's a need to create clear regulations that ensure victims of malfunctions or wrong decisions by AI systems can be adequately compensated without hindering innovation.

11.3 Feature Engineering

Let's now return to the technical aspects of the ML process. You've already made yourself familiar with the CRISP-DM model. Due to its enormous importance, we'll focus on the data preparation section. First, we have data available there in a wide variety of forms, but not always in such a way that we can use it directly for neural networks. This part of data preparation is also referred to as *feature engineering* and is an essential element of any ML project. Second, the training, which isn't dissimilar to sports training, must lead to success. This success must also be monitored and measured, and there are a number of challenges to consider. This section deals with both topics.

Most of the literature on neural networks (and this book is no exception) contains nice examples with clean, complete data. This is a good thing because it allows you to focus on the learning objective, which, in many cases, involves selecting the right model, training, and application.

We want an ML project to run as in Figure 11.6, with a nice raw dataset that can easily be converted into a dataset for neural networks.

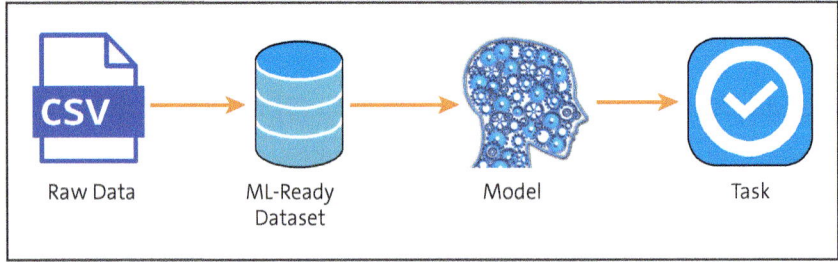

Figure 11.6 Desired Process of a Machine Learning Project

But the world is unfair—because the data is usually not well organized. In addition, if you ask a data scientist what their main task is, you often get the rhetorically exaggerated answer: "I spend 80% of my time preparing data." For this reason, Figure 11.7 reflects the real situation in a ML project much more accurately.

11.3 Feature Engineering

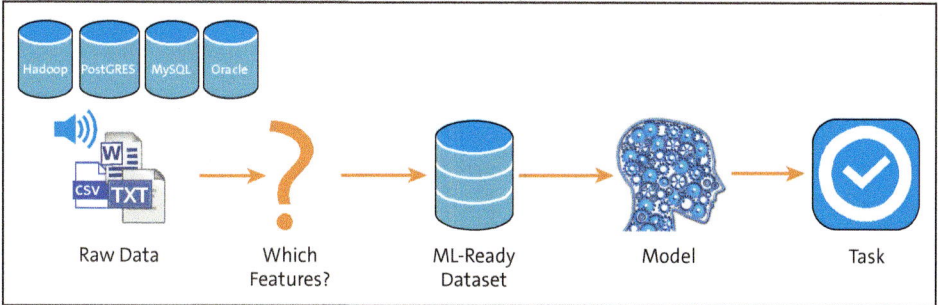

Figure 11.7 Reality of a Machine Learning Project

Andrew Ng, professor at Stanford University and chief scientist at Baidu, is one of the key figures in the development of the deep learning approach. The following quote reflects his opinion on feature engineering:

Coming up with features is difficult, time-consuming, requires expert knowledge. Applied Machine Learning is basically feature engineering.

By a dataset that has been prepared for ML, we mean a table of data in which each sample consists of different features (see Figure 11.8).

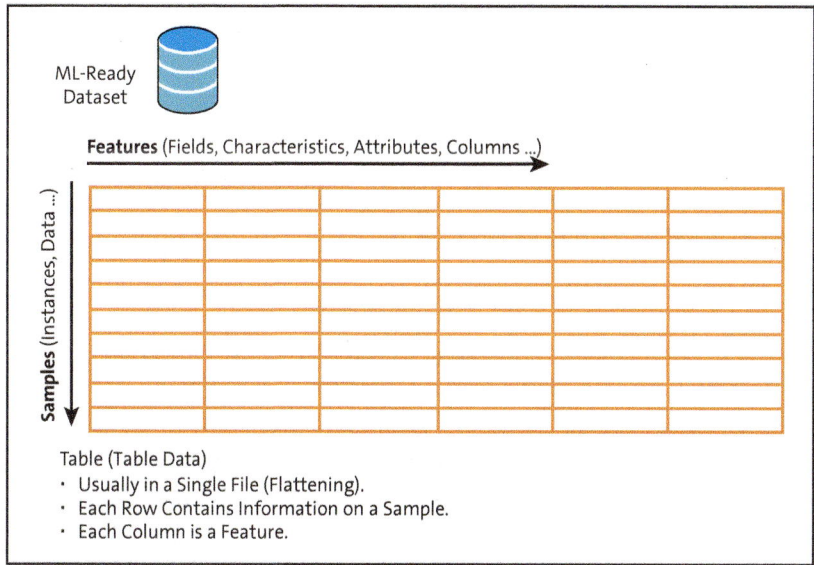

Figure 11.8 Dataset Prepared for Machine Learning

The tasks of feature engineering are very diverse. One challenge that shouldn't be underestimated is the extraction of data from databases, weblogs, signals, and images—that is, files or streams that are available in different formats. Every ML project requires numerical input values, but often the data is available in a different form and must be intelligently converted or encoded. In our example with the online newspaper articles (Section 11.1), the images for the article or the text itself could also serve as input data.

Data often contains hundreds of features and is also referred to as *high dimensionality data*. Methods for reducing this dimensionality are therefore part of every data scientist's toolkit. The uneven quantitative distribution of training datasets, that is, the underrepresented or overrepresentation of individual classes in the supervised classification, must also be taken into account.

11.3.1 Feature Coding

Neural networks need numerical values for the input vector $\vec{x}$. However, the raw data can be available in different formats. Even if this data is already available as numbers, a conversion may still be necessary. Figure 11.9 shows a classification of the features.

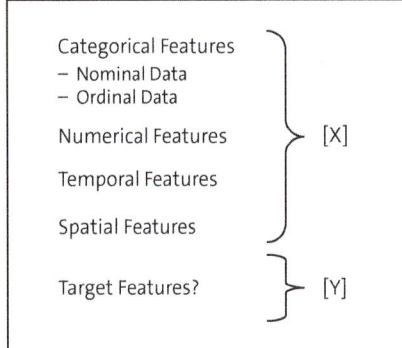

Figure 11.9 Classification of the Format of Raw Data

The conversion of raw data into ML-ready data is also known as *encoding*.

Let's take a look at the first category, the categorical features. *Categorical features* are usually in text format, for example, the hair color feature: black, blonde, brown, red. This is also an example of *nominal categorical features*, as there is no rating of the different features. Black is no better or worse than brown. Grades in school with the characteristics very good, good, satisfactory, sufficient, not sufficient, on the other hand, are an example of *ordinal categorical features* because there is a relationship between the values: *very good* is better than *good*, *good* is better than *satisfactory*, and so on.

Categorical raw data always requires some kind of pretreatment and therefore can't be transferred directly.

Categorical Features

The simplest form of encoding categorical features is the *label encoding*, which corresponds to a numbering of the individual characteristics of a feature (see Table 11.1).

Raw Data as Text	Assigned Label
very good	1
good	2
satisfactory	3
sufficient	4
not sufficient	5

Table 11.1 Example of Label Encoding (School Grades in Austria)

While this is perfectly suitable for ordinal data, an order is introduced for nominal data that can be problematic. In Listing 11.4, the hair color of students is encoded as a number, which automatically creates an order. Neural networks use the order of numbers, and this can have a negative impact on a model. However, the advantage of this form of encoding is that it doesn't increase the dimension or length of the input vectors.

For the following examples, you should open a new Jupyter Notebook and try out the listings yourself. Once again, we use some new Python libraries that need to be installed in advance. In Listing 11.4, this is sklearn, an ML library, and pandas, a very useful data analysis tool.

```python
# Import scikit-learn and pandas for label encoding
from sklearn import preprocessing
import pandas as pd

students_list = { 'Students' : [1, 2, 3, 4, 5, 6, 7, 8, 9, 10],
    'Grade': ["very good", "good", "good", "satisfactory", "sufficient", "not sufficient", "very good", "sufficient", "good", "good"],
    'Hair color': ["brown", "black", "blonde", "blue", "black", "brunette", "red", "brown", "black", "black"] }

df = pd.DataFrame(students_list, columns = ['Student', 'Grade', 'Hair Color'])

lab_enc = preprocessing.LabelEncoder()
lab_enc.fit(df['Grade'])
lab_enc.transform(df['Grade'])

# Output:
array([4, 2, 2, 0, 1, 3, 4, 1, 2, 2])
```

Listing 11.4 Sample Label Encoding of the Student's Hair Color Using sklearn

For nonordinal data, the *one-hot encoding* is recommended, in which each entry gets replaced by a binary vector. We've already used this type of encoding in Chapter 8. The resulting vector contains as many elements as there are different values. In the example in Listing 11.4, there are six different hair colors. If one-hot encoding is used, each occurrence is therefore represented by a six-digit binary vector. In this vector, a 1 is set at a certain point—depending on the hair color—while the other elements remain 0.

In Listing 11.5, we use the one-hot encoder from scikit-learn, which requires an array of integer values as input. For this reason, we convert the entries for the hair colors into integers using a `LabelEncoder` before we pass them on.

```
one_enc = preprocessing.OneHotEncoder(sparse=False)
help_enc = preprocessing.LabelEncoder()

# Conversion of the hair color column (= string data) to integer representation
help_col = help_enc.fit_transform(df['Hair Color'])
one_enc.fit_transform(help_col.reshape(len(df['Hair Color']),-1))
# Output:
array([[0., 0., 1., 0., 0., 0.],
       [0., 0., 0., 0., 0., 1.],
       [0., 1., 0., 0., 0., 0.],
       [1., 0., 0., 0., 0., 0.],
       [0., 0., 0., 0., 0., 1.],
       [0., 0., 0., 1., 0., 0.],
       [0., 0., 0., 0., 1., 0.],
       [0., 0., 1., 0., 0., 0.],
       [0., 0., 0., 0., 0., 1.],
       [0., 0., 0., 0., 0., 1.]])
```

Listing 11.5 One-Hot Encoding Using the scikit-learn Library

In this example, the hair color brown is therefore represented as vector [0, 0, 1, 0, 0, 0] (see the first line in the output of Listing 11.5).

As simple as this type of encoding is, it increases the dimensionality of the input data (the hair color is now represented by a six-dimensional vector), and this poses a problem because the search space for the optimization task gets significantly expanded.

The `category_encoders` library provides additional scikit-learn–compatible encoders for categorical variables, such as `HashingEncoder`, `PolynomialEncoder`, or `LeaveOneOutEncoder`. This library isn't part of the standard equipment and must be installed separately. The Jupyter Notebooks included in this chapter contain a notebook that accomplishes this task. Of course, you can also install this package via the Anaconda Navigator as before (see Chapter 2, Section 2.1.1).

In the example in Listing 11.6, we load the *Boston Housing Prices* dataset, which is provided by the scikit-learn library, and describe how you can use this library.

```
import category_encoders as ce
import pandas as pd


# Prepare the data
# The Boston_housing dataset is read from a csv file
# and prepared for scikit-learn
data = pd.read_csv('BostonHousing.csv')

# The medv column contains the target variable y - the median price of the
house
X = data.drop('medv', axis=1)
y = data['medv']
X.head(3)
```

Listing 11.6 Preparation of Data for category_encoders

```
# Output:
```

	CRIM	ZN	INDUS	CHAS	NOX	RM	AGE	DIS	RAD	TAX	PTRATIO	B	LSTAT
0	0.00632	18.0	2.31	0.0	0.538	6.575	65.2	4.0900	1.0	296.0	15.3	396.90	4.98
1	0.02731	0.0	7.07	0.0	0.469	6.421	78.9	4.9671	2.0	242.0	17.8	396.90	9.14
2	0.02729	0.0	7.07	0.0	0.469	7.185	61.1	4.9671	2.0	242.0	17.8	392.83	4.03

Figure 11.10 Output for Listing 11.6

> **The Boston Housing Prices dataset**
>
> The Boston Housing Prices dataset has long been used uncritically as a sample dataset in various statistics and ML books and courses. It was also included in the scikit-learn library as a teaching dataset up to version 1.2.
>
> However, this dataset has an ethical problem, which we address in Section 11.2.1: The creators of this dataset have created a *B* feature that falsely assumes that segregating people by ethnicity would have a positive impact on house prices. Originally, the dataset was supposed to examine the effects of air quality on prices.
>
> The developers of scikit-learn recommend that this dataset should no longer be used unless it's to learn about or discuss ethical issues in ML. It's precisely for this reason that we've left this dataset in this edition to initiate awareness of this topic.

In Listing 11.7, we use the `BinaryEncoder()` function for one-hot encoding. The function is called this way because it saves the result of one-hot encoding, a vector with 0 and 1, as a binary number. This saves a lot of storage space.

```python
# Use of binary encoding for two categorical features
encoded = ce.BinaryEncoder(cols=['CHAS', 'RAD']).fit(X, y)

# Transformation of the dataset
X_trans = encoded.transform(X)

# Display the result
X_trans.head(3)
```

Listing 11.7 Use of Additional Encoders

```
# Output:
```

	CRIM	ZN	INDUS	CHAS_0	CHAS_1	NOX	RM	AGE	DIS	RAD_0	RAD_1	RAD_2	RAD_3	RAD_4	TAX	PTRATIO	B	LSTAT
0	0.00632	18.0	2.31	0	1	0.538	6.575	65.2	4.0900	0	0	0	0	1	296.0	15.3	396.90	4.98
1	0.02731	0.0	7.07	0	1	0.469	6.421	78.9	4.9671	0	0	0	1	0	242.0	17.8	396.90	9.14
2	0.02729	0.0	7.07	0	1	0.469	7.185	61.1	4.9671	0	0	0	1	0	242.0	17.8	392.83	4.03

Figure 11.11 Output for Listing 11.7

Further information on the different variants of the category_encoders library can be found on the following documentation pages: *http://contrib.scikit-learn.org/category_encoders*.

Numerical Features

Even with numerical features, it can be an advantage if they are still being edited. Numerical values can come from sources as diverse as counts of user access, measurements from sensors, or from the geolocation of people or objects.

The size and distribution of the numerical features of our training datasets is crucial when you deal with these values. However, feature engineering doesn't only concern our input variables but can also be advantageous to edit the target variable, $\hat{y}$.

We'll introduce you to two methods: binning and feature scaling. *Binning* means that numerical values are converted into categories. Now you're probably wondering why this is done. Until now, we've always talked about converting categorical features into numerical values.

However, there are examples where the exact value is too precise. Let's assume that we're dealing with an economic issue involving people's purchasing behavior. Then, the *age* feature is an essential element, but too precise. Here, we would rather divide into age groups that roughly correlate with income.

Another example is user access or ratings of music, books, movies, and so on. Robots or unusual behavior on the part of users may result in numbers that differ by several orders of magnitude. The following examples in Listing 11.8 show how binning can be used in this case.

```python
import numpy as np
import matplotlib.pyplot as plt

# Generate 150 random age values in the range between 0 and 99
counter = np.random.randint(0,100,150)

# Assign the numerical values to the bins, in intervals of 10 (i.e. 0-9, 10-19,
etc.)
np.floor_divide(counter,10)

# The output as a graphic
plt.hist(counter, bins=10)
plt.ylabel('Frequency')
plt.xlabel('Age ranges')
plt.show()

# Show the first 10 values
print(counter[:10])

# Output:
[65 85  0 41 44 85 51 18 50 43]
```

Listing 11.8 Binning of Ages with Fixed Range Length

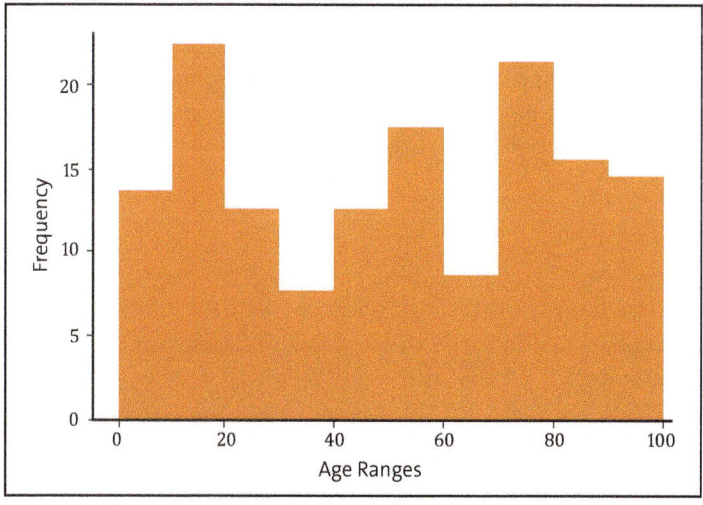

Figure 11.12 Output for Listing 11.8

Large datasets may also contain gaps in the number ranges, so that there are many bins without any content. However, the aim should be to obtain bins with an even number of elements. In this case, it's better to use the *quantile binning*. While you use a fixed division of the numerical range of the input values when binning with a fixed range length of the bin limits, quantile binning is based on the frequency of occurrence. The way it works with a 10-part binning (*percentile binning*) is that the first bin summarizes the first 10% of all occurring data, the second the next 10%, and so on. According to Listing 11.9, the limits of the first percentile are between 0.99 and 998.2, those of the second percentile between 998.2 and 2315.0, and so on.

```
import pandas as pd
# Generate arbitrary numbers
rand_elems = np.random.randint(0,10000,100)

# The pandas function qcut helps us to determine the percentile limits
perc_limits = pd.qcut(rand_elems, 10)
perc_limits
# Output:
[(5835.0, 6820.3], (998.2, 2315.0], (6820.3, 7949.4], (998.2, 2315.0], (5835.0,
6820.3], ..., (5835.0, 6820.3], (7949.4, 8530.6], (998.2, 2315.0], (8530.6,
9358.0], (5099.0, 5835.0]]
Length: 100
Categories (10, interval[float64]): [(0.999, 998.2] < (998.2, 2315.0] < (2315.0,
3483.6] < (3483.6, 4222.2] ... (5835.0, 6820.3] < (6820.3, 7949.4] < (7949.4,
8530.6] < (8530.6, 9358.0]]
```

Listing 11.9 Quantile Binning

But what do you do if the numerical values differ by several orders of magnitude? The *logarithmic binning* with a fixed range length is used here (see Listing 11.10), whereby the number groups are determined by powers of 10 (or by any power of an arbitrary integer): 0–9, 10–99, 100–999, 1000–9999, and so on.

```
import numpy as np
import matplotlib.pyplot as plt

# Generate 150 numbers with large numerical differences that are supposed to
# simulate web accesses
webaccesses = np.random.randint(0,1000000,150)

# Output of the logarithmic binning, in which the limits of the bins
# are logarithms of 10
plt.hist(webaccesses,bins=np.logspace(np.log10(0.1),np.log10(1000000.0),50))
plt.gca().set_xscale("log")
plt.ylabel('Frequency')
plt.xlabel('Web accesses')
plt.show()
```

```
print(webaccesses[:10])
```

```
# Output:
[3671012 2205678 9276288 3960628 8983584 4708590 7067269 9354739 6834788
 6917483]
```

Listing 11.10 Logarithmic Binning

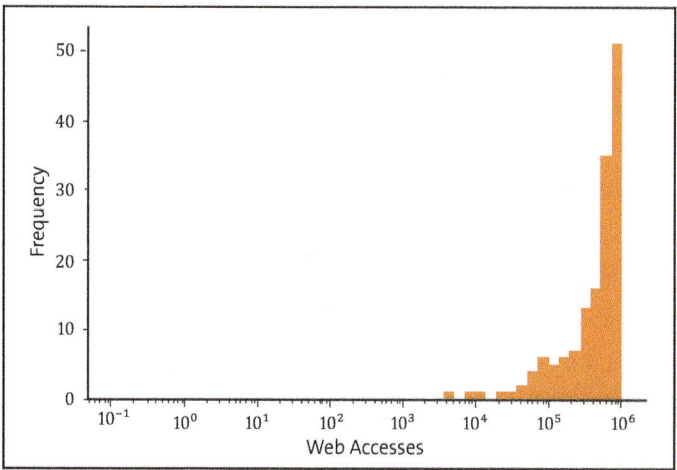

Figure 11.13 Image Output for Listing 11.10

The shape of the histogram in Figure 11.13 is easy to explain: our randomly selected web access numbers are generated using the `np.random.randint` function. In our example, this generates random numbers that are equally distributed over the number range 0, 1000000. However, because with logarithmic binning the limits of the bins become larger and larger (0–9, 10–99, 100–999, etc.), even exponentially larger, the bins include more and more values. The greater the web access figures, the more values go into the bins, which creates larger bins on the right-hand side.

Let's now look at the *scaling*. There are certain features that are restricted in the range of values, such as the body temperature, body weight, longitude, and latitude. Models such as decision trees aren't sensitive to the value range of the features, but neural networks react very strongly to them. *Feature scaling* (also known as *feature normalization*) allows for the transformation of the original data to a specific value range, typically a value range between 0 and 1, as shown in Figure 11.14.

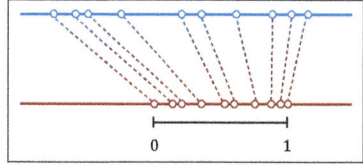

Figure 11.14 Sketch of Scaling in the [0,1] Value Range

The simplest way of scaling a value x_i is *min-max scaling*, whereby the minimum value of all data is mapped to 0 and the maximum to 1 using the following formula:

$$x\_minmax\_sc_i = \frac{x_i - \min(\vec{x})}{\max(\vec{x}) - \min(\vec{x})}$$

This formula is used for all elements x_i, where $\vec{x} = (x_1, x_2, \ldots, x_n)$.

The disadvantage, however, is that in the case of extreme minimum or maximum values, which in statistics are also referred to as *outliers*, the data is compressed to a very small range of values, thereby reducing differences. You could prevent this by not using the minimum or maximum value, but the 95% percentile limits. All values that are smaller than the lower percentile value are mapped to 0; all values that are larger than the upper percentile value are mapped to 1.

Another scaling option is *variance scaling*, which is performed using the following formula:

$$x\_var\_sc_i = \frac{x_i - \text{mean}(\vec{x})}{\sqrt{\text{var}(\vec{x})}}$$

A feature that is transformed in this way then has a mean value of 0 and a variance of 1.

Of course, there are already functions that perform these scalings. We first introduce a random vector rand_elems and then reshape it accordingly, as shown in Listing 11.11.

```
from sklearn.preprocessing import scale, minmax_scale
import matplotlib.pyplot as plt
import numpy as np

rand_elems = np.random.randint(0,10000,100).astype(np.float64)

x_minmax_sc = minmax_scale(rand_elems)
x_var_sc = scale(rand_elems,with_mean=True, with_std=True)
```

Listing 11.11 Min-Max and Variance Scaling Using sklearn

Temporal Features

Features that contain time information can be very tedious. On the one hand, you need to take time zones into account, for example, for user access data on websites. However, you must bear in mind that it's often not the absolute time that is important, but the position within a day.

Time binning is used for this, where the time information is divided into daily periods, as shown in Table 11.2.

11.3 Feature Engineering

Time of Day (Hour)	Category ID	Category Description
[5, 8]	1	early in the morning
[8, 11]	2	in the morning
[11, 14]	3	at noon
[14, 19]	4	in the afternoon
[19, 22]	5	in the evening
[22, 24) and [0, 5)	6	at night

Table 11.2 Time Binning of the Time of Day

Of course, you can also use larger binning periods, such as calendar week, month, quarter, or year.

It can also be useful not to take the time value itself, but the distance to certain events. When determining spending behavior, the time between the start of the vacations, Christmas, and major sporting events or the time of salary payment is more important than the time feature.

Spatial Features

Features that provide information about the location can be very useful. Just think of the use of credit cards in two very distant locations within an hour of each other, which is an important indicator of fraud detection due to the impossibly high speed of travel. Identifying attractive places for a tourist based on spatial distance can also provide valuable information.

In some datasets, the location description is available in the form of addresses, which of course can't be used directly in any ML project. Conversion to latitude and longitude is easy with the Python library, geopy. This library must be installed separately again, as shown in Listing 11.12.

```python
# Import the correct library
from geopy.geocoders import Nominatim

# Use of the Nominatim geocoder for OpenStreetMap data
# Nominatim needs a user_agent,
# which can have any name
geolocator = Nominatim(user_agent="My application")

# Enter the written address
location = geolocator.geocode("Rheinwerkallee 4, 52227 Bonn")
print("Address:", location.address)
```

```
print("LatLon:", location.latitude, location.longitude)
# Output:
Address Rheinwerkallee, Ramersdorf, Bonn-Beuel, Bonn, Regierungsbezirk Köln,
North Rhine-Westphalia, 53227, Germany
Longitude and latitude: 50.7172803 7.1538127
```

Listing 11.12 Converting an Address to Latitude and Longitude

Target Features

Up to now, we've always spoken of features, meaning all values of the input vector $\vec{x}$. However, the methods discussed can also affect the target vectors $\vec{y}$ and be applied in the same way.

11.3.2 Feature Extraction

It's difficult to predict where the raw data comes from, but essentially it comes from the following sources:

- Databases
- Text files
- Image data
- Audio data

Feature extraction refers to methods that transform raw data into an input vector $\vec{x}$. You could write a separate book about this topic alone, so we'll focus on the essential elements.

Databases

Naturally, a lot of data is collected in databases. For relational databases, there is a standardized language for defining, editing, and querying data. This database language, called *Structured Query Language* (SQL), is based on a relational algebra, and its structure is relatively simple and based on the English colloquial language.

> **SQL**
>
> There is an international standard that is defined by a joint committee of International Organization for Standardization (ISO) and International Electrotechnical Commission (IEC) with the participation of national standardization bodies. Although most databases support SQL, slightly modified dialects have developed that often don't cover the entire language range.

A relational database essentially consists of tables that are linked to each other; that is, they are in relation to each other. These tables and relations can be visualized in an entity relationship (ER) diagram, which is common in the database world (see Figure 11.15).

11.3 Feature Engineering

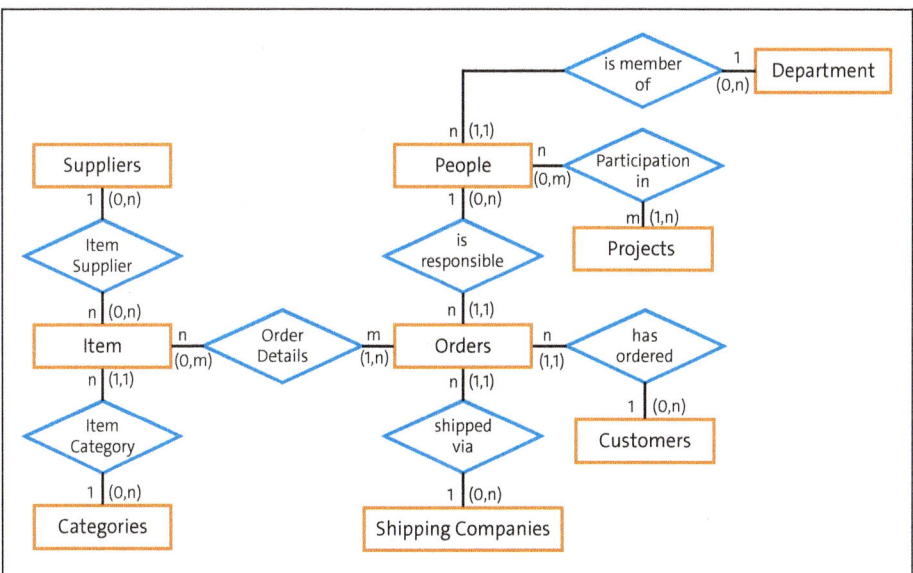

Figure 11.15 A "Simple" ER Diagram

> **Entity Relationship Diagram**
>
> The *ER diagram* is a way of representing or modeling logical database relationships. *Entities* (objects) and *relations* (relationships) are realized as tables in databases. In Figure 11.15, the entities are shown as rectangles, and the relations are shown as diamonds.

The representation of the ER diagram in Figure 11.15 shows the data model of a small database for a retail company's ordering system. The orange boxes represent tables. The blue diamonds and the connections represent relations between the tables.

This data model is of course highly simplified, as production systems usually have much more complex data models. But even this simple example shows that the creation of a database structure quickly becomes complicated due to the two-dimensionality of tables.

However, the training data for an ML task must be available as an ML-ready dataset, that is, in the form of a single table. This process of transforming an SQL database into a table is referred to as *data flattening*. Figure 11.16 shows an example.

With the advent of *big data*, more and more *Not Only SQL* (*NoSQL*) databases have become established. The name means that there are also other access languages that are oriented toward the special class of NoSQL database.

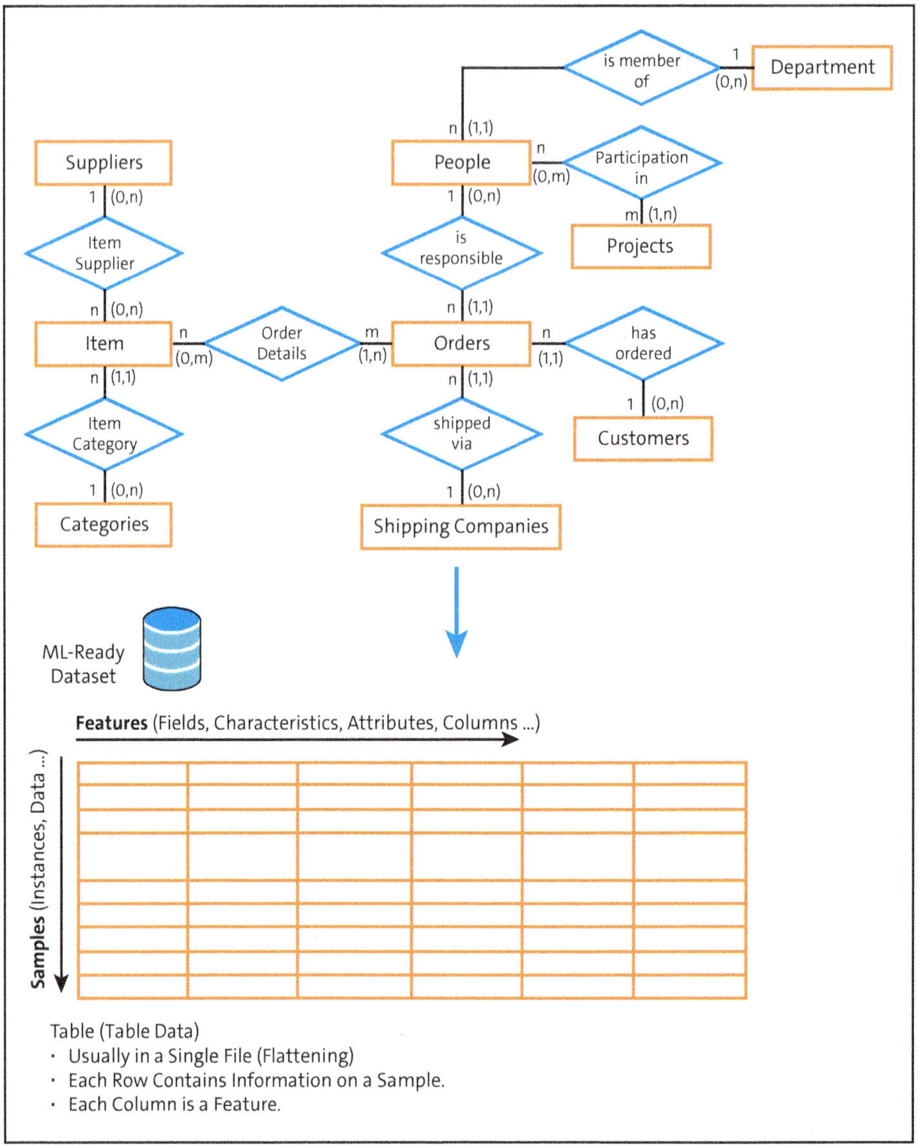

11 The Machine Learning Process

Figure 11.16 Data Flattening of a Database

Big Data and NoSQL

Big data is usually used as a collective term for technologies for high data volumes with rapid availability and high data type variability. Conventional database systems are unable to cope with the volume of data or the rapid growth in data. One feature of NoSQL databases is their simple scalability.

In Table 11.3, we compare a few typical SQL queries with a NoSQL query language.

SQL	NoSQL
SELECT firstname, age FROM user	db.user.find({}, {firstname:1, age:2})
SELECT * FROM user WHERE age=25	db.user.find({age:25})
SELECT * FROM user WHERE age=30 ORDER BY firstname	db.user.find({age:30}).sort({firstname:1})

Table 11.3 Examples of SQL vs. NoSQL Queries

The flexible data model of NoSQL databases is often cited as an advantage over relational databases. The aim is therefore to flatten the data available in complex structures. To do this, we very often have to create complex SQL queries that provide us with the data of our input vector, that is, our features.

Text Data

To analyze texts using neural networks, a text must somehow be transformed into a vector. Such texts are preprocessed is advance, so to speak. This preprocessing includes the following elements:

- Cleaning
- Removal
- Stemming, chopping
- Tokenization

There are many ways to clean up a text. Very often, the entire text is written in lowercase, numbers and punctuation marks are removed, or even words are corrected using grammar checkers. It's difficult to give some general advice here because this part depends very much on the specific task to be performed.

We'll look at these different steps using an example. For this purpose, we use a Python library called nltk, as shown in Listing 11.13. We no longer need to show you how to install this library, but you'll need to download some nltk data after the installation. These are text bodies, grammars, and so on.

```
import nltk

# nltk with GUI selection
nltk.download()
```

```
# nltk with direct download of popular text bodies
nltk.download("popular")
```

Listing 11.13 Command to Download the "nltk" Data

The `nltk.download()` function without parameters opens an interactive window (see Figure 11.17). There, you can see all the packages available in this library, select them manually, and then click on the **Download** button. The simpler way is of course to execute the `nltk.download('popular')` command to download some basic data and models.

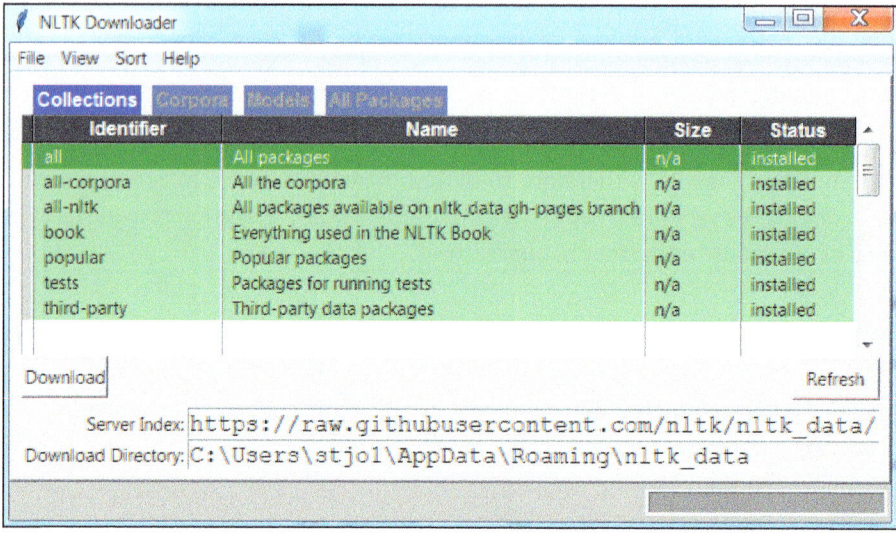

Figure 11.17 The Interactive "nltk" Download Window

In the first step, we now divide the input text into individual parts, as shown in Listing 11.14.

```
import nltk
from nltk.tokenize import word_tokenize
# Ensure that NLTK's tokenizer data is downloaded
nltk.download('punkt_tab')
myText = "Albert has had breakfast today, but he's really hungry again. It might be lunchtime again!"
tokens = word_tokenize(myText)
print(tokens)
# Output:
['Albert', 'has', 'had', 'breakfast', 'today', ',', 'but', 'he', "'s", 'really', 'hungry', 'again', '.', 'It', 'might', 'be', 'lunchtime', 'again', '!']
```

Listing 11.14 Tokenization of a Text

In addition to the individual words, this process also provides us with the punctuation, which doesn't contain any information. At the same time, we want to normalize everything to lowercase, as shown in Listing 11.15.

```
tokens = [w.lower() for w in tokens]
words = [w for w in tokens if w.isalpha()]
print(words)
# Output:
['albert', 'has', 'had', 'breakfast', 'today', 'but', 'he's', 'really',
'hungry'', 'again', 'it', 'might', 'be', 'lunchtime', 'again']
```

Listing 11.15 Now Lowercase Everything, and Remove Punctuation Marks

In our language, we use many words that don't contribute to the meaning of a phrase. Examples of this are definite and indefinite articles or verb forms of "be". These words are also known as *stop words*. The `nltk` library contains a list of stop words in different languages. These stop words must be actively loaded, as shown in Listing 11.16.

```
from random import shuffle
from nltk.corpus import stopwords
# Get the original list of stopwords
original_stopwords = stopwords.words('english')
# Assign to stopwords variable
stopwords = original_stopwords
# Shuffle the list and assign to a different variable name, i.e. shuffled_
stopwords
shuffled_stopwords = original_stopwords[:]  # Create a copy to avoid modifying
the original
shuffle(shuffled_stopwords)
# Now you can print both lists
print(stopwords[:10])
print(shuffled_stopwords[:10])
# Output:
['a', 'about', 'above', 'after', 'again', 'against', 'ain', 'all', 'am', 'an']
['mightn', 'be', 'they', 'myself', 'with', 'off', 'that', 'are', 'doing',
'before']
```

Listing 11.16 Random Selection from the List of English Stop Words

The list of stop words is of course sorted alphabetically. You can use the `shuffle()` function to shuffle the stop words; the output then shows a randomly selected list of elements.

Often, a word form doesn't necessarily provide relevant additional information. To take advantage of this, you can reduce terms to the word stem. To do this, you use *stemming* algorithms. Here too, the `nltk` library helps you with many stemming algorithms. Let's take a look at one of them, as shown in Listing 11.17.

```python
from nltk.stem.snowball import SnowballStemmer
snow = SnowballStemmer("english")
stemmed = [snow.stem(w) for w in tokens]
print(stemmed)
# Output:
['albert', 'has', 'had', 'breakfast', 'today', ',', 'but', 'he', "'s", 'realli',
'hungri', 'again', '.', 'it', 'might', 'be', 'lunchtim', 'again', '!']
```

Listing 11.17 Reduction to the Word Stem Using the SnowballStemmer

In most cases, the stemming algorithms already take care of lowercase and punctuation removal and contains a list of stop words. Note, however, that such stemming algorithms are algorithms and often produce unexpected results. Stemmers such as the *Hunspell stemmer*, which is based on a vocabulary, provide more precise results. Of course, there is also a Python library called hunspell, but that would go beyond the limits of this book.

We're still missing something essential: we need to convert this cleaned word list into a vector, as shown in Listing 11.18. There are methods based on the frequency of occurrence of words, such as *bag-of-words*. Here, each document is represented by a vector, with each coordinate of the vector representing a word. The value of the coordinate contains the occurrence of the word in the document text. The Python library scikit-learn (sklearn) is very helpful here.

```python
from sklearn.feature_extraction.text import CountVectorizer
myText = ["Albert has had breakfast today, but he's really hungry again. It
might be lunchtime again!"]

count_vector = CountVectorizer()
bow = count_vector.fit_transform(myText)
print(count_vector.vocabulary_)
# Output:
{'albert': 0, 'has': 5, 'had': 6, 'breakfast': 13, 'today': 4, 'but': 1, 'he's':
3, 'really': 11, 'hungry': 12, 'again': 2, 'it': 10, 'might': 9, 'be': 7,
'lunchtime': 8}
```

Listing 11.18 Bag-of-Words for Texts

You can see from the output of Listing 11.18 that tokenization has already been carried out.

With these functions, you can not only count the occurrence of words but also the occurrence of word combinations with two, three, or more words. These word combinations are also called *n-grams*, where the *n* stands for the length of the word combination. Of course, these parameters have a problem because frequently occurring words are overrated, and rare terms are underrated, although they can make a decisive contribution to the evaluation of the text.

For this reason, there is the *term frequency–inverse document frequency* (*TF-IDF*) indicator. *TF* is the number of the word *w* in a document divided by the total number of words in this document, and *IDF* is the log (number of all documents/number of all documents with word *w*). If TF and IDF are multiplied with each other, unusual terms get increased in value, as they can be decisive for the meaning of a text, and frequent terms are reduced.

But the best way to show this is to take a simple example and use the following sentences:

- Sentence 1: You drive the car on a highway.
- Sentence 2: You ride the bike on a road.

We calculate TF, IDF, and TF-IDF for these two sentences (= two documents) (see Figure 11.18).

Word	TF		IDF	TF*IDF	
	Sentence 1	Sentence 2		Sentence 1	Sentence 2
You	1/7	1/7	log(2/2)=0	0	0
the	1/7	1/7	log(2/2)=0	0	0
car	1/7	0	log(2/1)=0.3	0.043	0
bike	0	1/7	log(2/1)=0.3	0	0.043
on	1/7	1/7	log(2/2)=0	0	0
drive	1/7	0	log(2/1)=0.3	0.043	0
ride	0	1/7	log(2/1)=0.3	0	0.043
a	1/7	1/7	log(2/2)=0	0	0
road	0	1/7	log(2/1)=0.3	0	0.043
highway	1/7	0	log(2/1)=0.3	0.043	0

Figure 11.18 TF-IDF Example

All words with a TF-IDF of 0 aren't significant or are hardly significant for the respective document. Words such as "car," "bicycle," "road," and "highway" have a value other than 0 and contribute to the meaning of the document. The articles belong to the list of stop words and are therefore removed in advance. In a larger dataset of documents, the frequency of articles within a document will be very high, which increases their importance. However, because articles also occur very frequently across all documents, the articles lose importance and are given a very low value overall. This means that stop words wouldn't have to be removed in this case.

Image Data

Convolutional neural networks (CNNs) are already doing some of the work for us when it comes to image data. Automated feature engineering is carried out in the coding block of a CNN. The only preprocessing that is sometimes necessary is the scaling of images to a certain image dimension.

But what is of great importance for CNN is the amount of training data, and here we can help ourselves very easily. In general, this is referred to as *data augmentation*, when new and slightly modified data is generated based on existing data.

In the field of image processing, new images can be created using simple geometric transformations:

- Translation
- Rotation
- Scaling (enlarging, reducing)
- Shearing (scaling, but different in x and y directions)
- Horizontal and/or vertical mirroring

Applying these transformations changes the appearance, but usually doesn't change the class of the image, making it relatively easy to enlarge the existing dataset.

Because CNNs are ideal for image analysis, it's not surprising that the Python library tensorflow.keras provides us with functions that facilitate the creation of additional data. The code in Listing 11.19, as shown here, doesn't provide us with a result and only serves to demonstrate the capabilities of the ImageDataGenerator(). But in Chapter 13, we'll use it extensively.

```
from tensorflow.keras.preprocessing.image import ImageDataGenerator
datagenerator = ImageDataGenerator(
    rotation_range = 45,
    zoom_range=0.3,
    horizontal_flip=True,
    width_shift_range=0.3,
    height_shift_range=0.2,
    fill_mode='nearest')
```

Listing 11.19 Data Augmentation for Image Data

Here, we show just a few options for image data augmentation and take a closer look at what they mean:

- rotation_range specifies the range in degrees by which an image is randomly rotated.
- zoom_range is the range that is zoomed into, in this case, up to a factor of 0.3.
- width_shift_range and height_shift_range specify the range in which the images are randomly shifted in the x or y direction.

- `horizontal_flip` allows the image to be flipped randomly in a horizontal direction.
- `fill_mode` defines the method by which new pixels are created when rotating or shifting.

Figure 11.19 Sample Data Augmentation of Images (Source: Wikimedia Commons)

11.3.3 The Curse of Dimensionality

The *curse of dimensionality* describes the problem that occurs when the number of datasets and the dimension of the feature space don't match. This means that there are too few datasets. The number of datasets required for ML tasks increases exponentially with the number of features per dataset, which means the following:

- Tasks that are easy in low-dimensional space turn out to be much more difficult in high-dimensional space.
- Intuition (for two- or three-dimensional tasks) doesn't help in high-dimensional space; on the contrary, it can even be a hindrance.
- In high-dimensional spaces, the distances between datasets become greater.

For this reason, simple methods for converting categorical features, such as one-hot encoding, should be treated with caution, as they greatly increase dimensionality in one fell swoop. Figure 11.20 shows how to deal with the curse of dimensionality.

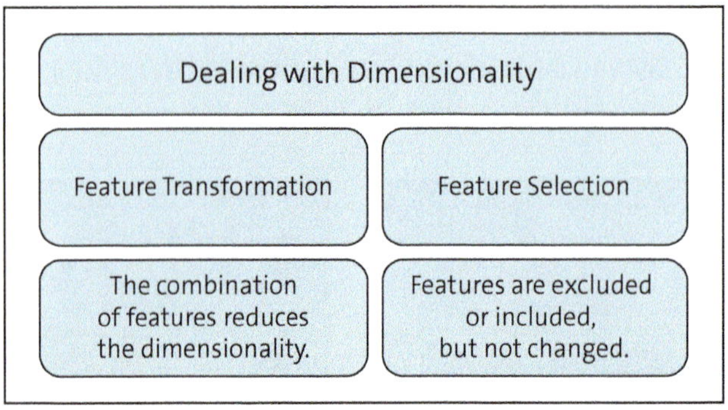

11 The Machine Learning Process

Figure 11.20 Dealing with Dimensionality

11.3.4 Feature Transformation

The idea of feature transformation is that the dimension is reduced by a combination of features.

Knowledge-Based

The knowledge-based feature transformation depends very much on the specific domain in which neural networks are used. To implement the knowledge-based feature transformation, you must therefore be well versed in this domain.

Simple examples of this include the body mass index (BMI), which is calculated using the following formula:

$$BMI = \frac{kg(\text{body weight})}{m(\text{body height})^2}$$

Here, the *body weight* and *height* features are replaced by a new combined feature: *BMI*.

An example from the field of satellite image remote sensing is the *Normalized Difference Vegetation Index* (NDVI). In the electromagnetic spectrum, satellite images provide the usual optical red, green, and blue reflections, but also the infrared, which is no longer visible to the eye. In ML tasks for vegetation questions, it's helpful to know that the difference between the red (R) and near infrared (NIR) bands is quite large. To reduce other influences when taking the satellite image, the NDVI is used to highlight the vegetation areas in the image, and the two features (R, NIR) are replaced by a new one (NDVI):

$$NDVI = \frac{(NIR - R)}{(NIR + R)}$$

Of course, you could say that this contradicts the idea of ML, as it involves algorithms that are supposed to learn from the data. However, there is nothing to be said against incorporating established, existing knowledge into this process.

Automatic

Another well-known method for reducing the dimensionality of a dataset is called *principal component analysis* (PCA). Our datasets consist of vectors with numbers that correspond to a specific coordinate system. The individual coordinates of the vector correspond to the characteristics of our features expressed in numbers. These features often correlate with each other; to put it simply, they are interdependent. You could now simply remove a feature *a* if it correlates with a feature *b* to reduce the dimensionality. However, even if features do correlate, this doesn't mean that these features don't contribute any information to the solution of our ML problem.

PCA is a method that creates a new coordinate system with the aim of obtaining as few correlated features as possible. These new uncorrelated features are referred to as *principal components*. These principal components can also be arranged according to their degree of uncorrelatedness, to put it somewhat casually. If you now only take the top part of the principal components, you've killed two birds with one stone, namely dimension reduction and new features with low correlation. Figure 11.21 shows how it works.

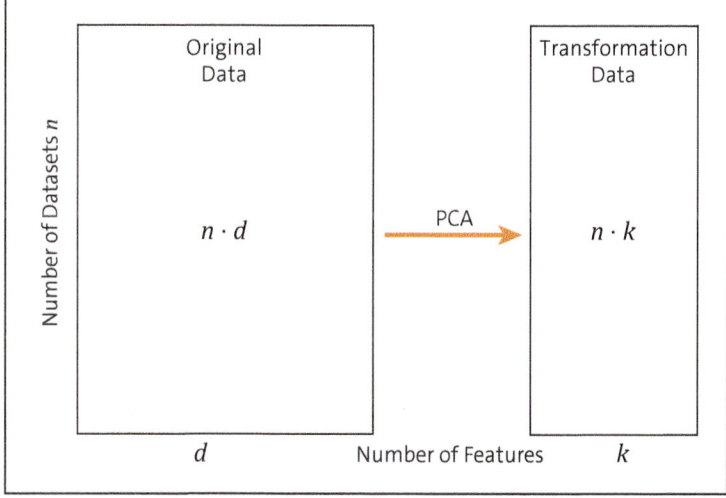

Figure 11.21 Functionality of the Principal Components Analysis

PCA uses an eigenvalue decomposition of the covariance matrix of the dataset. The covariance matrix *A* is a (*dxd*) matrix, where *d* corresponds to the number of features in our dataset, which represents the pairwise interaction of the features.

The exact derivation requires some linear algebra and wouldn't lead to much more information. The Python library scikit-learn provides a function that performs a PCA. Let's take a look at the consequences of PCA using the already familiar Iris dataset as an example (see Listing 11.20).

```python
from sklearn.datasets import load_iris
iris_data = load_iris()
iris_X, iris_y = iris_data.data, iris_data.target
```

Listing 11.20 Loading the Iris Dataset

Here, we load the Iris dataset provided by scikit-learn, as shown in Listing 11.21.

```python
from sklearn.decomposition import PCA
iris_pca = PCA(n_components=2)
```

Listing 11.21 PCA from scikit-learn

In Listing 11.21, we import the corresponding PCA function and specify how many new features we want. The n_components parameter therefore corresponds to our *k* from Figure 11.21. As shown in Listing 11.22, we've used the iris_pca.fit() function to determine the principal components (i.e., our eigenvectors) for transforming our data. The iris_pca.transform() function transforms our dataset into the new features.

```python
iris_pca.fit(iris_X)
iris_X_pca = iris_pca.transform(iris_X)

iris_ pca.components_
# Output:
array([[ 0.36158968, -0.08226889, 0.85657211, 0.35884393],
       [ 0.65653988, 0.72971237, -0.1757674 , -0.07470647]])
```

Listing 11.22 Determining the Principal Components

Let's now take a look at the effects of this transformation, as shown in Listing 11.23.

```python
import matplotlib.pyplot as plt
%matplotlib inline

for lab, mark, col in zip(range(3), ('^','s','o'), ('green','red','blue')):
    plt.scatter(x=iris_X[:,0].real[iris_y==lab],
                y=iris_X[:,1].real(iris_y==lab],
                color = col)
plt.title("Original iris data")
plt.xlabel("Sepal length (cm)")
plt.ylabel("Sepal width (cm)")
```

Listing 11.23 Code for a Scatterplot of the Iris Dataset

Figure 11.22 provides us with a scatterplot in which the sepal length is plotted against the sepal width for the individual iris classes, with the points for the *setosa* (green), *versicolor* (red), and *virginica* (blue) species colored according to their class. A typical characteristic of the correlation of the features is the relative proximity of the different cluster classes. The less correlated they are, the more clearly the class clusters are separated from each other.

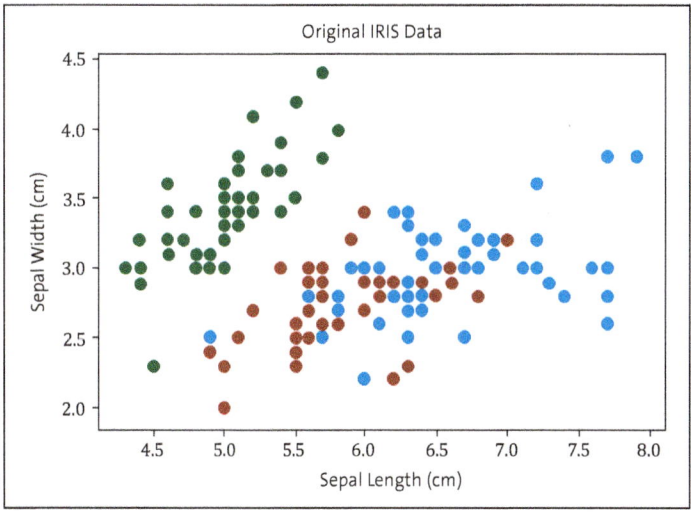

Figure 11.22 Scatterplot of the Original Iris Data

Task

Create the plot for the PCA-transformed iris data on the basis of Listing 11.23. Note that you must now use the transformed data (see Listing 11.22, `iris_X_pca`) instead of `iris_X`.

The solution to this task should provide the image shown in Figure 11.23.

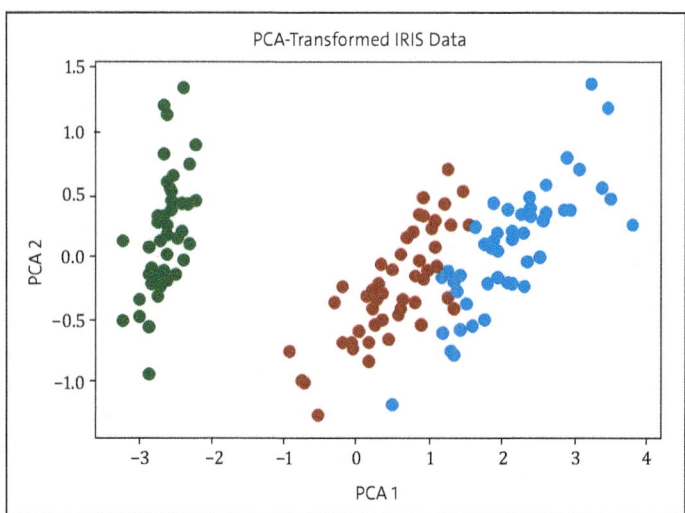

Figure 11.23 Scatterplot of the PCA-Transformed Iris Data

As you can see in Figure 11.23, the class clusters are more clearly separated, which is what we wanted to achieve.

11.3.5 Feature Selection

The datasets we have available may well contain features that contribute nothing or very little to our ML task or, in the worst case, have a negative impact on our results. Unnecessary features increase the training effort and reduce the generalization capability of the model.

The process of selecting a subset of relevant features can be done in the following way, where the program library scikit-learn provides us with a set of functions.

Filtering Methods

For filtering methods, a statistical measure is used to assign a value to each feature. The features are sorted by value, and the *k* best features are selected. Listing 11.24 outlines the procedure for this method.

```python
from sklearn.feature_selection import SelectKBest
from sklearn.feature_selection import chi2

X, y = iris_X, iris_y

test = SelectKBest(score_func=chi2, k=2)
myFilter = test.fit(X,y)
X_trans = myFilter.transform(X)
print(X_trans)
# Output:
[[1.4 0.2]
 [1.4 0.2]
 [1.3 0.2]
 ... ]
```

Listing 11.24 Filtering Method for Feature Selection for the X, y Dataset

The `SelectKBest()` function was selected in Listing 11.24; the *chi-square test* is used as the statistical measure, and the number of the best selected features of the dataset is specified with *k*.

Chi-Square Test

The chi-square test is a statistical measure that determines whether there is a correlation between any two (categorical) features. The name comes from the fact that a squared difference of expected and observed frequencies of features is used for the calculation.

Wrapper Methods

The selection of the feature set is regarded as an optimization task, whereby various feature combinations are prepared, evaluated, and compared with other combinations. A model is used to evaluate the combinations. The mode of operation is demonstrated in Listing 11.25.

```
X, y = iris_X, iris_y

from sklearn.feature_selection import RFE
from sklearn.linear_model import LogisticRegression

myModel = LogisticRegression()
test = RFE(myModel, n_features_to_select=3)
myFilter = test.fit(X,y)

X_fre = myFilter.transform(X)

print(X_fre)
# Output:
[[3.5 1.4 0.2]
 [3.  1.4 0.2]
 [3.2 1.3 0.2]
 ...]
```

Listing 11.25 Wrapper Method for Feature Selection for the X, y Dataset

Recursive feature elimination ((RFE())) is used here as a search method based on the *logistic regression* evaluation model (LogisticRegression()). The n_features_to_select parameter is again used to specify the number of features selected.

11.4 Summary

In the first part of this chapter, we looked at the ML process according to CRISP-DM. As AI applications using neural networks are becoming increasingly powerful and can also have an impact on people, we then turned our attention to the ethical and legal implications and challenges that need to be considered in the first phase of CRISP-DM, the business (process) understanding.

Feature engineering is part of the data preparation phase. In the section dedicated to it, we looked at the aspects of preparation for different ML methods, which of course include neural networks. Feature engineering is often underestimated, but the resources required for good data preparation are usually higher than the effort required for the actual model development.

11.5 Further Reading

- "AI4People's Ethical Framework for a Good AI Society: Opportunities, Risks, Principles, and Recommendations" by Luciano Floridi et al. (n.d., *https://ai4people.org/PDF/AI4People_Ethical_Framework_For_A_Good_AI_Society.pdf*)
- Coding of categorical variables (*https://contrib.scikit-learn.org/category_encoders/*)
- CRISP-DM (*https://en.wikipedia.org/wiki/Cross-industry_standard_process_for_data_mining*)
- "Interpretable Machine Learning: A Guide for Making Black Box Models Explainable" by Christian Molnar (2024, *https://christophm.github.io/interpretable-ml-book*)
- "OECD AI Principles Overview" (2024, *https://oecd.ai/en/ai-principles*)

Chapter 12
Learning Methods

As different as our brains are, so different are the learning processes.

In the heat of the moment, we almost overlooked the fact that there are other types of learning besides learning from input-output relationships. Consider, for example, a board placed on top of some bars—sturdy bars, mind you, to prevent the board from falling over. The board is attached to the bars at a height of approximately 3 feet. What would you call something like that?

What is going on in your mind right now is the search for similarities. This means that you try to assign this unknown thing to a group based on its property. This is a completely different way of learning than input-output learning.

Here's another example: Imagine ants walking on the ground in front of you, looking for material and food. On closer inspection, you notice that the ants walk as if on a road, almost as if there is an invisible traffic control center that tells the ants where to go. How do the ants do it? This is yet another way of learning. It's time for a classification of learning strategies, which we'll keep as simple as possible.

12.1 Learning Strategies

In Chapter 1, we've already dealt superficially with the different types of learning. We'll now delve deeper into the subject to develop a holistic view of learning.

A common classification of learning in a machine context distinguishes between four major approaches: *supervised*, *unsupervised*, *semi-supervised*, and *reinforcement* learning (see Figure 12.1). We'll take a closer look at these different approaches on the following pages and discuss simple examples of each. We've prepared a special treat for supervised learning: we'll implement the *Q-learning* algorithm that enabled *DeepMind* to teach a model to play Atari. We show the implementation both table-based and with the help of a neural network. And then there's a tasty piece in which we cluster data in the unsupervised learning style using Teuvo Kohonen's *self-organizing maps* (SOMs).

We begin our explanations with the most frequently used learning strategy: supervised learning.

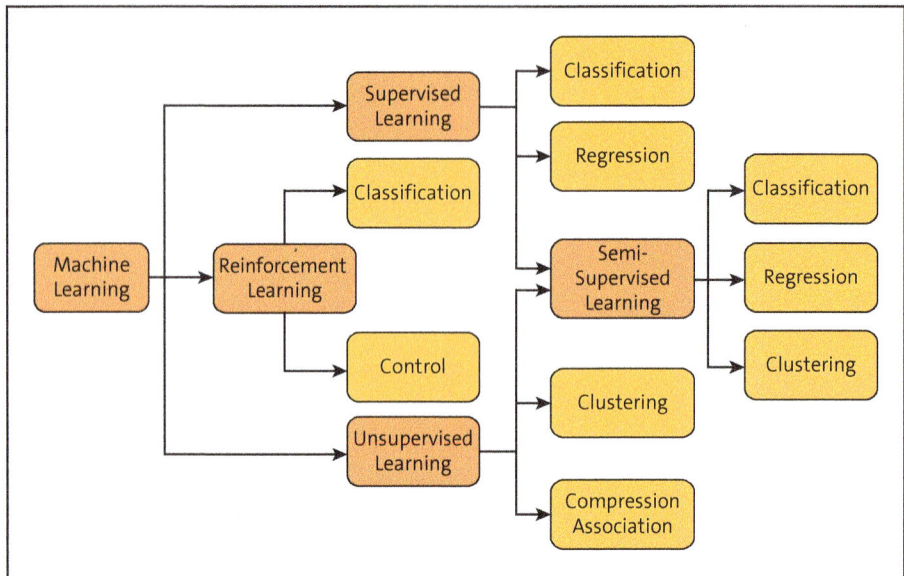

Figure 12.1 Learning Strategies and Their Possible Applications

12.1.1 Supervised Learning

The task in supervised learning might be "Develop a prediction model based on input and provided output" or "Learn the assignment of an input to a given class." As the name suggests, someone monitors the learning process. This is almost like trying to teach another person something by giving them an example with the desired result, that is, an image of a cat entitled "This is a cat." (The provided name of the image is also referred to as a *label* in this context.)

When you hear, "No, that's not a dog, that's a cat!" this means that an *association* is learned between things (input/output). A popular application of this learning technique is *classification*, although this is only one example of a technique. Another area of application is *regression*, also known as *prediction*. Let's now take a closer look at the areas of application.

Classification

As an example of *classification* (see Figure 12.2), imagine the assignment of books to a grouping, possibly only to two categories: entertainment books and nonfiction books. Where would you place the book on programming neural networks that you're currently reading? And what example would you have for the other category? Where would you place *The Little Prince*, for example?

Another example of classification is the sorting into good and poor production quality in the context of a snack production. For example, the quality is good if the packaging's welds are flawless.

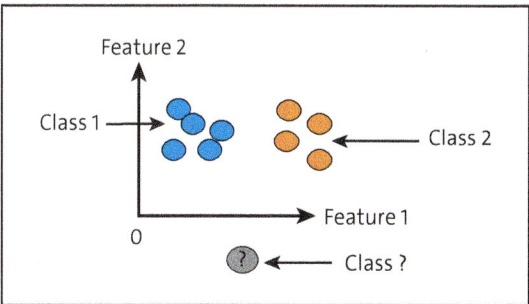

Figure 12.2 Classification

In general, classification is based on using different *characteristics*, also known as *features*, as input. The length of the *feature vector* depends on the number of features used as input. A square, for example, can be characterized by the single characteristic side length, whereas a rectangle requires two characteristics: length and width.

Regression

The other area of application of supervised learning concerns *regression*, which is used to make *predictions*. If you take values that are related to each other, networks can learn the relationships and apply them to as yet unknown values. This could be described as the *ability to abstract*. A tangible example is to determine whether the weather has an influence on the quality of the wine produced or whether the number of listeners at an electronic dance festival remains the same for the next few years.

In (linear) regression, the accuracy of the prediction depends on the (linear) correlation of the variables. The variables are referred to as *independent* and *dependent variables*. For example, the winemaker wants to have predicted the dependent variable "quality" for the independent variable "weather". The more strongly the two variables are related, the higher their correlation. *Correlation* is the word for the relationship with regard to an assumption, such as that "weather" and "quality" can be represented as a line (see Figure 12.3).

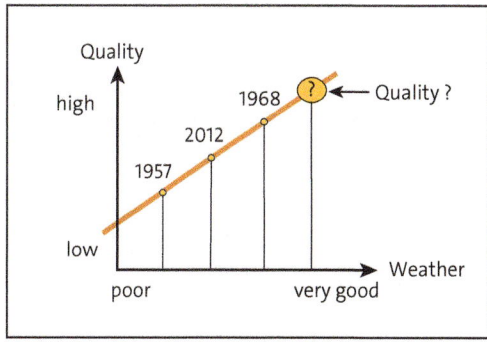

Figure 12.3 Wine-Quality Regression

12 Learning Methods

Various tools are available for the analysis, for example, a *scatterplot* like the one shown in Figure 12.3. This is a wonderful illustration of the perfect linear relationship between weather and quality. The more combinations of values that fulfill the relationship, the more certain we can be that we've found a valid description of the relationship.

But the reality isn't as ideal as shown in our example. Realistically, the points wobble and form point clouds around the ideal. In Figure 12.4, you can see a concrete example of this wobbling around the ideal model: specifically, the development of the temperature over a certain period of time.

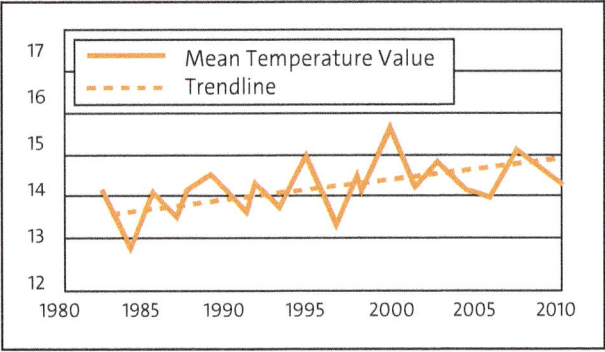

Figure 12.4 Schematic Representation of the Trend Analysis for a Mean Temperature Value

The question that now arises is how exactly the line can be laid through this cloud. There are, of course, many options; it's therefore necessary to define a criterion for this. One option is that the sum of the distances between the points y and the points on the line $\hat{y}$ is as small as possible. To ensure that both positive and negative deviations from the line—the point can lie above or below the line—are included in the calculation, increasing the error, the square of the distance is used. This results in the formula below for the error E, where n is the number of points measured or the number of points on the line:

$$E = \frac{1}{2}\sum_{i=1}^{n}(y_i - \hat{y}_i)^2$$

You already know this equation, and you've seen that it plays a decisive role in error minimization, that is, learning, in the Adaline and backpropagation algorithms.

Time Series

A special aspect of regression is its application to *time series*. The values are *ordered* along the time axis. A value is available at all times, for example, sales figures for a specific item, possibly a chocolate bar. Our goal could be to derive the demand for chocolate bars from past data and predict it for the next day. Three data-based methods have evolved:

- Time series predictions
- Causal predictions
- Combined predictions

The *time series prediction* uses the ordered sequence of input values with a specific time window size (see Figure 12.5). In our example, the window size depends on the number of input neurons. The window is moved over the time series starting from index 0.

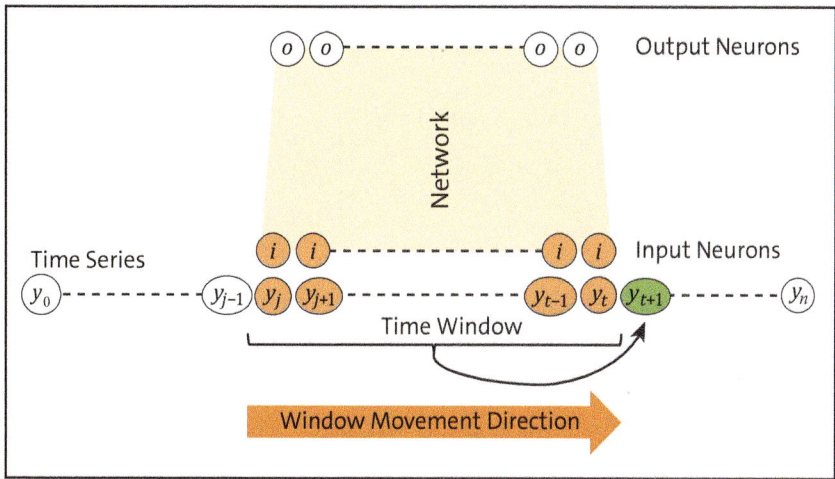

Figure 12.5 Time Series with Network

A value is supposed to be predicted from the values in the time window, whereby we've plotted the value in the next index with y_{t+1}, but it can also be located further in the future. It should be noted that there is a functional relationship between the calculated value and the values in the past.

The *causal prediction* doesn't use the time series of y_t, but takes into account other independent influences, for example, the day of the week, the weather, or the Super Bowl. The influences are also referred to as *independent/influencing* or *explanatory variables*. This means that the value to be determined is functionally dependent on influencing variables.

As its name suggests, the *combined prediction model*, also known as *dynamic regression*, uses both of the previous methods for making predictions.

Learning Algorithms

In Table 12.1, we've summarized known algorithms for supervised learning. However, we've also included the neural networks here.

Algorithm	Area of Use
Classification rule learners	Classification
Decision trees	Classification
Naive Bayes	Classification
Nearest neighbor	Classification
Linear regression	Regression
Model trees	Regression
Regression trees	Regression
Neural networks	*Classification/regression*
Support vector machines	Classification/regression

Table 12.1 Learning Algorithms for Supervised Learning

12.1.2 Unsupervised Learning

In *unsupervised learning*, the task is as follows: "Discover an internal representation of the data using only the input."

This means if there are no output categories or labels the algorithm can use to establish a relationship, *inherent* and possibly still unknown properties must be discovered to determine similarities.

A (cluster) model is created for the data without us actually knowing what to call the individual clusters. Nevertheless, these groupings do exist. It's almost as if you've seen thousands of images of chairs but haven't yet heard the term for chair. At some point, you'll invent the term "chair" out of convenience to communicate more easily and efficiently. You then use the term to represent an element from the group of chair objects. This learning strategy therefore generates an internal representation of the data and can handle new, untrained data. The main techniques used in unsupervised learning are clustering and *association*.

Clustering

The task of the clustering algorithms is to divide inputs into groups, categories, or classes only on the basis of the information contained in the inputs, without any further additional information (see Figure 12.6). In contrast to classification, where the output values are predefined, this information isn't available for clustering.

Among other things, clustering is used for *segmentation analysis*, which identifies groups of individuals with similar behavior or demographic information to better target advertising campaigns. It's important to note at this point that although the algorithms can perform clustering, humans are still needed to evaluate and interpret these clusters.

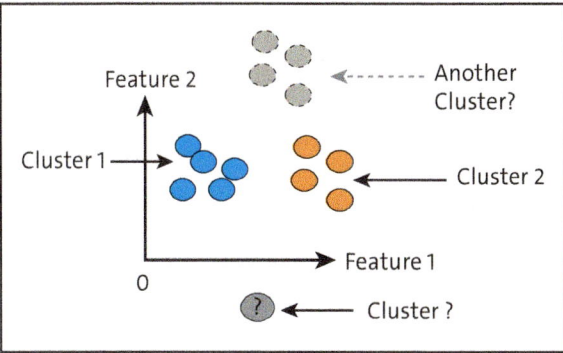

Figure 12.6 Cluster

Association

Learning algorithms from this category extract rules and patterns from the datasets, explain the relationships between the variables, and show frequencies (repetitions) and patterns. These rules in turn enable organizations to gain deeper insights into their data. A simple example is the analysis of shopping baskets and derivations from the compositions, for example:

{Cheese,Cold Cuts} ⟹ {Bread}

This can be read as follows: "When customers buy cheese and cold cuts, they also buy bread."

Learning Algorithms

In Table 12.2, we've collected well-known algorithms for unsupervised learning, in addition to the neural networks that we've also included. (Algorithms shown in italics belong to the group of neural networks.)

Algorithm	Area of Use
k-means	Clustering
Gaussian mixture	Clustering
Hierarchical clustering	Clustering
PCA principal component analysis (dimensional reduction)	Clustering
Hidden Markov model	Clustering
Neural networks	*Clustering*

Table 12.2 Learning Algorithms for Unsupervised Learning

Algorithm	Area of Use
Hopfield networks	*Association*
Self-organizing map (SOM)	*Clustering*
Association rules	Pattern discovery

Table 12.2 Learning Algorithms for Unsupervised Learning (Cont.)

Application area

One use case for the descriptive approach to data is *pattern discovery*, which discovers associations in the data. It's used to carry out market basket analyses with the aim of determining which products are bought together and should therefore be suggested to the customer when buying online or should be placed differently in the store.

Example: Self-Organizing Maps

With Teuvo Kohonen's *self-organizing maps* (SOM) and the *adaptive resonance theory* (ART), two types of networks are available for unsupervised learning. We now want to introduce you to SOM so you can get a little hands-on experience and explore this versatile approach.

A SOM has two central components: the *input layer*, as you know it from other network types, and the *map*. In this map, nodes/neurons are connected to each other like in a fishing net: Each input neuron is connected to each node *v* in the map. Each node *v* in the map receives input from all input neurons (see Figure 12.7).

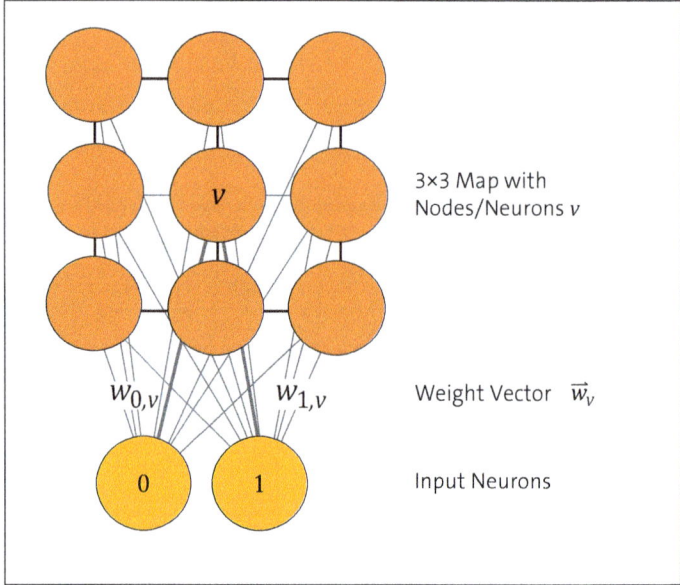

Figure 12.7 SOM with Two Input Neurons

The trick is that inputs that are in close proximity are also mapped to neighboring nodes in the map. For this reason, we could also say that the SOMs represent a *topographical organization* in which nearby places in the map represent inputs with similar properties.

Error correction isn't used for learning, as is the case with backpropagation, but *competitive learning*, which is a form of unsupervised learning. In this learning strategy, a node "fights" to be allowed to respond to an input. As indicated in Figure 12.7, each node v in the map has a weight vector $\vec{w}_v$, where each node is connected to each input neuron.

Algorithm

With this background information, we can take a closer look at the learning algorithm:

1. Initialize the weights of the weight vectors $\vec{w}_v$ for each node v with small random values from the interval [0,1].
2. Repeat the following for a certain number of iterations *I*:
   - Randomly select an input vector $\vec{x}(s)$ from the set of training examples *X*, where s is the iteration step.
   - Calculate the distance *d* from the input vector to the weight vectors of all nodes with $d(\vec{x}(s), \vec{w}(s))$.
   - Determine the node u with the smallest distance. This is also referred to as the *best matching unit* (BMU).
   - Adjust the BMU weight vector $\vec{w}_u(s)$ and that of the node neighbors in the direction of the input vector. The adjustment is made using the following formula:

   $$\vec{w}_v(s+1) = \vec{w}_v(s) + \theta(u,v,s) \cdot \alpha(s) \cdot (\vec{x}(s) - \vec{w}_v(s))$$

The different elements of the formula have the following meaning:

- $\vec{w}_v(s)$
  The weight vector for nodes with index v for iteration step s.

- s
  The iteration step.

- u
  The index for the BMU with regard to the input vector $\vec{x}(s)$.

- $\vec{x}(s)$
  The input vector.

- $\alpha(s)$
  A learning rate that decreases over time.

- $\theta(u,v,s)$
  The neighborhood function to determine the neighbors with index v for BMU and with index u for iteration step s. With increasing iteration step s, the neighborhood shrinks and gets values in the interval [0,1].

We still need to take a closer look at some aspects, such as the calculation of the *distance* between two vectors, for example, the input vector $\vec{x}(s)$ and a weight vector $\vec{w}_v(s)$. This distance is defined as the *Euclidean distance* with the following:

$$d(\vec{x}, \vec{w}) = \|\vec{x} - \vec{w}\|$$
$$= \sqrt{(x_1 - w_1)^2 + (x_2 - w_2)^2 + \ldots} = \sqrt{\sum_i (x_i - w_i)^2}$$

This is the way mathematicians put it. Import the math module to translate it into readable Python code, as shown in Listing 12.1.

```
def distance(self, vec1, vec2):
    return math.sqrt((vec1[0]-vec2[0])**2+(vec1[1]-vec2[1])**2)
```

Listing 12.1 Euclidean Distance between the Two 2D Vectors "vec1" and "vec2"

You can also visualize the distance by taking a closer look at the geometric representation in Figure 12.8. We've inserted sample values for the distance to practice the calculation.

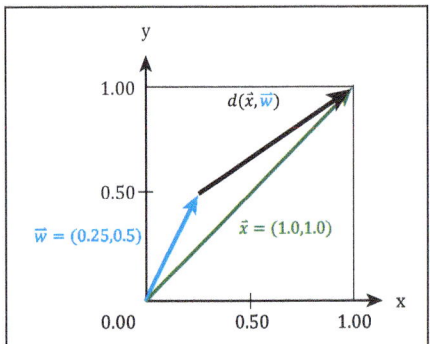

Figure 12.8 Euclidean Distance of Two Vectors

Let's assume we have the two vectors $\vec{w} = (0.25, 0.5)$ and $\vec{x} = (1.0, 1.0)$. If we insert them into the formula for the Euclidean distance, we get the following result:

$$d(\vec{x}, \vec{w}) = \|\vec{x} - \vec{w}\|$$
$$= \sqrt{(1.0 - 0.25)^2 + (1.0 - 0.5)^2}$$
$$= \sqrt{0.5625 + 0.25}$$
$$\cong 0.9$$

Great, now you can calculate the distance between two vectors. If you will, this is the basic arithmetic operation for the SOM algorithm.

Various functions can be used to define the *learning rate*. An easy-to-implement function that decreases evenly can be defined as follows, depending on the iteration step s and the initial learning rate $\alpha(0)$:

$$\alpha(s) = \alpha(0) \cdot \frac{1}{s}$$

This corresponds to the requirement that strong changes are desired at the beginning of the learning process, and then stabilization sets in over time (see Figure 12.9).

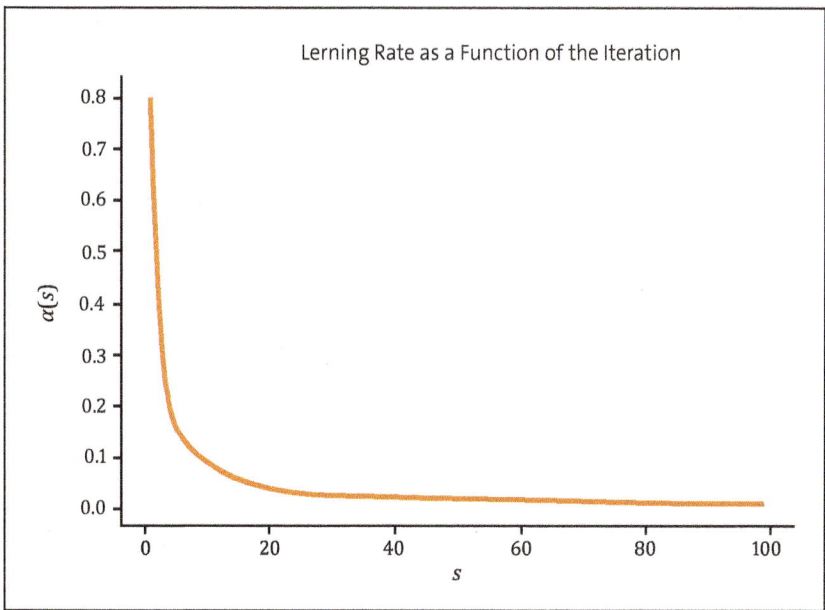

Figure 12.9 Decrease in the Learning Rate as the Number of Iterations Increases

As an example, we've set the learning rate to $\alpha(0) = 0.80$ and the number of iterations to 100. For iteration step 10, the learning rate would then be the following:

$$\alpha(10) = \frac{\alpha(0)}{10} = \frac{0.80}{10} = 0.08$$

An alternative function, which we can also use in the implementation for the learning rate, ensures a linear reduction during the I iterations:

$$\alpha(s, I) = \alpha(0) \cdot (1 - \frac{s}{I})$$

The next aspect we look at is the *neighborhood* of a vector. In Figure 12.10, you can see a neighborhood on the left-hand side that is defined by the radius. Whenever an intersection occurs within the circle, this means that a node from the SOM is in the neighborhood. The radius should decrease with the iterations.

On the right-hand side, you can see two different options to take into account the distance from the BMU to the neighbor. In the upper case, the rating is always the same, no matter how far away the neighbor is—as long as it's still within the radius. In the lower case, a *bell curve* is used for the evaluation, which doesn't always evaluate the value for the distance to the BMU equally.

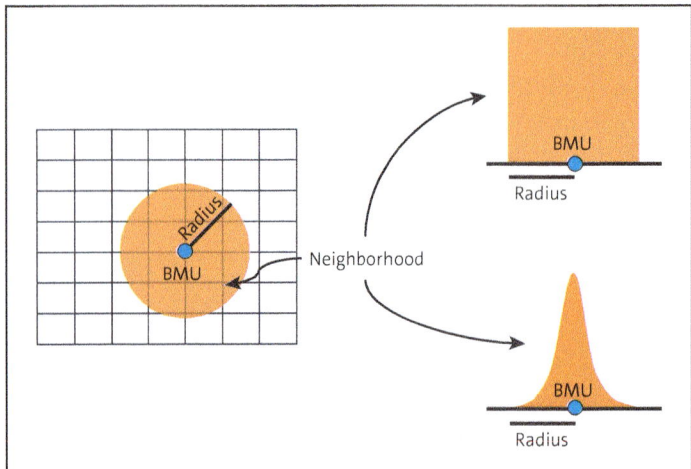

Figure 12.10 Different Neighborhood Functions

We then cast these facts into a calculable mold. For the calculation that we'll use for the *neighborhood* between nodes *u* and *v*, the following formula applies:

$$\theta(u, v, s) = \begin{cases} h_{u,v}(s), & \text{if } d(u,v) \leq r(s,I) \\ 0, & \text{else} \end{cases}$$

This means that if a node with index *v* lies within the distance *r(s)* of the BMU *u*, the neighborhood is evaluated with function $h_{u,v}(s)$ at time *s*. For the value of the neighborhood, you can use any function that loses value over time so that the neighborhood becomes smaller over time, for example, $\alpha(s)$, as defined previously.

The *radius r(s,I)* is the last element we still need to define. There are, of course, plenty of options available for this too, but it should be defined in such a way that it becomes smaller and smaller with each iteration step *s*, for example, like this:

$$r(s, I) = r(0) \cdot e^{\left(-\frac{s}{I}\right)}$$

In Figure 12.11, you can see an example of the development of the radius with the iterations. It starts at 6 and decreases to a value below 2 over the course of the 100 iterations.

Before we move on to the implementation, let's take another look at the learning formula for the SOM with what we've heard so far:

$$\vec{w}_v(s+1) = \vec{w}_v(s) + \underbrace{\theta(u,v,s)}_{\text{less than 1}} \cdot \underbrace{\alpha(s)}_{\text{less than 1}} \cdot \underbrace{(\vec{x}(s) - \vec{w}_v(s))}_{\text{Vector for BMU}}$$

We can now summarize the formula in words as follows: "Change the weight of a map node in the direction of the BMU per iteration step."

The direction from the weight vector to the BMU results from the difference between the vectors $(\vec{x}(s) - \vec{w}_v(s))$, which is explained in detail in Appendix B.

One more thing to note: if the difference vector is added to vector $\vec{w}_v(s)$, the new vector $\vec{w}_v(s+1)$ points more in the direction of the input vector.

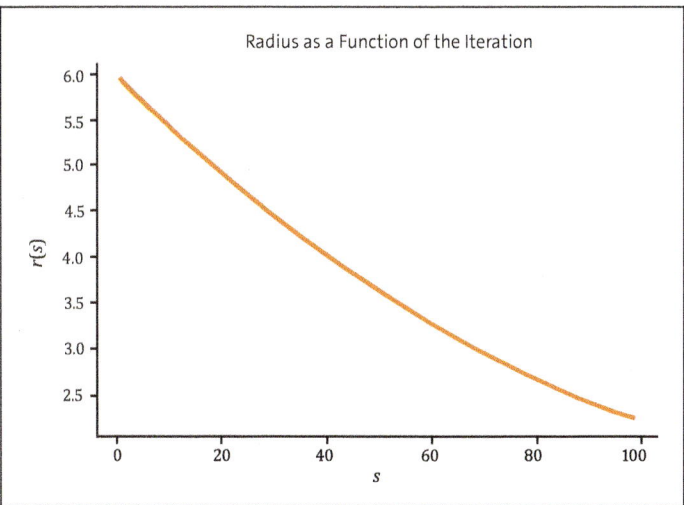

Figure 12.11 Radius as a Function of Iteration

The Code

We're now ready for the implementation and will take a look at the Python code. Of course, we want to encapsulate the functionalities cleanly, not attach too much importance to performance, and also use object orientation. The basic components we discussed in the introduction to SOM are inputs, map with nodes, and weights (see Figure 12.12).

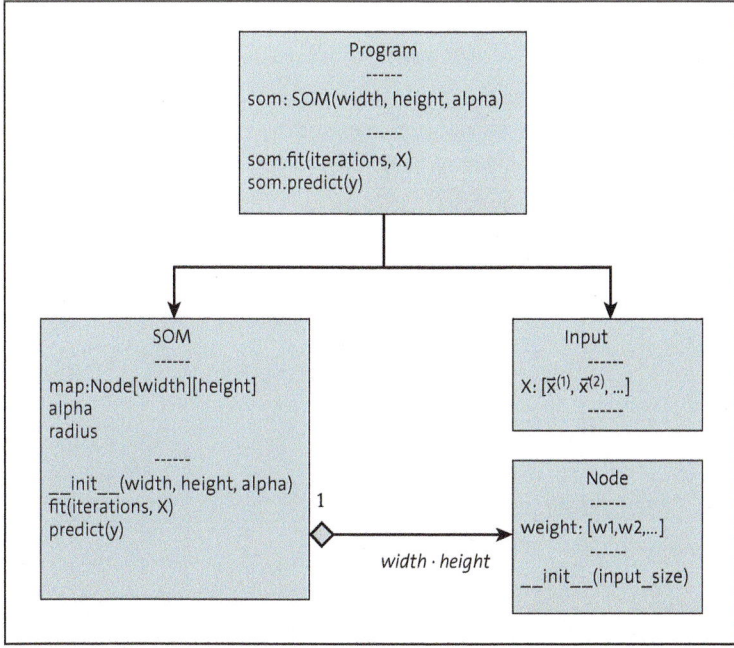

Figure 12.12 The Components of SOM and Their Interrelationships

As always, the main task is performed by the `fit()` method in the SOM class. The weight adjustments are carried out there; that is, learning is implemented as described at the beginning. We've described and commented on the code in detail, so it shouldn't be too difficult for you to understand it. We've again used our popular XOR problem as an example, as shown in Listing 12.2.

```python
# Teuvo Kohonen's SOM
import random
import math
import numpy as np
import matplotlib.pyplot as plt
import matplotlib.colors as col

# Slight overkill, as we only use weights, but expandable
class Node(object):
    def __init__(self, input_size=2):
        self.input_size = input_size
        self.weight = np.array([random.random() \
                                for e in range(self.input_size)])
# This is the SOM
class SOM(object):
    def __init__(self, map_width=10, map_height=10, alpha=0.005):
        """Initialization of the SOM
        """
        self.map_width = map_width
        self.map_height = map_height
        self.radius = 0.6 # Radius
        self.alpha = alpha # Learning rate
        self.map = [[Node() for j in range(self.map_width)] \
                    for i in range(self.map_height)] # The map with the nodes

    def fit(self, iterations, X):
        """ Competitive learning for the SOM
        """
        for s in range(1,iterations+1):
            radius_s = self.radius * math.exp(-1.0*s/iterations)
            # Alpha: V1
            # alpha_s = self.alpha / s
            # Alpha: V2
            alpha_s = self.alpha*(1.0 - s/iterations)
            # Select random input vector
            x = X[random.randint(0,X.shape[0]-1)]
            distances = np.empty((self.map_width, self.map_height))
            # Calculate all distances
            for i in range(self.map_width):
```

```python
            for j in range(self.map_height):
                distances[i][j] = self.distance(x,self.map[i][j].weight)
        # Find best matching unit index
        bmu_index = np.unravel_index(np.argmin(distances, axis=None),\
                distances.shape)
        for i in range(self.map_width):
            for j in range(self.map_height):
                v=self.map[i][j].weight # Weight vector of a node
                u=self.map[bmu_index[0]][bmu_index[1]].weight # BMU
                distance=self.distance(u,v) # Distance between BMU and
weight vector
                if distance <= radius_s:
                    neighborhood = radius_s # neighborhood value
                    self.map[i][j].weight += \
                      neighborhood * alpha_s * (x - v) # weight adjustment
        # Draw weight vectors in 2-dimensional plot
        self.plot_weights(x)

    # Which vector is closest?
    def predict(self, y):
        distances = np.empty((self.map_width, self.map_height))
        # Calculate all distances
        for i in range(self.map_width):
            for j in range(self.map_height):
                distances[i][j] = self.distance(y,self.map[i][j].weight)
        # Node with smallest distance between weight vector and target y
        min_dist_index = np.unravel_index(np.argmin(distances, axis=None),\
                distances.shape)
        # Return of the weight vector
        return self.map[min_dist_index[0]][min_dist_index[1]].weight

    # Distance between two vectors, calculation optimized
    def distance(self, u, v):
        return np.linalg.norm(u-v)

    # Output weights as a plot
    def plot_weights(self,x):
        # For output
        fig, ax = plt.subplots()
        weights_x = [] # x-coordinates
        weights_y = [] # y-coordinates
        # All weights
        for i in range(self.map_width):
            for j in range(self.map_height):
```

```python
                weights_x.append(self.map[i][j].weight[0])
                weights_y.append(self.map[i][j].weight[1])
        ax.scatter(weights_x, weights_y, color='b')
        plt.title('Weight vectors - Iteration ' + str(i))
        plt.xlabel('x')
        plt.ylabel('y')
        xticks=np.arange(0,1,0.1)
        yticks=np.arange(0,1,0.1)
        plt.yticks(yticks)
        plt.xticks(xticks)
        plt.show()
########### Let the SOM begin #################
###############################################
# Control of SOM training and evaluation
# Width and height of the map
map_width = 4
map_height = 4
# Learning rate
alpha = 0.8
# Iterations
iterations = 100
# Random
random.seed(1)
# Instantiate the SOM
som=SOM(map_width,map_height,alpha)
# Training
# Initialization of the training examples
X=np.array([[0.0,0.0],[0.0,1.0],[1.0,0.0],[1.0,1.0]])
print("Train the XOR function")
som.fit(iterations,X)
# Evaluation
# Output three decimal places
np.set_printoptions(formatter={'float': lambda x: "{0:0.3f}".format(x)})
print("Statements on the XOR function")
print("Statement 0 0, {}".format(som.predict([0,0])))
print("Statement 1 0, {}".format(som.predict([1,0])))
print("Statement 0 1, {}".format(som.predict([0,1])))
print("Statement 1 1, {}".format(som.predict([1,1])))
```

Listing 12.2 SOM Implementation

If you run the code, the `plot_weights()` method is called in the `fit()` method after each iteration step. This ensures that a diagram is drawn with the weight vectors and we can discover the migration of the weights toward the BMUs and the formation of clusters.

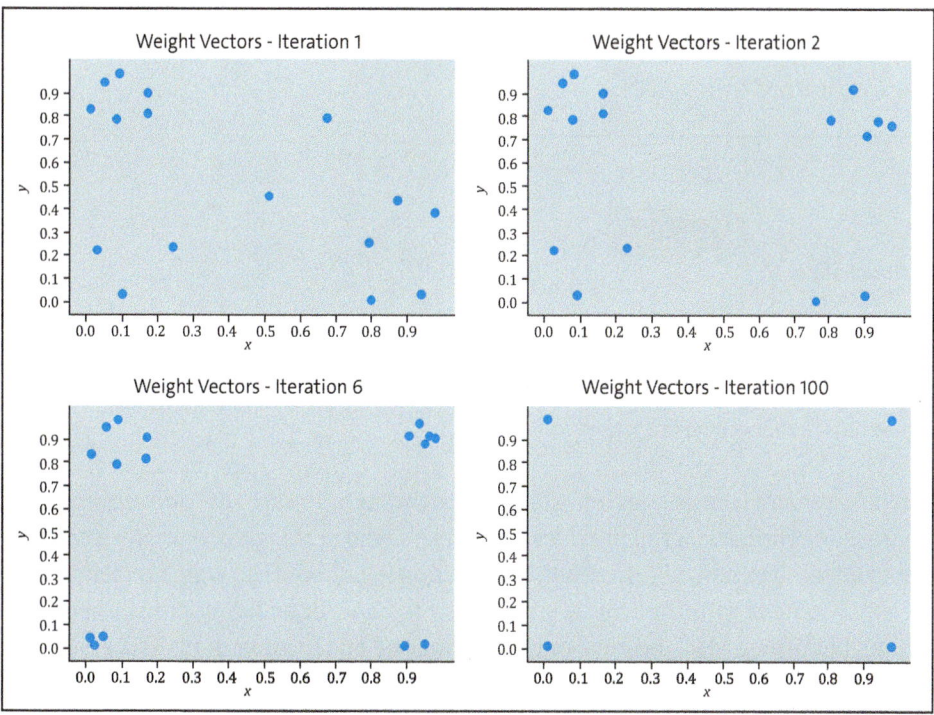

Figure 12.13 SOM and Cluster Formation during the Learning Process

In Figure 12.13, we've extracted some of the iteration steps to show you the evolution of the clusters, which ultimately leads to an impressive classification accuracy, as shown in Listing 12.3.

```
# Output:
Statements on the XOR function
Statement 0 0, [0.001 0.000]
Statement 1 0, [0.999 0.000]
Statement 0 1, [0.001 0.999]
Statement 1 1, [0.999 1.000]
```

Listing 12.3 Result of the XOR Problem Evaluation

We hope you enjoy adapting the code to your needs and tasks. With the preceding implementation and the theory discussed, you get a powerful tool that you can use in many different ways. But unfortunately, we have to move on to the next topic, reinforcement learning.

12.1.3 Reinforcement Learning

Reinforcement learning represents the third major category of learning approaches alongside supervised and unsupervised learning and is motivated by approaches from psychology. As described previously, supervised learning is characterized by the fact that the artificial neural network (ANN) is presented with value pairs (x, y) whose mappings are learned by the ANN.

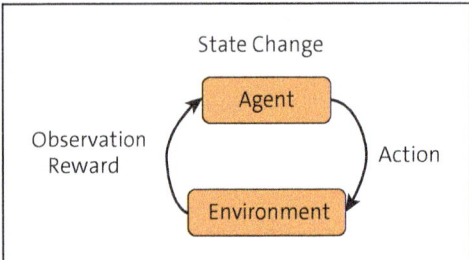

Figure 12.14 Reinforcement System

In reinforcement learning, learning takes place through "reward" or "punishment" (see Figure 12.14). The basic setup provides for an *agent* who performs *actions* in an *environment* and receives *rewards* as a result. Information is gathered through interaction with the environment with the aim of being rewarded and/or avoiding punishment. The actions performed lead the environment from one state to the next. As a result, the agent learns a tactic or a strategy that enables it to react in situations that are partly known and partly unknown. The approach has a high *gamification* component, as the agent seeks to maximize the reward.

The agent doesn't know the environment in advance, but explores it step-by-step (incrementally). So, if the agent performs an action a_t at the time t and environment state s_t, the state of the environment changes to s_{t+1}, and the agent receives the reward r_t.

The agent will choose an action in such a way that it can maximize its expected profit at time t and in the future. Let's simply call the profit R_t at a point in time t. Then, you can see the future profit as the sum of the individual profits in each step, with the future going T steps further:

$$G_t = R_{t+1} + R_{t+2} + R_{t+3} + \cdots + R_{t+T}$$

For example, in the next step, the agent drives into the wall and receives a reward of -10, drives into a free space to get +3 points, visits a previously visited space again to receive 1 point, and so on. If we insert this into the previous formula, the result is

$$G_t = -10 + 3 + 1 + \cdots$$

This formula isn't yet elegant enough; we can write it more simply:

$$G_t = \sum_{k=1}^{T} R_{t+k}$$

If we now also want to take into account that future rewards are subject to a discount because it's not yet certain whether we'll receive the reward and the further into the future it is, the more uncertain this is, then we have to make a small change to the formula: the *discount factor* γ (gamma), as shown here:

$$G_t = \sum_{k=1}^{T} \gamma^k \cdot r_{t+k} \text{ with } 0 \leq \gamma < 1$$

So that's it, this formula will be the basis for a famous reinforcement algorithm, which we'll explore in depth together with you: *Q-learning* (by Chris Watkins in 1989; see Section 12.4). You'll use this algorithm to teach a small agent to move optimally in an unknown terrain and to develop a strategy: not to fall into a hole, but to avoid walls!

Theory: What Is Q-Learning?

Q-learning is a technique based on a state-action-value (*reward*) function ($Q: S \times A \rightarrow \mathbb{R}$) that indicates the future return (i.e., the reward) an agent can expect if it performs action a in state s and then follows the best possible strategy.

Unlike model-based methods, where the agent tries to develop a precise idea of the environment, Q-learning learns directly through interaction with the environment.

The Q function, which is often referred to as $Q(s,a)$, receives the current state and a possible action as input and calculates the potential reward for this action and all subsequent actions. More precisely, this is referred to as *temporal difference learning*, which adjusts the estimate of the future total reward G_t for each step. Let's take a look at the formula for Q-learning, also known as the *Bellman equation*, in all its beauty and without shying away from mathematical notation:

$$Q^{\text{new}}(s_t, a_t) \leftarrow Q^{\text{old}}(s_t, a_t) + \underbrace{\alpha}_{\text{LR}} \cdot \left(r_t + \underbrace{\gamma}_{\text{DF}} \cdot \underbrace{\max_{a \in A} Q(s_{t+1}, a)}_{\text{FV}} - Q^{\text{alt}}(s_t, a_t) \right)$$

Let's look at this one thing at a time:

- The old value $Q^{\text{old}}(s_t, a_t)$ is the profit for the state s_t and the action a_t, to which an estimated future value is added. It starts with a random value defined by the developer.
- The learning rate (LR α) controls the influence of the future value. If α is 0, for example, then the future value has no influence whatsoever.
- The reward r_t was obtained by using the action a_t in the state s_t.
- The discount factor (DF γ) ensures that older findings are valued higher than more recent ones. To start with, it can simply be set to 1; as a rule, it's in the value range of $0 \leq \gamma \leq 1$.
- The estimated future value (FV) $\max_{a \in A} Q(s_{t+1}, a)$ represents the maximum of all actions in the next step.

At the beginning, choosing the action isn't easy, as nothing has been learned yet: Should the agent wander around and discover a lot or fall back on what has already been discovered? This dilemma can also be found in the literature under the term *exploration-exploitation dilemma*.

One possible solution is a *selection procedure* for actions that also takes the learning time into account. For this purpose, the ε limit (pronounced *Epsilon*, that is, E as in *exploration* or *exploitation*) is defined, which can assume a value between 0 and 1 (e.g., 0.2), as shown in Figure 12.15.

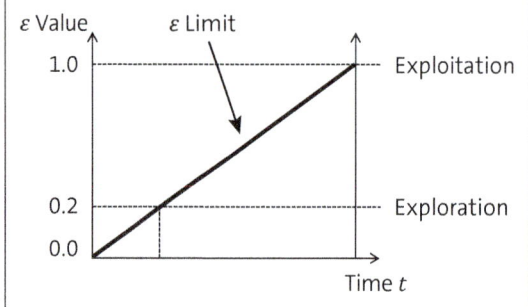

Figure 12.15 Exploration vs. Exploitation

Here is how the action gets selected:

1. The action selection generates a random value between 0 and 1.
2. If this value is greater than the ε limit, a random action is determined.
3. If the value is less than the ε limit, the action with the maximum Q value is used.
4. As the number of iterations increases, the value of the limit increases, while the probability of a random selection of an action decreases.

This takes into account the fact that the agent learns over time, which is sometimes even assumed for humans. Now let's move on to the real world.

Real-Life Example: The Robot in the Maze

To put theory into practice, let's look at a simple example: the robot in the *maze*. Imagine a simple, square maze consisting of a 5x5 grid area. Some of the fields represent obstacles the robot can't pass, while one field marks the exit. The robot always starts in the lower-left corner of the maze (*position (0,0)*), and the goal is to reach the exit in the upper-right corner (*position (4,4)*).

> **Note**
> You'll already know this anyway, but computer scientists start counting from 0.

The robot has no information about the maze at the beginning. It doesn't know which paths are passable and which are not. Each time it performs an action, it receives a reward:

- A small negative reward (-1) for each movement to prevent aimless wandering
- A larger negative reward (-10) if it hits a wall or obstacle
- A high positive reward (+100) when it reaches the exit

In each *episode* the robot starts again in the lower-left corner and tries to find the exit. Initially, the robot will make many mistakes and often run into obstacles. But over time and through many repetitions, it will learn which paths lead to the goal and which do not. Each action the robot performs will make it a little "smarter" by adjusting the Q values for this action and the corresponding state.

Over the course of many episodes, the robot will eventually (hopefully) find the best way through the maze—according to the Q-learning formula: the chosen path minimizes the negative rewards and maximizes the positive rewards by choosing the shortest and safest route to the exit. This is the path we refer to as the *optimal path*.

To implement Q-learning, we can use either the *tabular Q-learning* or the *deep Q-network* (DQN) approach:

- *Tabular Q-learning* is well suited for small state spaces in which all possible states and actions can be explicitly stored. However, this method becomes inefficient for larger problems such as games with complex environments because the memory and computing effort increases exponentially with the size of the state space. The deep Q-network approach is used to work around this problem.

- The *deep Q-network* approach uses neural networks to approximate the Q-values instead of storing them in a table. Due to its ability to generalize, a neural network can also make effective predictions in large or continuous state spaces. It learns from the data and adjusts the weightings in the network to recognize patterns and approximate optimal Q values.

Implementation: Q-Learning with a Neural Network

Now that we've looked at the theory and the real-life example, let's move on to the implementation of the Q-learning algorithm using a neural network. The model learns to predict Q values—and we'll take a look at how this works in Listing 12.4. We've peppered the code with comments and thus incorporated the explanations into the code.

```python
# For the mathematical operations
import numpy as np
# Our neural network framework
import tensorflow as tf
from tensorflow import keras
from tensorflow.keras import layers
# Plot
import matplotlib.pyplot as plt
# Data container - double ended queue
from collections import deque
```

12 Learning Methods

```python
# The coincidence
import random

# Parameters
alpha = 0.001  # learning rate
gamma = 0.9    # discount factor
epsilon = 0.1  # exploration rate
epsilon_decay = 0.995 # From exploit to explore
epsilon_min = 0.01
episodes = 10  # Number of episodes
batch_size = 32 # Number of learning examples
memory = deque(maxlen=2000)

# Maze setup
maze_size = (5, 5)
start_position = (0, 0)
goal_position = (4, 4)
walls = [(1, 1), (2, 2), (3, 3)]

# Helper functions
def is_valid_move(position):
    x, y = position
    return 0 <= x < maze_size[0] and 0 <= y < maze_size[1] and position not in walls

def get_next_position(position, action):
    x, y = position
    if action == 0: return (x - 1, y) # top
    if action == 1: return (x, y + 1) # right
    if action == 2: return (x + 1, y) # bottom
    if action == 3: return (x, y - 1) # left

def get_reward(position):
    if position == goal_position:
        return 100
    elif position in walls:
        return -10
    else:
        return -1

# Converts the state (position) into an input vector for the network
def state_to_input(state):
    state_input = np.zeros(maze_size)
    state_input[state] = 1
```

```python
    return state_input.reshape(1, -1)

# Define Q-network (neural network)
model = keras.Sequential([
    layers.Input(shape=(maze_size[0] * maze_size[1],)),  # Input layer with the correct input shape
    layers.Dense(24, activation='relu'),  # First hidden layer
    layers.Dense(24, activation='relu'),  # Second hidden layer
    layers.Dense(4, activation='linear')  # Output layer with 4 possible actions
])

# Adam = Stochastic Gradient Descent Method
# mse = Mean Square Error
model.compile(optimizer=keras.optimizers.Adam(learning_rate=alpha), loss='mse')

# Function for selecting the action (epsilon limit)
def choose_action(state):
    if np.random.rand() <= epsilon:
        return np.random.randint(4)  # Random action
    q_values = model.predict(state_to_input(state), verbose=0)
    return np.argmax(q_values[0])  # Action with the highest Q value

# Stores experiences in the replay memory
def remember(state, action, reward, next_state, done):
    memory.append((state, action, reward, next_state, done))

# Trains the network with experience from the replay memory
def replay():
    global epsilon
    if len(memory) < batch_size:
        return
    minibatch = random.sample(memory, batch_size)
    for state, action, reward, next_state, done in minibatch:
        target = reward
        if not done:
            target = reward + gamma * np.amax(model.predict(state_to_input(next_state), verbose=0)[0])
        target_f = model.predict(state_to_input(state), verbose=0)
        target_f[0][action] = target
        model.fit(state_to_input(state), target_f, epochs=1, verbose=0)
    if epsilon > epsilon_min:
        epsilon *= epsilon_decay
```

```python
# Q-learning algorithm with neural networks
for episode in range(episodes):
    state = start_position
    done = False

    step = 0
    while not done:
        step += 1
        print(f"Step: {episode}/{step}")
        action = choose_action(state)
        next_state = get_next_position(state, action)
        if not is_valid_move(next_state):
            next_state = state
            reward = -10
        else:
            reward = get_reward(next_state)

        done = next_state == goal_position
        remember(state, action, reward, next_state, done)
        state = next_state

        replay()

# Calculate the optimal path
optimal_path = []
state = start_position
while state != goal_position:
    optimal_path.append(state)
    action = choose_action(state)
    next_state = get_next_position(state, action)
    if not is_valid_move(next_state):
        break
    state = next_state

optimal_path.append(goal_position)

# Plot the maze
fig, ax = plt.subplots(figsize=(6, 6))

# Draw the walls (obstacles)
for wall in walls:
    rect = plt.Rectangle(wall, 1, 1, facecolor='black')
    ax.add_patch(rect)
```

```
# Draw the optimal path
for i in range(len(optimal_path) - 1):
    start = optimal_path[i]
    end = optimal_path[i + 1]
    ax.arrow(start[1] + 0.5, start[0] + 0.5, end[1] - start[1], end[0] - start[
0], head_width=0.2, head_length=0.2, fc='blue', ec='blue')

# Mark the starting point
ax.plot(start_position[1] + 0.5, start_position[0] + 0.5, 'go', markersize=10)
# Green for the start

# Mark the end point
ax.plot(goal_position[1] + 0.5, goal_position[0] + 0.5, 'rs', markersize=10)  #
Red for the goal

# Configuration of the plot
ax.set_xlim(0, maze_size[1])
ax.set_ylim(0, maze_size[0])
ax.set_xticks(np.arange(maze_size[1] + 1))
ax.set_yticks(np.arange(maze_size[0] + 1))
ax.grid(True)
ax.set_aspect('equal')

plt.show()
```

Listing 12.4 Q-Learning with Neural Networks

Here's an explanation of the code:

- **Q-network**
  We use a neural network implemented with Keras. The network has two hidden layers with 24 neurons each and uses a *rectified linear unit* (ReLU) as an activation function. The output layer has 4 neurons, each representing the Q values for the four possible actions.

- **Epsilon limit (epsilon-greedy strategy)**
  This strategy is used to either select a random action (exploration) or to execute the action with the highest Q value (exploitation). The epsilon value is gradually reduced to promote learning.

- **Replay memory**
  Experience (i.e., transitions between states) are stored in a replay memory to make the network training process more stable.

12 Learning Methods

- **Replay**
  The neural network is periodically trained on a random batch of experiences to implement the Q-learning algorithm.

- **Visualization**
  After training, the path is determined on the basis of the neural network and displayed visually, for example, as shown for the optimal path in Figure 12.16.

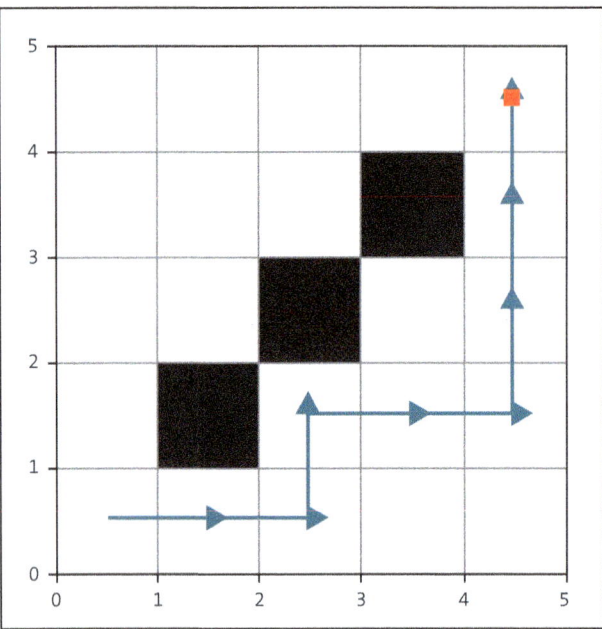

Figure 12.16 The Optimal Path for the Robot through the Maze, Learned from Deep-Q Networks

Conclusion

By using a neural network, we've realized the agent's ability to learn in more complex environments. The neural network learns to approximate the Q values and enables the robot to learn efficiently even if the state space is large or changing. This is a first step in the direction of how advanced applications of reinforcement learning work, as can be seen in the work of Richard S. Sutton and Andrew Barto or in the famous DeepMind applications.

This concludes our little trip into the world of reinforcement. Now we want to look at another learning strategy: semi-supervised learning.

12.1.4 Semi-Supervised Learning

In the case of supervised learning, the desired output values may not be known for each dataset or the effort required to provide the inputs with the desired output is too high. For this mixed scenario in which the desired outputs are only partially available, *semi-supervised learning* is a useful option.

As an example, imagine a set of data that doesn't have a label (see Figure 12.17): For example, by way of clustering, two clusters are created that don't have a label. If data is added that does have a label, it's possible to use these labels as representatives for the cluster.

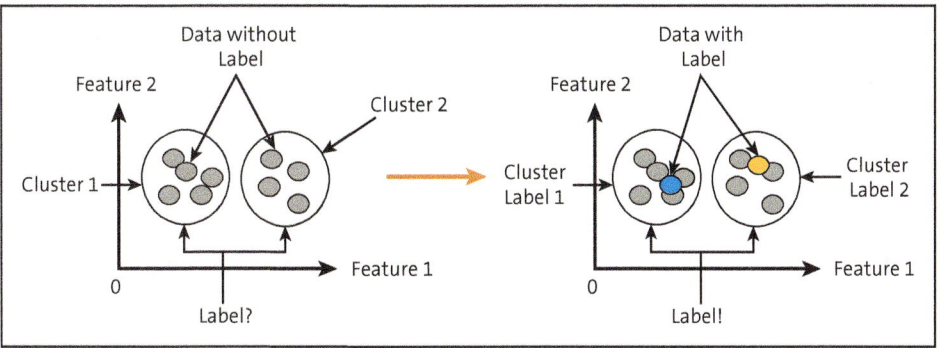

Figure 12.17 Semi-Supervised Learning

The approach just described is known as *semi-supervised classification* and has recently gained popularity as it allows unlabeled data, which is very easy to collect, to be put to good use. If you tried to label every date, this wouldn't be possible due to the time, complexity and cost involved. Here, semi-supervised learning helps, as it uses the large amount of unlabeled data and labeled data together to create better classifiers. In addition to semi-supervised classification, there is also *semi-supervised clustering* and *semi-supervised regression*.

Now that we know all about the learning strategies, we'll discuss a few more tools used to determine the classification quality or to enable comparisons of models.

12.2 Tools

How can the quality of ANNs be assessed? Or how can models be compared with each other? Answering this question will be our task in the final sections of this chapter.

12.2.1 Confusion Matrix

One tool that helps us evaluate the quality is the *confusion matrix*. It explains the classification behavior of a network. You can see an example of a confusion matrix in Table 12.3.

12 Learning Methods

		Actual Class		
		A	Not A	Total
Predicted Class	A	True Positive (TP)	False Positive (FP)	TP + FP
	Not A	False Negative (FN)	True Negative (TN)	FN + TN
	Total	TP + FN	FP + TN	

Table 12.3 Confusion Matrix

In the confusion matrix, you can see how well the neural network has learned to classify, for example, to name the stop traffic signs in a picture. In Table 12.3, this is referred to as "Class A". Depending on the prediction, there are different results; in the best case, the network can correctly classify a stop sign (true positive) or also correctly name a not-stop sign (true negative). However, it's also likely to happen that the network doesn't recognize stop signs as such (false negatives) or designates not-stop signs as stop signs (false positives).

Based on the individual evaluation fields, key figures can be derived that provide information about the classification quality. We'll now discuss three of these key figures in more detail and reuse two of them in the following. As a simple example with numbers plucked out of thin air and calculated results that are as nice as possible, we take the results from Table 12.4 for a stop sign classifier.

		Actual Class		
		A	Not A	Total
Predicted Class	A	9 (TP)	3 (FP)	12 (TP + FP)
	Not A	6 (FN)	9 (TN)	6 (FN + TN)
	Total	6 (TP + FN)	12 (FP + TN)	27

Table 12.4 Confusion Matrix for a Stop Sign Classifier

The *accuracy* as the first simple example of a key figure gives the proportion of events classified as correct (*TP*) in relation to the total number of events classified as class "A" (*TP* + *FP*), that is, as follows:

$$Accuracy = \frac{TP}{TP + FP}$$

This results in the *Accuracy* = $\frac{9}{12}$ = 0.75.

The *sensitivity* is the proportion of events (*TP*) classified as correct in relation to the total of all events to be classified as class "A" or actually belonging to class "A" (*TP* + *FN*):

$$Sensitivity = \frac{TP}{TP + FN}$$

This results in the *Sensitivity* = $\frac{9}{15}$ = 0.6.

The *specificity* is the proportion of correctly classified not-stop signs (*TN*) in relation to the total of all events classified as class "not A" (*FP* + *TN*):

$$Specificity = \frac{TN}{FP + TN}$$

This then results in the *Specificity* = $\frac{9}{12}$ = 0.75.

With these figures as a basis, let's take a look at the next tool that is enjoying great popularity in the machine learning community: ROC curves.

12.2.2 Receiver Operating Characteristic Curves

The *receiver operating characteristic* (ROC) analysis evaluates the quality of the diagnosis of a (binary) classification model by graphically and numerically illustrating the relationship between the *sensitivity* and *specificity* for this model. Furthermore, the area under the ROC curve (*AUROC* = area under ROC curve) can be used to determine the advantage of using a particular classification model compared to another model.

The ROCs were developed in the United States in 1941 after the attack on Pearl Harbor. The original task was to find out why the receiver operators of the US radar were able to overlook the Japanese aircraft. The ROCs were then quickly introduced into other scientific disciplines and used accordingly, for example, in psychology or medicine. So, to build a ROC curve for a classification ANN, the following steps are involved:

1. We need an ANN that fulfills a classification task in the sense of the previous section (e.g., stop signs vs. not-stop signs).
2. This is followed by determining the sensitivity and specificity for different threshold values for the classification of the networks in the output layer (e.g., *softmax*) plus the calculation of *1-specificity* (i.e., the false positive rate). The latter is the proportion of false positives classified as "not A"—that is, images classified as stop sign—in the set of all images not classified as stop sign.
3. The ROC curve can be drawn using these x/y value pairs (1-specificity, sensitivity).

A small numerical example will help demonstrate these steps in practice. We start with the values in Table 12.5, in which we've compiled fictitious values for 1-specificity and sensitivity.

Threshold Value	1-Specificity	Sensitivity
0.00	0.00	0.00
0.10	0.00	0.13
0.20	0.01	0.37
...	...	...
0.50	0.25	0.60
...		
1.00	1.00	1.00

Table 12.5 Specificity and Sensitivity for Determining the ROC Curve

The *threshold value* refers to the value in the output layer of an ANN. Remember that a feed-forward network is trained to categorize inputs into class "A" or "B". If the result at the output is 0, then the input is assigned to class "A"; and if the result at the output is 1, then the input is assigned to class "B". In real life, however, there will be no 0 or 1 at the output, but a number in the interval [0,1]. This is why you're now faced with the difficult task of determining where to draw the line. This means that you must set the threshold value between class "A" and class "B". Depending on the threshold value selected, the quality of the classification will change.

Let's continue our example and draw the ROC curve using the specificity and sensitivity values determined from Table 12.5.

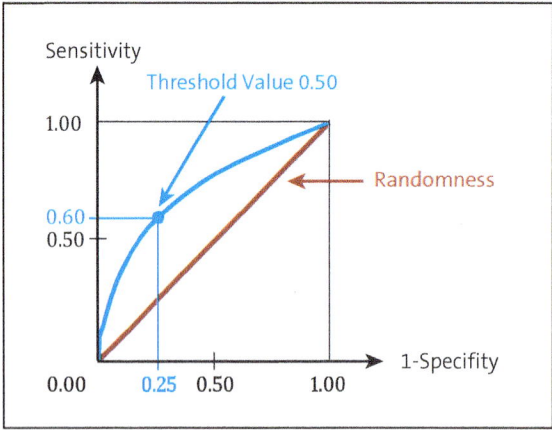

Figure 12.18 ROC Curve

In Figure 12.18, we've taken the points from the table and created the blue curve from them. We've highlighted the point for the 0.50 threshold so that you can see the application of the points more clearly. In more general terms, this means that the closer a point is to the top-left corner, that is, (0.0, 1.0), the more suitable the threshold value is for classification. If the point at the top-left corner were actually reached, the classes "A" and "not-A" could be distinguished with 100% sensitivity and specificity—a value that is hardly achievable in reality.

If the measurements belonging to class "A" and "not A" overlap completely, there will be no ROC curve, but an ROC line, which we've drawn in red in Figure 12.18. In this case, complete coincidence is the order of the day. That would be like flipping a coin to make a decision.

There is one more aspect of the ROC curves we want to explain: if you want to compare two classification models with each other, you can use the ROC curve or the AUROC. For example, two ANN classification models differ in the different number of hidden units (see Figure 12.19).

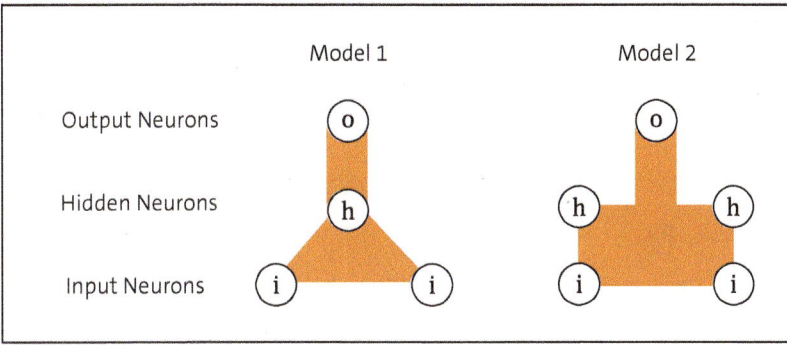

Figure 12.19 Model Comparison

By varying the threshold values for the output unit for model 1 or model 2, two ROC curves can be created. These can look like Figure 12.20, for example.

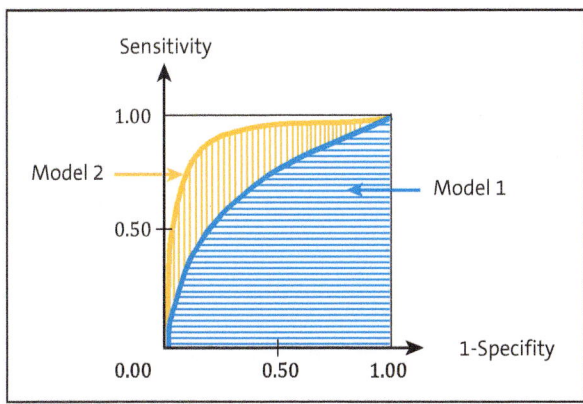

Figure 12.20 Two ROC Curves and Their AUROC

A simple statement for comparing the models is that the further the curve extends to the top left, the more efficient the model. In our case, model 2 is definitely better than model 1.

12.3 Summary

That's it for the additional learning considerations. We've discussed the main categories of learning and implemented two algorithms for unsupervised and reinforcement learning. The implementations have shown possible applications of networks in Q-learning on the one hand and completely new learning strategies in self-organizing maps (SOMs) on the other.

The topic of SOMs in particular has shown that it's very useful to know the development history of ANNs, as even the supposedly old approaches still have very useful aspects for current tasks. Now that we're talking about areas of application, you should go straight on to the next chapter, in which we'll introduce the practical side and areas of application of ANNs.

12.4 Further Reading

- Adventures in Machine Learning (*http://adventuresinmachinelearning.com/reinforcement-learning-tutorial-python-keras*)
- Farama Foundation Gymnasium (*https://github.com/Farama-Foundation/Gymnasium*)
- "Machine Learning for Humans, Part 5: Reinforcement Learning" by Vishal Maini (2017, *https://medium.com/machine-learning-for-humans/reinforcement-learning-6eacf258b265*)
- OpenAI Gym Beta (*https://openai.com/index/openai-gym-beta*)
- Reinforcement learning (*https://en.wikipedia.org/wiki/Reinforcement_learning*)
- "Reinforcement Learning" by Chris Watkins (n.d., *www.cs.rhul.ac.uk/~chrisw*)

Chapter 13
Areas of Application and Real-Life Examples

Application of neural networks in the practical world

With deep neural networks, we've developed a tool that can be used in many different ways. In this chapter, we want to show you a few examples that impressively demonstrate how powerful neural networks can be. We start with a simple regression task.

Then, we describe examples of image classification with a custom convolutional neural network (CNN) and a pretrained CNN. Finally, we let neural networks dream and make an already trained network available as a product via a web application.

In the course of the chapter, we'll also use functions from Python libraries that haven't yet been used. The installation of libraries is described in Chapter 2. In addition, the Jupyter Notebooks in the download area already contain cells with the installation commands.

13.1 Warm-Up

Let's get started with a very simple example to remind ourselves of the essential elements of classification and regression. We use the US Census Service's California Housing dataset, which is a few years old but will suffice for our warm-up. The exact description can be found at *www.kaggle.com/datasets/camnugent/california-housing-prices*.

The task is to predict the value of a building based on data such as crime rate, average number of rooms per building, tax rate, student/teacher ratio, age, and nitrogen oxide concentration. We'll solve this task using a network that isn't too deep, as shown in Listing 13.1.

```
import os
from time import time
from sklearn.datasets import fetch_california_housing
import tensorflow
from tensorflow.keras import models
from tensorflow.keras import layers
```

Listing 13.1 Importing the Major Libraries

We use the `keras` library integrated in TensorFlow, which makes life easier for us with neural networks. It's also convenient that `keras`, like many other machine learning (ML) libraries, provides us with the dataset at the same time.

Then, we split our dataset into a training dataset and a test dataset and display some more information about this dataset, as shown in Listing 13.2.

```
data = fetch_california_housing()
X, y = data.data, data.target
print("Number of input features:", X.shape[1])
print("The first 5 house prices:",y[0:5]*10000)
print("Median house price in dollars:",y.mean()*10000)
# Output:
Number of input features: 8
The first 5 house prices: [45260. 35850. 35210. 34130. 34220.]
Median house price in dollars: 20685.58169089147
```

Listing 13.2 Loading the California Housing Dataset

In Listing 13.2, you can see that we have 8 input features as expected. On average, the median house prices to be estimated with the neural network are around $20,685.

We define our not very deep neural network in Listing 13.3. Because it's so important, we repeat at this point that only the training dataset may be used for training. The test dataset is only there to evaluate our trained network!

```
# Split data into training and test dataset
X_train, X_test, y_train, y_test = train_test_split(X, y, test_size=0.2,
random_state=42)

model = models.Sequential()
model.add(layers.Dense(32, activation='relu',
                       input_shape=(X_train.shape[1],)))

model.add(layers.Dense(64, activation='relu'))
model.add(layers.Dense(1))
model.compile(optimizer='rmsprop', loss='mse', metrics=['mae'])
```

Listing 13.3 Creating the Neural Network

We've selected the *RMSprop* method as the optimization method. This method isn't dissimilar to the Adam method from Chapter 7. It takes into account past gradients from the optimization process. We select the *mean squared error* (mse) as the `loss` function. Furthermore, we choose the *mean absolute error* (mae) as a measure of accuracy.

As a result, we have a neural network with 13 input features. The first layer consists of 32 neurons, the second layer of 64 neurons, and exactly 1 output neuron provides us with the estimated median house value (see Figure 13.1).

```
Layer (type)                 Output Shape              Param #
=================================================================
dense_3 (Dense)              (None, 32)                448
_____
dense_4 (Dense)              (None, 64)                2112
_____
dense_5 (Dense)              (None, 1)                 65
=================================================================
Total params: 2,625
Trainable params: 2,625
Non-trainable params: 0
```

Figure 13.1 Network Model for the California Housing Dataset

It's always advisable to save the network model (model.save(), see Listing 13.4) so that it can be compared with other models later. Using model.summary() we can take a closer look at the structure.

```
model.summary()
model.save('california_housing.keras')
```

Listing 13.4 Storage and Output of the Network Model

How can we interpret the result from Listing 13.5? First, we must point out that you'll most likely get similar, but not exactly the same results. The MSE is approximately 27.6, and the MAE is approximately 3.4. If we multiply the MAE by 1,000, we get the error as a dollar amount. So we're off by about 3,400 dollars, based on the absolute error. With a median house price of around 23,000 dollars, this isn't a bad result.

```
model.fit(X_train, y_train, epochs=500, batch_size=16, verbose=0)

test_mse_score, test_mae_score = model.evaluate(X_test, y_test)
print(test_mse_score,test_mae_score)
# Output:
27.615190393784466 3.3850744
```

Listing 13.5 Training and Evaluation of the Neural Network

> **Task**
>
> Experiment with additional layers with different numbers of neurons. To do this, you must add the model.add(layers.Dense(#NumberNeurons, activation='relu')) function to Listing 13.3. Try to achieve a better result.

13.2 Image Classification

Before we turn to the example of image classification, let's first establish some order in the terminology: Are we talking about image classification, object recognition, object identification, or even object segmentation?

These terms aren't used very consistently in the literature, which is why we're now defining an interpretation for ourselves. Because we're in the field of image analysis, the best way to explain this is with images.

13.2.1 Definitions

In *image classification*, an image is assigned to a class. The class typically denotes the (individual) object that is displayed in the image, that is, the "Scissors" class in Figure 13.2.

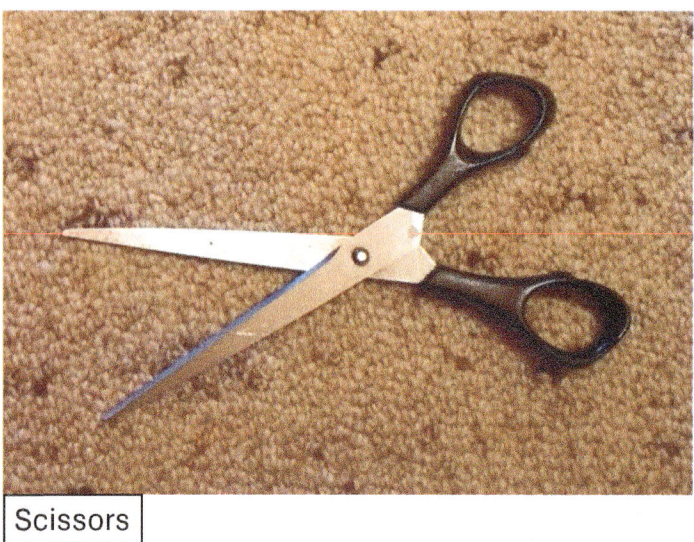

Figure 13.2 Example of Image Classification

Object recognition is about recognizing or detecting one or more objects. This often involves determining an axis-parallel *bounding box* and thus the size and position of the object in the image. That is useful if you want to detect objects in a video to be able to react to the movement or size of the object.

Objects are detected *and* classified for *object identification* (see Figure 13.3). In most cases, not only the class itself is specified but also the classification accuracy.

Object segmentation goes one step further than object identification and determines the pixels belonging to the object. In Figure 13.4, the segments are displayed in color. This allows you to separate the object relatively sharply from the background and other objects.

13.2 Image Classification

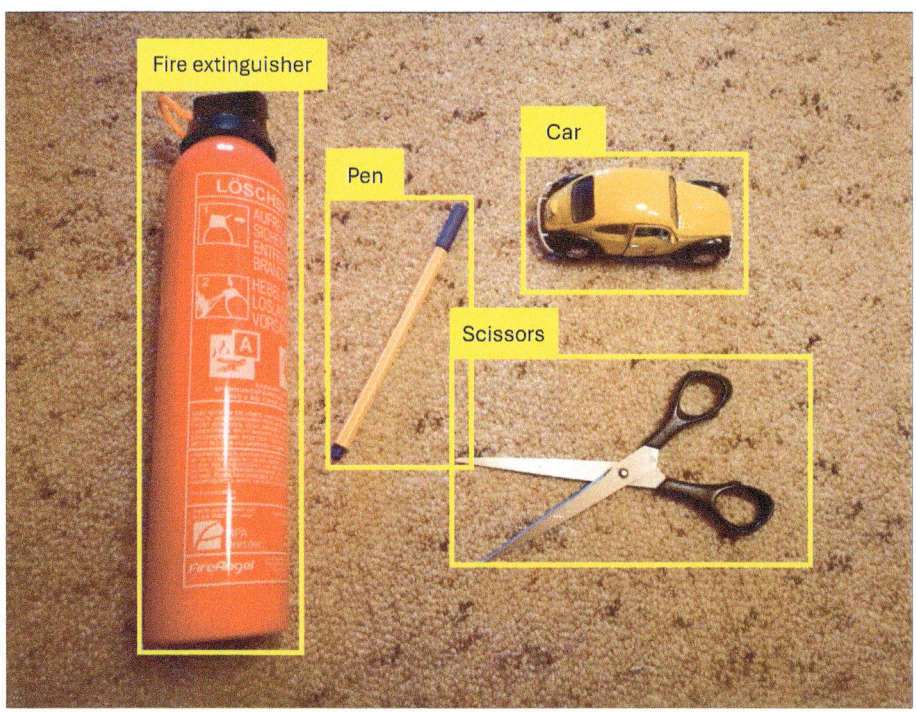

Figure 13.3 Example of Object Identification

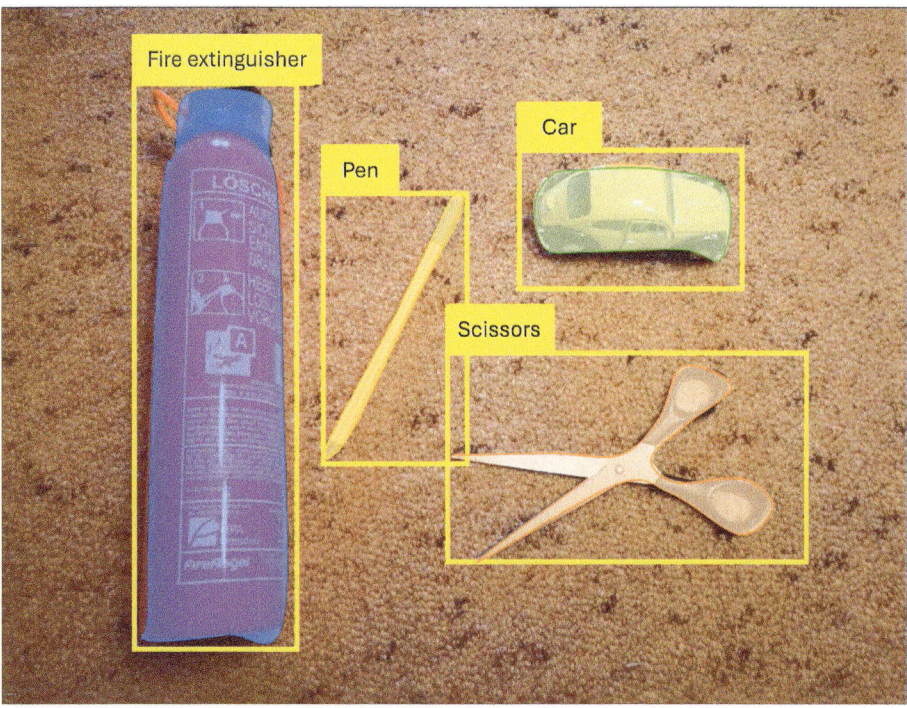

Figure 13.4 Example of Object Segmentation

13.2.2 On Bees and Bumblebees

We started the book with bees and end it in the last chapter with these interesting and useful insects. We'll try to train a network that can distinguish images of bees and bumblebees.

The first step consists of getting the training data. We're taking longer to show that the tedious part takes place before the actual training. As we mentioned in Chapter 11, the preparation and extraction of data involves a lot of work. But enough whining, let's get started!

Data Preparation

We use a Naive Bees dataset that comes from an ML competition by Driven Data Labs. This dataset consists of approximately 4,000 images of bumblebees and bees. About a quarter of the pictures are bee pictures, while the majority are bumblebee pictures. We're therefore dealing with a binary classification with an unbalanced dataset. At first glance, the number of images seems large, but there aren't too many for a CNN. However, we also know how to help ourselves with data augmentation.

Implementation

As usual, we start by importing the required libraries, as shown in Listing 13.6.

```
import pandas as pd
import urllib
import os
import PIL
import shutil
import numpy as np
%matplotlib inline
```

Listing 13.6 Libraries for Data Preparation

We use `pandas` to load a *.csv* file into a `DataFrame` and to edit the `DataFrame` object. We don't need the `PIL` library directly, but functions from `keras` for image processing use elements from it. Here, it serves as an indication that this library must be installed in advance, as usual.

`urllib` is a program library that provides us with functions for downloading data. `shutil`, on the other hand, is needed to copy the downloaded files into a specific file structure.

Let's now turn to the base directory and the file structure, as shown in Listing 13.7.

```
basedir = r'bee_bumblebee'
dir_dict = {
    'download': os.path.join(basedir,'download'),
    'train':    os.path.join(basedir,'train'),
    'test':     os.path.join(basedir,'test'),
    'valid':    os.path.join(basedir,'valid'),
    'train_bee':    os.path.join(basedir,'train','bee'),
    'train_bumble': os.path.join(basedir,'train','bumblebee'),
    'test_bee':     os.path.join(basedir,'test','bee'),
    'test_bumble':  os.path.join(basedir,'test','bumblebee'),
    'valid_bee':    os.path.join(basedir,'valid','bee'),
    'valid_bumble': os.path.join(basedir,'valid','bumblebee')}

# Create the directories (only if they do not yet exist)
for key in dir_dict:
    if os.path.exists()=False: os.makedirs(dir_dict[key])
```

Listing 13.7 Base Directory and File Structure as Dictionary

The base directory basedir is specified as bee_bumblebee. You must of course create your own directory and assign the basedir variable accordingly. The preceding r before 'bee_bumblebee' isn't a typo, but means for Python to take the string as it is, that is, not to interpret anything within the string. This is particularly important if slashes, especially backslashes, occur within the string. These could otherwise be interpreted as the start of an escape sequence.

> **Escape Sequences**
>
> An *escape sequence* is a combination of characters that doesn't represent text, but a type of function, for example, \n for a line feed or \t for a tab character.

Let's take a closer look at the necessary directory structure created by Listing 13.7 in Figure 13.5.

Figure 13.5 Directory Structure for the Naive Bees Dataset

We see a *download* directory and a separate directory for each of the training, test, and validation datasets. Except for the *download* directory, separate subdirectories named *bee* and *bumblebee* are created for each class.

Let's first download the images and the *train_labels.csv* file from a Git repository, as shown in Listing 13.8.

GitHub repository

GitHub is a very popular, freely available version control system (*https://github.com*). You can use it to manage data and its versioning. It's primarily used for joint program development.

In simple terms, a *repository* is a directory in which program-specific data such as program code, documentation, and other information is stored.

```
url = 'https://raw.githubusercontent.com/dionhagan/naive-bees/master/train_
labels.csv'
urlretrieve(url, os.path.join(dir_dict['download'],'train_labels.csv'))
```

Listing 13.8 Downloading the train_labels.csv File

Using `urlretrieve()`, the file is downloaded via the web address and copied into our *download* directory. If you think this takes too long, you can also get the data from the download area for this chapter.

Let's take a look at what the file looks like, as shown in Listing 13.9.

```
df = pd.read_csv(os.path.join(dir_dict['download'],'train_labels.csv'))
df.info()
print(df.head(5))
# Output:
<class 'pandas.core.frame.DataFrame'>
RangeIndex: 3969 entries, 0 to 3968
Data columns (total 2 columns):
id       3969 non-null int64
genus    3969 non-null float64
dtypes: float64(1), int64(1)
memory usage: 62.1KB
      id  genus
0    520    1.0
1   3800    1.0
2   3289    1.0
3   2695    1.0
4   4922    1.0
```

Listing 13.9 Transferring the File to a DataFrame

13.2 Image Classification

We can use the `df.info()` and `df.head()` functions to view the properties and content of the `DataFrame` element. It consists of 3,969 entries and an `id` column, which is also a component of the filename of the associated image, as well as a `genus` column, which tells us the class. The bumblebee images have the value 1, while the bee images have the value 0.

We modify this `DataFrame` to make copying into our directory structure a little easier. The changes in Listing 13.10 will sort by the `id` and add another column that shows us the class name.

```
df = df.sort_values(['id'])
df['class'] = np.where(df['genus']==1,'_bumble','_bee')
print(df.head(5))
# Output:
      id  genus   class
2681   1    1.0  _bumble
1056   2    1.0  _bumble
2745   3    1.0  _bumble
2234   4    0.0     _bee
2645   5    0.0     _bee
```

Listing 13.10 Modification of the DataFrame

If you looked carefully, you'll also notice that we're using `np.where()`, a function of the `numpy` library, to fill the `class` column depending on the value of the `genus` column.

Now, we get the 4,000 or so images and save them in the *download* directory (see Listing 13.11). Be patient; this process will take a few minutes. Incidentally, the filename of an image consists of the ID and the *jpg* file extension, that is, *4.jpg* for the image with ID 4 (incidentally a bee image, as shown earlier in Listing 13.10).

```
for index, row in df.iterrows():
    fname = "{:.0f}.jpg".format(row['id'])
    url = 'https://raw.githubusercontent.com/dionhagan/naive-bees/master/images/train/'+fname
    print(url)
    urllib.request.urlretrieve(url, os.path.join(dir_dict['download'],fname))
```

Listing 13.11 Downloading the Images of the Naive Bees Dataset

To check this, we have an image displayed, as shown in Listing 13.12 and Figure 13.6.

```
from IPython.display import Image
Image(os.path.join(dir_dict['download'],'8.jpg'))
```

Listing 13.12 Code for Displaying the 8.jpg Image

Figure 13.6 Image Number 8 (Bee or Bumblebee? Blurring Adds to the Difficulty in Identification)

Incidentally, the images have a format of 200×200 pixels, which is important for the definition of our network, as we have to specify the input size.

Now we distribute the images from the *download* directory to our directory structure according to Figure 13.5. We use the first 3,000 images as training data and the next 500 as test data. The rest, that is, fewer than 500 images, serve as a validation dataset, as shown in Listing 13.13.

```
for index, row in df.iterrows():
    fname = "{:.0f}.jpg".format(row['id'])

    classname = row['class']
    if row['id']<3000:
        dest_dir = os.path.join(dir_dict['train'+classname])
    elif row['id']<3500:
        dest_dir = os.path.join(dir_dict['test'+classname])
    else:
        dest_dir = os.path.join(dir_dict['valid'+classname])

    src = os.path.join(dir_dict['download'],fname)
    dest = os.path.join(dest_dir,fname)
    shutil.copyfile(src, dest)
```

Listing 13.13 Distribution of Image Data to the Network Structure

Now we get to defining our own CNN, as shown in Listing 13.14.

```
from tensorflow.keras import layers
from tensorflow.keras import models
from tensorflow.keras import optimizers
```

```python
model = models.Sequential()
model.add(layers.Conv2D(32, (3, 3), activation='relu',
                        input_shape=(200, 200, 3)))
model.add(layers.MaxPooling2D((2, 2)))
model.add(layers.Conv2D(64, (3, 3), activation='relu'))
model.add(layers.MaxPooling2D((2, 2)))
model.add(layers.Conv2D(128, (3, 3), activation='relu'))
model.add(layers.MaxPooling2D((2, 2)))
model.add(layers.Conv2D(128, (3, 3), activation='relu'))
model.add(layers.MaxPooling2D((2, 2)))
model.add(layers.Flatten())
model.add(layers.Dense(512, activation='relu'))
model.add(layers.Dense(1, activation='sigmoid'))

model.compile(loss='binary_crossentropy',
              optimizer=optimizers.RMSprop(lr=1e-4),
              metrics=['acc'])

model.summary()
```

Listing 13.14 Code for the Model Structure of the CNN

At the first level, we specify the image data format. We have color images with the format of 200×200 pixels. An image consists of three color channels—red, green, and blue—which together form a color image. This is why the input to the network ultimately reads (200, 200, 3). The coding block of this CNN goes up to the `Flatten()` layer and contains the typical sequence of convolutional blocks with a convolution layer (here, `Conv2D()`) and pooling layers (here, `MaxPooling()`). The prediction block consists of a fully connected neural network with a hidden layer containing 512 neurons (see Figure 13.7).

With the few lines from Listing 13.14, we've created a model with a total of almost 7 million parameters, which is pretty remarkable. This network has exactly one output neuron that contains the class. We use `model.compile()` to define further important parameters for the training. The `loss` function is `binary_crossentropy`, which is often used for binary classification tasks, that is, tasks with only two classes.

We chose RMSprop as the optimization method because it's particularly suitable for neural networks. It was developed and published by Geoffrey Hinton. The learning rate `lr` is set relatively low here at 0.0001.

Now, we're almost ready for the training phase; we just have to tell the model how to get the images. Moreover, we face two challenges: the training dataset isn't particularly large, and we have an unbalanced dataset as there are much fewer bee images than bumblebee images (with a ratio of 1:4).

```
Layer (type)                 Output Shape              Param #
=================================================================
conv2d (Conv2D)              (None, 198, 198, 32)      896
_____
max_pooling2d (MaxPooling2D) (None, 99, 99, 32)        0
_____
conv2d_1 (Conv2D)            (None, 97, 97, 64)        18496
_____
max_pooling2d_1 (MaxPooling2 (None, 48, 48, 64)        0
_____
conv2d_2 (Conv2D)            (None, 46, 46, 128)       73856
_____
max_pooling2d_2 (MaxPooling2 (None, 23, 23, 128)       0
_____
conv2d_3 (Conv2D)            (None, 21, 21, 128)       147584
_____
max_pooling2d_3 (MaxPooling2 (None, 10, 10, 128)       0
_____
flatten (Flatten)            (None, 12800)             0
_____
dense (Dense)                (None, 512)               6554112
_____
dense_1 (Dense)              (None, 1)                 513
=================================================================
Total params: 6,795,457
Trainable params: 6,795,457
Non-trainable params: 0
```

Figure 13.7 Layers of the Model

Fortunately, Keras again helps us immensely with the `ImageDataGenerator()` function (see Listing 13.15) you already know from Chapter 11. This function allows you to simultaneously scale the pixel values from the numerical range 0–255 to the numerical range 0–1 and create additional images by rotating, moving, zooming, and shearing. The `flow_from_directory()` function allows us to establish the connection between the image files and the necessary input for the CNN. This function contains a `target_size` parameter, the effects of which are more powerful than you might think at first glance. In our example, all images have the same format—(200, 200, 3). If our images have different sizes, they will be scaled to the size specified by `target_size`.

```python
from tensorflow.keras.preprocessing.image import ImageDataGenerator

train_datagen = ImageDataGenerator(
    rescale=1./255,
    rotation_range=40,
    width_shift_range=0.2,
    height_shift_range=0.2,
    shear_range=0.2,
    zoom_range=0.2,
    horizontal_flip=True,)
```

13.2 Image Classification

```
test_datagen = ImageDataGenerator(rescale=1./255)

train_generator = train_datagen.flow_from_directory(
    dir_dict['train'],
    target_size=(200, 200),
    batch_size=20,
    class_mode='binary')

validation_generator = test_datagen.flow_from_directory(
    dir_dict['valid'],
    target_size=(200, 200),
    batch_size=20,
    class_mode='binary')
```

Listing 13.15 Generating Training, Test, and Validation Data from the Image Data

This is where our directory structure comes in handy, because Keras automatically assigns the images to the respective class, depending on whether an image is in the *bee* or *bumblebee* directory, as shown in Listing 13.16.

```
from sklearn.utils import class_weight
import numpy as np

class_weights = class_weight.compute_class_weight(
            'balanced',
            np.unique(train_generator.classes),
            train_generator.classes)

print(class_weights)
# Output:
[2.39547038 0.63189338]
```

Listing 13.16 Determination of Class Weights for the Unbalanced Dataset

These class weights are used in the `loss` function in such a way that in the training phase misclassifications of the underrepresented class (in our case, the bees) contribute more to the error value, that is, are penalized. That is why it's also referred to as a *penalty term*.

Now we're ready to train the network, as shown in Listing 13.17.

```
history = model.fit_generator(
    train_generator,
    steps_per_epoch=100,
    epochs=30,
    validation_data=validation_generator,
```

```
        validation_steps=50,
        class_weight = class_weights)
# Output:
Epoch 1/30
100/100 [==============================] - 383s 4s/step - loss: 0.5191 - acc:
0.7921 - val_loss: 0.6164 - val_acc: 0.7660
Epoch 2/30
100/100 [==============================] - 364s 4s/step - loss: 0.4958 - acc:
0.7944 - val_loss: 0.5100 - val_acc: 0.7720
...
```

Listing 13.17 Starting the Training

Depending on the available computing capacity, training can take a few minutes (GPU machine) to a couple of hours (standard CPU computer).

Finally, we look at the development of the accuracy or error (loss) over the course of the training, which we create via Listing 13.18. This code uses the history variable from Listing 13.17, which stores training history information.

```
import matplotlib.pyplot as plt
acc = history.history['acc']
val_acc = history.history['val_acc']
loss = history.history['loss']
val_loss = history.history['val_loss']

epochs = range(len(acc))

plt.plot(epochs, acc, 'bo', label='Training accuracy')
plt.plot(epochs, val_acc, 'b', label='Validation accuracy')
plt.title('Training and validation accuracy')
plt.legend()

plt.figure()

plt.plot(epochs, loss, 'bo', label='Training loss')
plt.plot(epochs, val_loss, 'b', label='Validation loss')
plt.title('Training and validation loss')
plt.legend()

plt.show()
```

Listing 13.18 Creation of the Progression Graphs

You can see the two progression curves in Figure 13.8 and Figure 13.9.

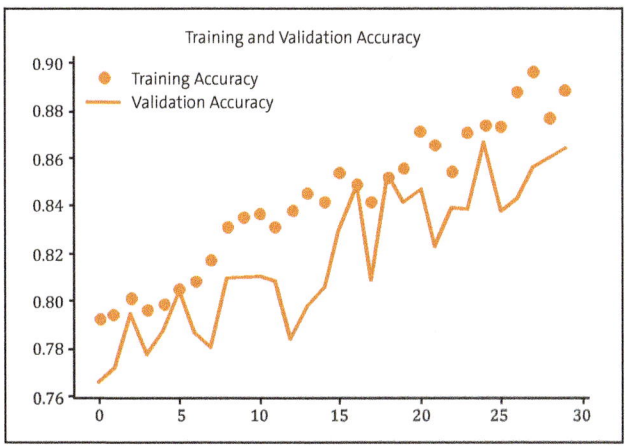

Figure 13.8 Progression Curve for Training and Validation Accuracy

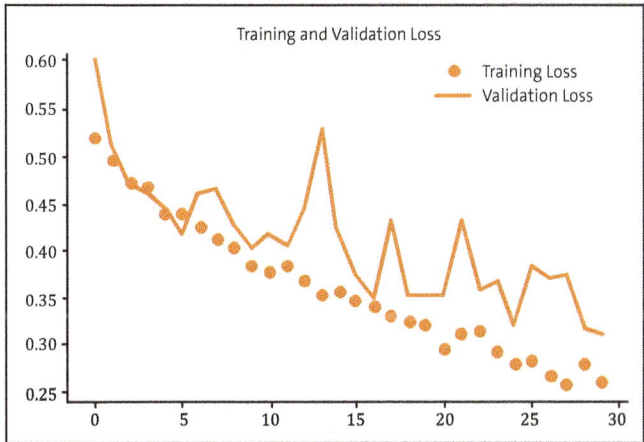

Figure 13.9 Progression Curve for the Training and Validation Loss

Thus, we achieve an accuracy of approximately 86%, which isn't a bad result for our relatively simple CNN without much fine-tuning. Typically, results from approximately 95% onward should be the goal. Accuracy values of up to 99% were achieved in a competition (*https://github.com/drivendataorg/naive-bees-classifier*), although this involved working with combinations of pretrained networks and a great deal of fine-tuning of parameters.

> **Task**
>
> Experiment with this dataset by changing the model from Listing 13.14, shown earlier. You can add a convolution block or a dense layer in the prediction block of the model or use other optimization methods. Can you achieve more than 90% accuracy?

13.2.3 Pretrained Networks

In Chapter 8, you've already learned how to use pretrained networks by classifying arbitrary images. There, we got by without any training. Now, we want to use pretrained networks and the already trained coding block for our new images. We're only training a new prediction block that we create ourselves. In addition, you already know that the use of pretrained deep neural networks is referred to as *transfer learning*.

For this example, we use a popular image dataset with cats and dogs. The task is similar to the previous example—here, instead of bees and bumblebees, the task is to distinguish between images of cats and dogs.

Data preparation

We download the dataset in the conventional way, that is, with a simple download, via the following link. You can find the dataset on the Kaggle platform at *www.kaggle.com/chetankv/dogs-cats-images*. You'll receive a compressed file in ZIP format which, after decompression, conveniently provides the data in the directory structure required for the Keras `ImageDataGenerator()`.

Kaggle

Kaggle is an online community for data scientists and ML enthusiasts. It allows users to find datasets for ML tasks and take part in a competition. It's also possible to upload datasets yourself to find suitable solutions. Many of the competitions are also endowed with prizes. Kaggle was acquired by Google in 2017.

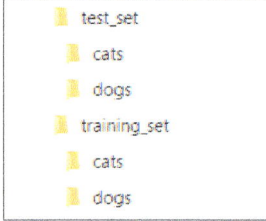

Figure 13.10 The Directory Structure of the Cats and Dogs Dataset

The training dataset contains 4,000 images of cats and 4,000 images of dogs, and the test dataset contains 1,000 cat and 1,000 dog images. This means we're dealing with a balanced dataset. However, the amount of training data is relatively small for a deep learning task. But pretrained CNNs in which the coding block has already been trained to recognize patterns are ideal for precisely this case.

Implementation

As usual, we start by importing the required libraries, as shown in Listing 13.19.

```python
import pandas as pd
import os
import PIL
import shutil
import numpy as np
```

Listing 13.19 Importing the Libraries

Similar to Listing 13.7, we again create a dictionary that stores the base directory and its subdirectories, as shown in Listing 13.20.

```python
basedir = r'cat_dog'
dir_dict = {
    'train': os.path.join(basedir,'training_set'),
    'test': os.path.join(basedir,'test_set'),
    'train_cats': os.path.join(basedir,'training_set','cats'),
    'train_dogs': os.path.join(basedir,'training_set','dogs'),
    'test_cats': os.path.join(basedir,'test_set','cats'),
    'test_dogs': os.path.join(basedir,'test_set','dogs')
        }

# Create the directories (only if they do not yet exist)
for key in dir_dict:
    if not os.path.exists(dir_dict[key]):
        os.makedirs(dir_dict[key])
```

Listing 13.20 Dictionary for our Directories

Once again, we check whether there are really 4,000 images per class for the training dataset and 1,000 images for the test dataset by outputting the number of files in the respective directory, as shown in Listing 13.21.

```python
print("Number of cat images (training):",len(os.listdir(dir_dict['train_cats'])))
print("Number of dog images (training):",len(os.listdir(dir_dict['train_dogs'])))
print("Number of cat images (test):",len(os.listdir(dir_dict['test_cats'])))
print("Number of dog images (test):",len(os.listdir(dir_dict['test_dogs'])))
# Output:
Number of cat images (training): 4000
Number of dog images (training): 4000
Number of cat images (test): 1000
Number of dog images (test): 1000
```

Listing 13.21 Checking the Number of Images

A small selection of the test images shows that these are quite complex. Figure 13.11 and Figure 13.12 show that the images can also contain writing, people, or other elements. This makes CNNs' ability to distinguish between such images all the more remarkable.

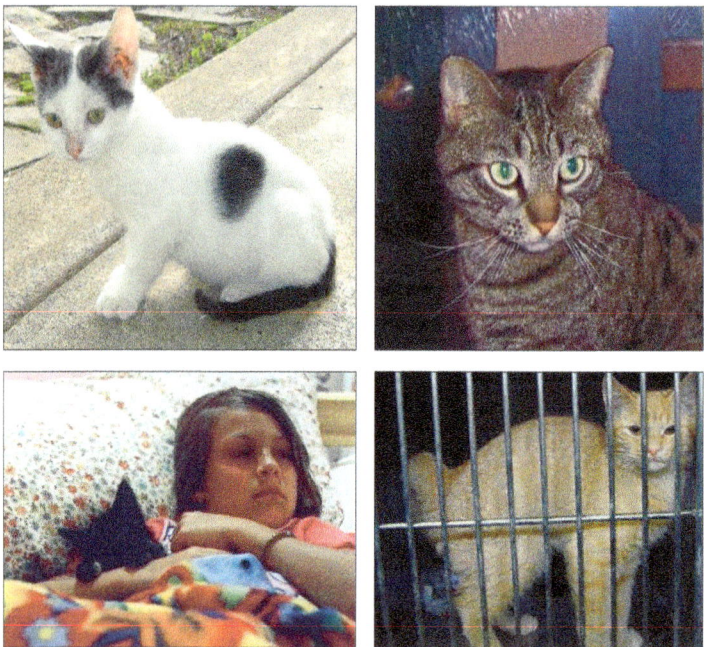

Figure 13.11 Extract from the Test Dataset for Cats

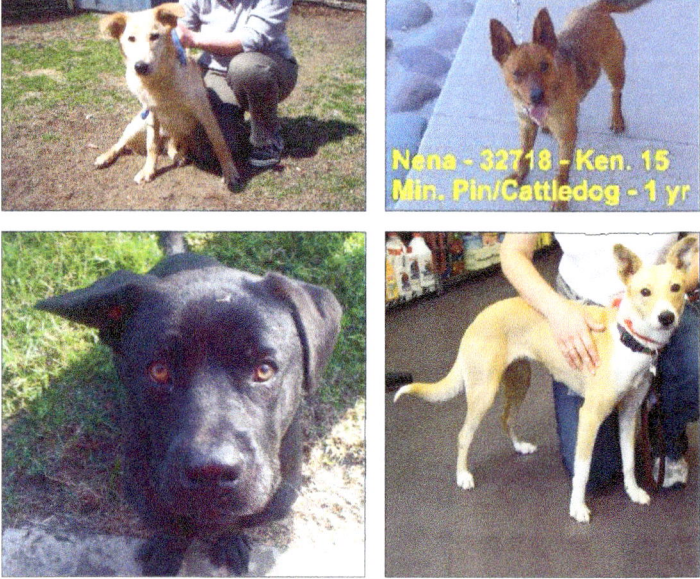

Figure 13.12 Extract from the Test Dataset for Dogs

Now we're able to put the pieces of our network together and start importing the `tensorflow.keras` libraries, as shown in Listing 13.22.

```
from tensorflow.keras import layers
from tensorflow.keras import models
from tensorflow.keras.applications.inception_v3 import InceptionV3
from tensorflow.keras.applications.inception_v3 import preprocess_input,
decode_predictions
from tensorflow.keras.preprocessing import image

model = models.Sequential()
base_model = InceptionV3(weights='imagenet',include_top=False,input_shape=(200,
200, 3))
model.add(base_model)
model.add(layers.Flatten())
model.add(layers.Dense(256, activation='relu'))
model.add(layers.Dense(512, activation='relu'))
model.add(layers.Dense(1, activation='sigmoid'))
```

Listing 13.22 Import and Setup of the Pretrained CNN

The pretrained *Inception-v3* network is already provided to us by `tensorflow.keras`. The `base_model` variable contains the structure of Inception-v3. Using the `weights='imagenet'` parameter, we not only transfer the structure but also the trained weights based on the `imagenet` dataset. As we only need the coding block, we pass the `include_top=False` parameters during the initialization of Inception-v3. This means we dispense with the prediction block of this CNN. That's good because we want to train our own prediction block based on our Dogs and Cats dataset.

Our network model starts with the `base_model`, and we add our fully connected neural network, which consists of two layers with 512 and 256 neurons, respectively.

When we train our network model, however, we no longer want to change the weights of our coding block. It has already been pretrained and has the ability to recognize corresponding patterns. Fortunately, fixing the weights for our `base_model`, Inception-v3, is quite simple, as you can see in Listing 13.23.

```
base_model.trainable = False
model.summary()
```

Listing 13.23 Fixing the Weights of the Coding Block and Outputting the Model

Figure 13.13 shows us that around two-thirds of the parameters can't be trained. This is a consequence of fixing the weights of our pretrained network. But we still have almost 9 million parameters left.

```
Layer (type)                    Output Shape              Param #
=================================================================
inception_v3 (Model)            (None, 4, 4, 2048)        21802784
_____
flatten (Flatten)               (None, 32768)             0
_____
dense (Dense)                   (None, 256)               8388864
_____
dense_1 (Dense)                 (None, 512)               131584
_____
dense_2 (Dense)                 (None, 1)                 513
=================================================================
Total params: 30,323,745
Trainable params: 8,520,961
Non-trainable params: 21,802,784
```

Figure 13.13 Output of model.summary() for Our Network

What we're still missing is the definition of the `loss` function, the optimization method, and the accuracy metric, as shown in Listing 13.24.

```
from tensorflow.keras import optimizers

model.compile(loss='binary_crossentropy',
              optimizer=optimizers.RMSprop(lr=1e-4),
              metrics=['acc'])
```

Listing 13.24 Definition of Loss Function and Optimization Method

Because we're again dealing with a binary classification problem, we use `binary_crossentropy` and `RMSprop` as optimization methods in Listing 13.24, as we did for the bees and bumblebees.

In Listing 13.25, we again use the `ImageDataGenerator()` from `tensorflow.keras` to prepare the data accordingly. Because, as already mentioned, we have relatively little training data, we generate additional training data by rotating, moving, and turning the images. Again, the transformation to a pixel value range from [0,255] to [0,1] and the required input size of the image of (200,200,3) are crucial.

```
from tensorflow.keras.preprocessing.image import ImageDataGenerator

train_datagen = ImageDataGenerator(
    rescale=1./255,
    rotation_range=40,
    width_shift_range=0.2,
    height_shift_range=0.2,
    shear_range=0.2,
    zoom_range=0.2,
    horizontal_flip=True,)
```

```
test_datagen = ImageDataGenerator(rescale=1./255)

train_generator = train_datagen.flow_from_directory(
    dir_dict['train'],
    target_size=(200, 200),
    batch_size=20,
    class_mode='binary')

validation_generator = test_datagen.flow_from_directory(
    dir_dict['test'],
    target_size=(200, 200),
    batch_size=20,
    class_mode='binary')
# Output:
Found 8000 images belonging to 2 classes.
Found 2000 images belonging to 2 classes.
```

Listing 13.25 Preparing the Image Data

We're now ready to send our network to training camp. Listing 13.26 will again require some patience if you're not using a GPU computer.

```
history = model.fit_generator(
    train_generator,
    steps_per_epoch=100,
    epochs=40,
    validation_data=validation_generator,
    validation_steps=50)
```

Listing 13.26 Training Our Model

The trained model can be saved again to compare different network structures, as shown in Listing 13.27.

```
model.save(os.path.join(basedir,'cats_dogs_1.h5'))
```

Listing 13.27 Saving the Trained Net

The training process is again displayed as graphs that show the accuracy of the training and validation dataset (see Figure 13.14) and the progression of the loss function (*Loss*, see Figure 13.15).

The pretrained network already gives us a very high validation accuracy of approximately 97% in the first epochs, while the training accuracy slowly but steadily increases to approximately 90%. This isn't unusual when using pretrained networks, as long as the trained images correspond approximately to the training dataset of the pretrained dataset. The ImageNet dataset also contains dogs and cats as a class and is therefore ideally suited.

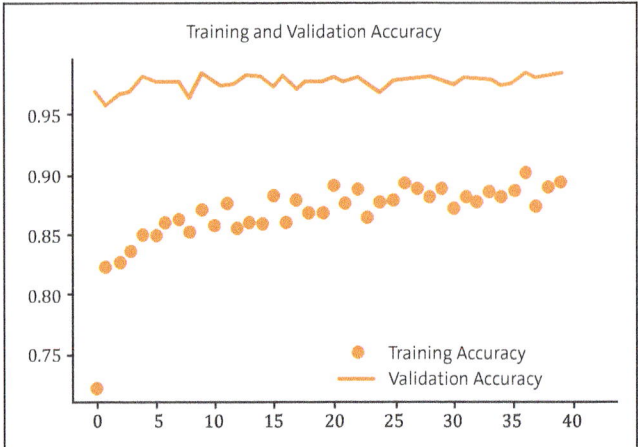

Figure 13.14 Accuracy Development of More Than 40 Training Epochs

The course of the loss (Loss) is similar to the development of the accuracy, as shown in Figure 13.15. These graphics were created with Listing 13.18, which you can use unchanged.

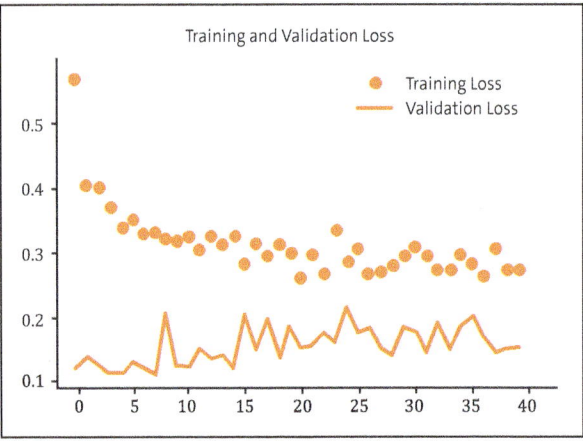

Figure 13.15 Loss Development of More Than 40 Training Epochs

This means that we already have a powerful network for classifying cat and dog images.

To achieve greater accuracy, further fine-tuning of the network is required, that is, a refinement of the hyperparameters. Unfortunately, there is no silver bullet for fine-tuning. Experience in dealing with different datasets and procedures plays a major role here, and you'll have to experiment with different network structures.

> **Task**
>
> Try using another pretrained network instead of *Inception-v3*, for example, *ResNet50*. Listing 13.22 must be changed so that ResNet50 gets imported:
>
> from tensorflow.keras.applications.resnet50 import ResNet50
>
> The initialization is the same except for the input dimension of the image (224,224,3). In addition, make sure that you import the correct image preparation function: preprocess_input. Of course, you can and should also experiment with the prediction block and change the hyperparameters here too.

13.3 Dreamed Images

In 2015, Google presented a method known as *DeepDream*. Alexander Mordvintsev had developed this method, and its publication initiated a veritable internet hype. Researchers in the field of deep learning wanted to know what their CNNs are dreaming about.

Figure 13.16 shows an example of a landscape in the fall on the left and the image on the right after applying a DeepDream algorithm, that is, the dreamed version.

Figure 13.16 Image of a Fall Landscape and the Dream Version (Source: Wikimedia Commons)

13.3.1 The Algorithm

How does this DeepDream algorithm work? Briefly summarized, it consists of three steps:

1. Start with an input image.
2. Process the input image with different resolutions (levels).
3. For each stage, maximize the activation of a set of layers, and mix the results.

Unlike other applications, we don't have to train a network or modify weights through training. On the contrary, we leave the weights as they are and modify the input screen instead. The modified input image is fed back to the CNN in a different resolution—so we run the CNN in reverse. More precisely, we create a *feedback loop*, in which the input image modified with each stage is fed to the CNN.

Feedback Loop

There is usually one input and one output in a process. The input is modified (e.g., sent through a neural network), and an output is obtained. If this process is run through again and the output of the previous stage is used as input for the next stage, a *feedback loop* is obtained, as outlined in Figure 13.17.

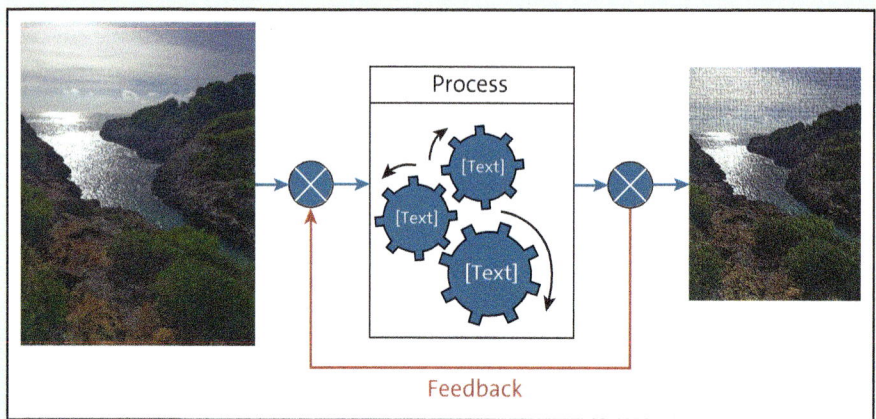

Figure 13.17 Diagram of a Feedback Loop

We're again using a pretrained network—Inception-v3. All of them are suitable for implementation, but of course they influence the results: CNNs trained on the ImageNet dataset contain more elements of dogs, cats, and birds. For this reason, the DeepDream images contain elements of these objects.

For the actual algorithm, we need to create an image pyramid by reducing the original image by a certain factor. In our sample program, we use factor 2. We don't do this indefinitely, but we choose a certain number of levels, usually 5 to 6, that the image gets reduced, as shown in Figure 13.18.

We start the algorithm with the smallest image and let it "dream" in the CNN by applying a gradient ascent method to the input for the optimization, which we still have to select. At the same time, we try to maximize the activation of entire layers. We choose a `Loss` function in which a selected number of layers make a certain contribution. The selection of weights controls the dreaming a little; the first layers lead to geometric patterns, and the later layers lead to representations in which complex objects (e.g., a dog's head, a bird) are depicted in the result.

Figure 13.18 Five-Level Image Pyramid

As shown in Figure 13.19, the result is scaled up in both dimensions (i.e., height and width) after a dream stage. As a result, we naturally lose a lot of details. But we can help ourselves by combining the dream result with the matching original image from the image pyramid we created at the start to obtain details. We then send this new image combination into the dream world again.

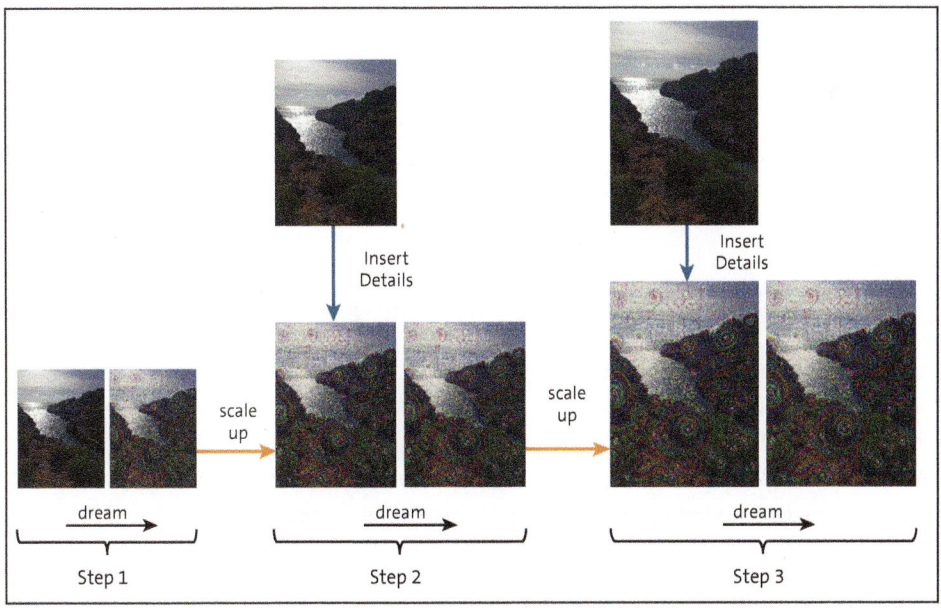

Figure 13.19 Visualization of the Stepwise DeepDream Algorithm

13.3.2 Implementation

First, as usual, we import all the required libraries from which we need functions or classes for programming, as shown in Listing 13.28.

```
import numpy as np

# IPython.display is needed to delete the output of a
# cell - very useful
from IPython.display import clear_output

import matplotlib as mpl
from matplotlib import pyplot as plt

from tensorflow.keras.preprocessing import image
import tensorflow as tf
```

Listing 13.28 Importing Libraries and Functions

We can use any pretrained CNN again, and this time we use Inception-v3.

Then, we define some useful functions that display images in the Jupyter Notebook, convert the data types of the image from float to integer and vice versa, and ensure the correct input format for our Inception-v3 network architecture (see Listing 13.29).

```
# Normalize the image
def deprocess(img):
    img = 255*(img + 1.0)/2.0
    return tf.cast(img, tf.uint8)

# Image display
def show(img):
    plt.figure(figsize=(12,12))
    plt.grid(False)
    plt.axis('off')
    plt.imshow(img)
    plt.show()
```

Listing 13.29 Useful Image Functions

The show_img() function is only used to display images in Jupyter Notebook and uses the matplotlib library. The deprocess() function transforms the pixel values of the resulting image into the range (0, 255) so that it can be output accordingly.

As further preparation, we load an image of a beach scene from Majorca, which is then "dreamed up" in the neural network (as if the beach itself wasn't dreamlike enough already), as shown in Listing 13.30.

13.3 Dreamed Images

```python
import os

basedir = 'deepdream'
original_image = os.path.join(basedir,'mallorca.jpg')
final_image = os.path.join(basedir,'mallorca_dream.jpg')

original_img = np.array(original_img)

show(original_img)
```
Listing 13.30 Loading the Original Image and Displaying the Loaded Image

You've already seen the beach scene from Majorca in Figure 13.18. You can also use your own images and adjust the `basedir` variable according to your own directory structure.

Now we're going to load the network we dream of. This network structure, in this case, Inception-v3, is assigned weights resulting from training with the ImageNet dataset, as shown in Listing 13.31.

```python
base_model = tf.keras.applications.InceptionV3(include_top=False, weights='imagenet')
base_model.summary()
```
Listing 13.31 Loading the InceptionV3 Model

Figure 13.20 shows an excerpt from its structure.

There are several ways to control dreaming. In Listing 13.32, we define the layers we use for dreaming and their weighting. The higher the weight, the more the layer contributes to learning.

```python
# The aim is to maximize activation in these layers
names = ['mixed3', 'mixed5']
layers = [base_model.get_layer(name).output for name in names]

# Our dream model is based on Inception_v3
dream_model = tf.keras.Model(inputs=base_model.input, outputs=layers)
```
Listing 13.32 Layers That Are Used for Dreaming

The name of the layer must be taken directly from the structure. In Figure 13.20, you can see the name of the network layer in the left-hand column. If you use a different model for dreaming, you must adopt the corresponding designation.

```
Layer (type)                    Output Shape              Param #
=================================================================
input_3 (InputLayer)            (None, None, None, 3)     0
block1_conv1 (Conv2D)           (None, None, None, 64)    1792
block1_conv2 (Conv2D)           (None, None, None, 64)    36928
block1_pool (MaxPooling2D)      (None, None, None, 64)    0
block2_conv1 (Conv2D)           (None, None, None, 128)   73856
block2_conv2 (Conv2D)           (None, None, None, 128)   147584
block2_pool (MaxPooling2D)      (None, None, None, 128)   0
block3_conv1 (Conv2D)           (None, None, None, 256)   295168
block3_conv2 (Conv2D)           (None, None, None, 256)   590080
block3_conv3 (Conv2D)           (None, None, None, 256)   590080
block3_conv4 (Conv2D)           (None, None, None, 256)   590080
block3_pool (MaxPooling2D)      (None, None, None, 256)   0
block4_conv1 (Conv2D)           (None, None, None, 512)   1180160
block4_conv2 (Conv2D)           (None, None, None, 512)   2359808
block4_conv3 (Conv2D)           (None, None, None, 512)   2359808
block4_conv4 (Conv2D)           (None, None, None, 512)   2359808
block4_pool (MaxPooling2D)      (None, None, None, 512)   0
block5_conv1 (Conv2D)           (None, None, None, 512)   2359808
block5_conv2 (Conv2D)           (None, None, None, 512)   2359808
block5_conv3 (Conv2D)           (None, None, None, 512)   2359808
block5_conv4 (Conv2D)           (None, None, None, 512)   2359808
block5_pool (MaxPooling2D)      (None, None, None, 512)   0
=================================================================
Total params: 20,024,384
Trainable params: 20,024,384
Non-trainable params: 0
```

Figure 13.20 A Section of the Network Structure of Inception-v3 (Output of model.summary())

Then, we define functions that allow us to get the loss and gradient value of our image (see Listing 13.33). Remember, the aim is to maximize the activation of entire layers.

```python
def calculate_loss(img, model):
    # Forward pass of the image through Inception_v3 to obtain the
    # activation
    img_batch = tf.expand_dims(img, axis=0)
    layer_activations = model(img_batch)

    losses = []
    for act in layer_activations:
```

```
    loss = tf.math.reduce_mean(act)
    losses.append(loss)

  return tf.reduce_sum(losses)
```

Listing 13.33 Loss and Gradient Function for the Optimization Process

For each optimization task, we have to calculate the `loss` function, which we've done in Listing 13.33. Note that we're using the layers we defined in Listing 13.32.

This leaves us to define a function that performs the optimization itself, using `tf.GradientTape()`—a new feature in TensorFlow 2 that allows us to easily calculate the gradients, as shown in Listing 13.34.

```
@tf.function
def deepdream(model, img, step_size):
    with tf.GradientTape() as tape:
      # We need gradients relative to `img`
      tape.watch(img)
      loss = calculate_loss(img, model)

    # Calculate the gradient of the loss function in relation to the image
values of the input image
    gradients = tape.gradient(loss, img)

    # Normalize the values of the gradient
    gradients /= tf.math.reduce_std(gradients) + 1e-8

    # In the gradient ascent, the loss is maximized so that the input image
    # "stimulates" the layers more and more
    # The image is modified by adding the gradient
    img = img + gradients*step_size
    img = tf.clip_by_value(img, -1, 1)

    return loss, img
```

Listing 13.34 Optimization Function for Dreaming

> **What Is @tf.function?**
>
> `@tf.function` is a *decorator* function that is applied to an original function to decorate it (in our case, function `deepdream()`), as the name suggests. The easiest way to explain a decorator is to use an example. We now use Listing 13.35 to define a decorator function, `@timeit`.

```
import time

# This is our decorator function
def timeit(original_fn):
    def decorator_fn(*args, **kwargs):
        start = time.time()
        res = original_fn(*args, **kwargs)
        end = time.time()
        print('func:%r args:[%r, %r] Result: %r Time: %2.6f sec' % (original_
fn.__name__,args, kwargs, res, end - start))

        return decorator_fn

@timeit
def add(x, y):
    return x + y

add(3,4)
#Output:
func: 'add' args:[(3, 4), {}] Result: 7 Time: 0.000000 sec
```

Listing 13.35 The @timeit Decorator

The original add() function is embellished by the @timeit decorator in such a way that the execution time of the original function is measured and output in addition to the result for the execution of the add() function.

The @tf.function decoration is a bit more complex and transforms a Python function into a TensorFlow graph representation for optimized calculation.

In the deepdream() function, we use the calculate_loss() loss function to calculate the activation on the selected layers and tf.GradientTape() to calculate the gradient.

Now that we have all the necessary functions together, we can start our feedback dream loop using the function shown in Listing 13.36.

```
def run_deepdream_feedback(model, img, steps=100, step_size=0.01):
  # Conversion of uint8 to the range expected by Inception_v3.
  img = tf.keras.applications.inception_v3.preprocess_input(img)

  for step in range(steps):
    loss, img = deepdream(model, img, step_size)

    if step % 100 == 0:
      clear_output(wait=True)
```

```
    show(deprocess(img))
    print ("Step {}, loss {}".format(step, loss))

result = deprocess(img)
clear_output(wait=True)
show(result)

return result
```

Listing 13.36 The Dream Process

Up to this point, we've now sent our image through our pretrained network multiple times, which has tried to discover any patterns. If it detects even the slightest pattern, we reinforce it and feed it back in the feedback loop. Now we still have to create our image pyramid (i.e., octaves) to obtain a higher-resolution image, as shown in Listing 13.37.

```
OCTAVE_SCALE = 1.3

img = tf.constant(np.array(original_img))
base_shape = tf.cast(tf.shape(img)[:-1], tf.float32)

for n in range(5):
  new_shape = tf.cast(base_shape*(OCTAVE_SCALE**n), tf.int32)

  img = tf.image.resize(img, new_shape).numpy()

  img = run_deepdream_feedback(model=dream_model, img=img, steps=500, step_size=0.001)

show(img)
```

Listing 13.37 Run through the Pyramid and Output of the Result Image (see Figure 13.21)

> **Task**
>
> Use this code to dream up different images. Vary the parameters, for example, the layers weights in Listing 13.32, or use a pyramid with more levels or different scaling (refer to Listing 13.37) and increase the number of iterations in the run_deepdream_feedback() function (refer to Listing 13.36). However, you can also use another of the known pretrained networks instead of Inception-v3, whereby you must ensure that you use the correct layer designations.

Figure 13.21 Original and Dream Image Based on Inception-v3

13.4 Deployment with Pretrained Networks

In the previous examples, we've dealt with neural networks, their structure, training, and representation. Now we want to show a simple way to make a trained network available as a product via a web application. To do this, we again use the libraries provided by the Hugging Face platform (see Chapter 8, Section 8.3).

The first example uses a pretrained neural network that expects an image as input and outputs a corresponding textual description of the image, that is, an image-to-text task. We want to make this network available via a web application so that a user can upload an image and receive a corresponding image description. To create the web application, we use gradio, a library we can install using the !pip install gradio command (either in a code cell or via the Anaconda environment).

Gradio

Gradio is an open-source library that enables developers to quickly create user-friendly web interfaces for ML models and other applications. With Gradio, models can be tested and used interactively by presenting them in a web application without the need for extensive programming skills. Gradio supports various input and output components such as images, text, and audio, facilitating the visualization and use of ML models in real time. Gradio is particularly useful for creating prototypes and demo applications, and for collaboration.

13.4.1 A Web Application for a Neural Network to Generate Image Descriptions

You already know how easy it is to create a `pipeline` using the Transformer library, but before we start, we need to install the `gradio` library. For this purpose, we chose a corresponding model for the image-to-text task, `Salesforce/blip-image-captioning-large`, as shown in Listing 13.38.

```
from transformers import pipeline

mypipe = pipeline("image-to-text", model=" Salesforce/blip-image-captioning-large", device="cuda")
mypipe("https://huggingface.co/datasets/Narsil/image_dummy/resolve/main/parrots.png")
# Output:
[{'generated_text': 'there are two parrots that are standing next to each other'}]
```

Listing 13.38 Image Description Using the Transformer Library

In this program code, we define the task and the associated model and pass an image to the `mypipe` pipeline—in this case, the image of a parrot, which is made available via Hugging Face. Let's check the accuracy of the output using Figure 13.22.

Figure 13.22 Image of Two Parrots

But how do we make this available as a web application? If you're expecting a long program, we must disappoint you (see Listing 13.39).

```
import gradio as gr

gr.Interface.from_pipeline(mypipe).launch(share=True)
# Output:
Colab notebook detected. To show errors in colab notebook, set debug=True in launch()
```

```
Running on public URL: https://94b05d4907012a73df.gradio.live
```

This share link expires in 72 hours. For free permanent hosting and GPU upgrades, run `gradio deploy` from Terminal to deploy to Spaces (https://huggingface.co/spaces)

Listing 13.39 Web Application via Gradio

The URL *https://94b05d4907012a73df.gradio.live* is now valid for 72 hours, but it can be created permanently via Hugging Face. This is done using the !gradio deploy command. Of course, this URL will look slightly different for you.

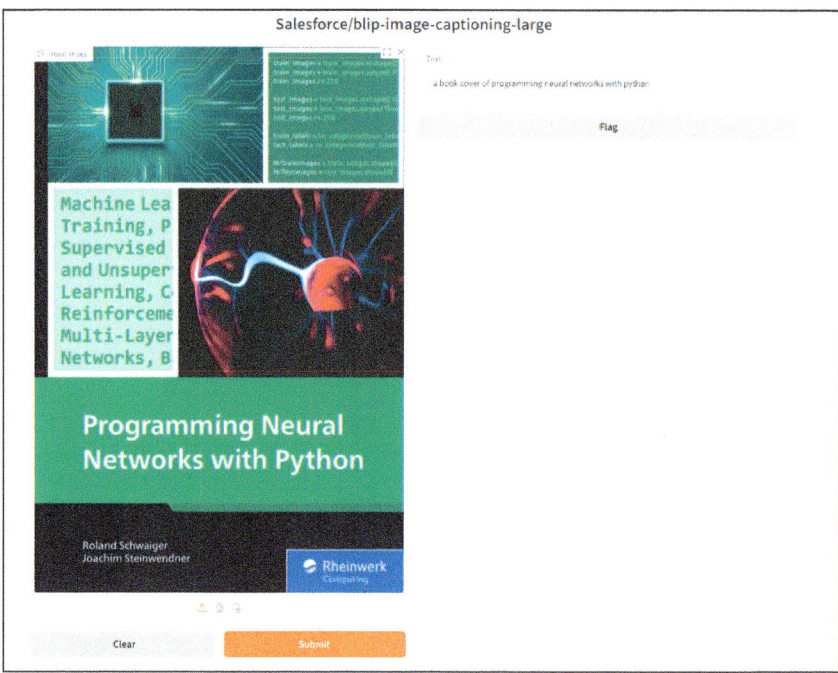

Figure 13.23 Web Application with Gradio for Uploading Custom Images

For the image in Figure 13.23, the web application displayed the description, "a book cover of programming neural networks with python." That's not quite perfect because it doesn't recognize a book, but it's certainly appropriate.

13.4.2 A Web Application for Image Generation

Generative AI, that is, models that can create text, images, audio, and videos, have become increasingly important in recent years. In this section, we want to describe a *stable diffusion model* that is able to generate an image from a text (aka *prompt*). We proceed in a similar way to Section 13.4.1, but define the necessary entries for generating the web application via Gradio itself (see Listing 13.40).

13.4 Deployment with Pretrained Networks

```python
from diffusers import StableDiffusionPipeline
import torch

model_id = "runwayml/stable-diffusion-v1-5"
SDpipe = StableDiffusionPipeline.from_pretrained(model_id, torch_dtype=
torch.float16)
SDpipe = SDpipe.to("cuda")
```

Listing 13.40 A Stable Diffusion Model for Image Generation

Here, we use the `diffusers` library which was developed by Hugging Face specifically for the use of diffusion models for image generation, and `torch`, a neural network library that provides similar performance to the `tensorflow` library we're most familiar with.

For the web application, we now define the `inputs`, `outputs`, and a function that converts the input into the output (`fn`), as shown in Listing 13.41. This function, we call it `get_completion`, is then passed to the Gradio interface and expects a prompt. It then returns the model generated using the `runwayml/stable-diffusion-v1-5` model.

```python
import gradio as gr

def get_completion(prompt):
    negative_prompt = """
    simple background, duplicate, low quality, lowest quality,
    bad anatomy, bad proportions, extra digits, lowres, username,
    artist name, error, duplicate, watermark, signature, text,
    extra digit, fewer digits, worst quality, jpeg artifacts, blurry
    """
    return SDpipe(prompt, negative_prompt=negative_prompt).images[0]


genai_app = gr.Interface(fn=get_completion,
    inputs=[gr.Textbox(label="Your prompt")],
    outputs=[gr.Image(label="Result")],
    title="Create cool images",
    description="Create arbitrary images with stable diffusion",
    allow_flagging="never",
    examples=["astronaut, riding a horse, on mars",
    "cargo ship, flying, in space"])
genai_app.launch()
```

Listing 13.41 Web Application for Stable Diffusion

This code is now used to create a Gradio app that allows us to enter any prompts and generate an image, as shown in Figure 13.24.

13 Areas of Application and Real-Life Examples

Figure 13.24 Web Application for Image Generation

13.5 Summary

In this chapter, we've shown some sample applications for regression and image classification tasks. We created and trained our own CNN using `tensorflow.keras`. The use of pretrained networks such as Inception-v3, ResNet50, or VGG19 facilitates image classification with a small amount of image data training. To see what information is contained in a deep neural network, the DeepDream algorithm was implemented—the dreaming of a deep network belongs to the class of generative networks. To take a step toward deployment, we've used the Transformer library and Gradio to show how you can make pretrained networks available as a web application.

13.6 Further Reading

- California Housing dataset (*www.kaggle.com/datasets/camnugent/california-housing-prices*)
- "Hedonic Housing Prices and the Demand for Clean Air" by David Harrison Jr. and Daniel L. Rubinfeld (1978, *https://lawcat.berkeley.edu/record/1111234/files/fulltext.pdf%E2%80%8B*)
- DeepDream by Alexander Mordvintsev (*https://znah.net*)
- Gradio (*www.gradio.app/*)
- Naive Bees dataset (*www.drivendata.org/competitions/8/naive-bees-classifier/page/31*)
- "Keras: The High-Level API for TensorFlow" (*www.tensorflow.org/guide/keras*)

Appendices

A Python in Brief 389

B Mathematics in Brief 417

C TensorFlow 2 and Keras 435

Appendix A
Python in Brief

Life is like a roller coaster: when we go downhill, we start screaming; when we go uphill, we admire the view. Programming is a similar adventure.

The activity of programming is its own emotional roller coaster with constant change between elation (when the program does what it's supposed to do) and deep sadness (when you pore over the program for hours and still haven't discovered the error).

In addition, terms such as *object-oriented*, *recursive*, *lambda*, and *dynamic typing* mystify the world of programming and make it difficult for the inexperienced to take their first steps. Fortunately for us, a programming language called Python has become established in the field of neural networks, which provides a number of advantages:

- Python is managed in an open, community-based development model (open source by the Python Software Foundation). This means that the language is freely available under a specific license, and the source code for Python is also open to everyone.
- Python has a concise, easy-to-read syntax, which results in simple and understandable programming code. This is achieved by the fact that the language has relatively few different language constructs and very often uses indentations instead of parentheses.
- Python has a very large number of standard libraries. These are also called *modules* or *packages* and are essentially ready-made programs that you can use without having to program them yourself. In addition, there are powerful third-party packages for machine learning (ML) and neural networks, such as scikit-learn, SciPy, NumPy, TensorFlow, and Keras.

But before we get started, we should say that programming is exciting, challenging, and fun—especially with Python!

> **Naming**
>
> The Python programming language was developed by Guido van Rossum at the Centrum Wiskunde & Informatica in Amsterdam. The language owes its name not to the giant snake, but to the well-known BBC comedy series *Monty Python's Flying Circus*.

A Python in Brief

Figure A.1 The Python Roller Coaster in Efteling, Netherlands: Where People Seem to Know What Programming Can Feel Like

This appendix provides a short introduction to Python, which admittedly can only cover a small part of all the capabilities of this programming language. Our goal is to provide you with a small backpack of skills that will facilitate your journey into the world of neural network programming.

> **Warning**
>
> To be able to use the program examples here, you need an appropriate development environment. In this book, we use Anaconda to manage Python and its packages and to launch Jupyter Notebook as well as Google Colab, which is our programming environment. You can find out more about installing these tools in Chapter 2, which you should work through first.
>
> To reproduce the code examples, start Jupyter Notebook, and type the individual listings into a cell. Alternatively, you can make it easy for yourself and use the listings from the download material at *www.rheinwerk-computing.com/6059*.

A.1 The First Climb: Data Types, Variables, and Values

The seat and safety belt have been adjusted and the Python roller-coaster ride starts with the following question: What exactly is a computer program? A *computer program* is a sequence of exact statements that are implemented by the computer, for example, adding up numbers, calculating the sine of an angle, outputting text on the screen, printing a page, or deleting files. The results of these statements naturally always depend on data that is fed into a program via the keyboard, files, as data from sensors, and so on. A neural network is also a computer program, but it also stores weights and settings. These weights control the output of a neural network and are learned from a set of (training) data; the learning is in turn a sequence of statements that changes the weights. The value of ML—and neural networks are part of the family of methods—is that new data constantly changes or adapts the settings and thus the output of the computer program.

The statements are created by you in a high-level language such as Python in the form of program code. This program code is in turn translated into a *machine language* that can be read by the computer. This machine language essentially consists of a (very long) sequence of zeros and ones.

A.1.1 Variables Are Valuable

One of the simplest statements is to assign values (e.g., a number, a text, a yes/no value) to a variable. A variable in Python is like an empty container that can have different contents. In the following example, the container named pot is assigned the number 12345, and then the text "This is a text" and the contents are displayed by specifying the variable name pot. A special feature of Python is that the pot variable can be assigned a number, a text, a matrix, or any other object. This is known as *dynamic typing*. In many other programming languages, the data type (e.g., an integer, floating point number, or text) of a variable is defined in advance, and it can then only contain values of this specific data type. Python is more flexible, as you can see in Listing A.1.

```
pot = 12345
pot
# Output:
12345

pot = 'This is a text'
pot
# Output:
This is a text
```

Listing A.1 Assignment of Values

In an assignment, the name of the variable (or container) to be filled is always on the left-hand side, followed by an equal sign =. On the right, there is always the filling (referred to as *expression*), which can be quite complicated: numbers can be added (e.g., 3 + 5; the plus signs, minus signs, etc. are also referred to as *operators*).

The excerpts in Listing A.2 also contain a *comment* on the program code. This comment is always preceded by a # character in Python, which means that the subsequent text isn't regarded as a statement, but is ignored by the computer. Comments are used in programming to describe the program code and are important because the logic and structure of a program can become very complex. Even experienced programmers need this information to find their way around their own program code later on.

```python
# This is a first comment in the program code
a = 3 + 5 # results in 8 (comment after statement possible!)
b = 5 - 8 # results in -3

# The variables can also be used on the right-hand side
# The "filling" of the variable is then simply used for the calculation
c = (1 + 4) * a # should result in 40

# ... and not so well-known operators
# rounded division
d = 7 // 4  # results in 1
e = -7 // 5

# modulo operation: Remainder of a division
f = 11 % 3 # results in 2
```

Listing A.2 Examples of Simple Data Types

The `print()` function deserves a little more attention (see Listing A.3); it's generally used for the output of text, variables, and so on. `print()` is also interesting because you can pass it almost any number of parameters, which are then output accordingly. Most functions, on the other hand, expect a fixed number of parameters.

```python
# here the output gets generated using the print() function
print("a =", a)
print("b =", b)
print("c =", c)
print("d =", d)
print("e =", e)
print("f =", f)
# Output:
a = 8
b = -3
c = 40
d = 1
e = -2
c = 2
```

Listing A.3 The print() Function

A.1.2 1 and 0 Equals Boole

Because the computer ultimately only has to be able to handle 0 and 1, it's also logical that there must be a special data type that can only contain 0 or 1. Variables with this data type are also called *Boolean variables*, named after the English mathematician, logician and philosopher George Boole. These variables can only have two values, which we interpret as yes or no, 1 or 0, or True or False. This data type is also important because, together with control structures (discussed in Section A.2), the flow of a program can be controlled very well, as shown in Listing A.4.

```
# Boolean values
True # means yes, 1, on ... True
False # means no, 0, off ... False

# You can also negate the Boolean values
not True # corresponds to False
not False # corresponds to True

# Boolean operators are very important, such as
# the query for equality ==
0 == 1 # results in False
2 == 2 # results in True

# the query for inequality !=
0 != 1 # results in True
2 != 2 # results in False

# more comparisons: less than <, greater than >,
# less than or equal to <= , greater than or equal to >=
1 < 100  # results in True
20 < 10 # results in False
3 <= 8   # results in True
11 >= 11 # results in True
# Output:
True
```

Listing A.4 Boolean Values and Operators

The output of this listing only shows the result of the last comparison. If you don't believe the comments, you can surround the expressions with print()—for example, print(2 !=2).

A.1.3 A Text Is a String

Of course, in addition to the Boolean type and numbers, there's also text. In programming jargon, texts are usually referred to as *strings*. The term *string* stands for *sequence*, *succession*, or *chain*; in our case, it refers to a chain of consecutive characters (these can be letters, numbers, or even special characters).

But now it gets a bit difficult because the entire program code is actually a sequence of characters. To distinguish a text as the value of a variable from the rest of the program code, it's enclosed in single or double quotation marks in Python, as shown in Listing A.5.

```python
# Strings are surrounded by either
# single ' quotation marks or
# double " quotation
# marks
myText = "This is my first Python string"
# or
myText = 'This is my second Python string'

# Strings can even be "added", or more precisely joined together
addText = "Part1 and " + "Part2"

# There is also a function that calculates the number of characters
# in a string. This is also known as the length of a string
lenText = len("This text has how many characters?")

# For the composition of strings there is an
# interesting function available: .format()
formatText = "The {} is called {}.".format("Hund","Bello")
print(formatText)
# Output: 'The dog is called Cooper.'

# A string is a sequence of characters
# Python allows you to access individual characters with square brackets []
theABC = 'abcdefghijklmnopqrstuvwxyz'
firstLetter = theABC[0] # WARNING: The numbering starts at 0
lastLetter = theABC[25] # Therefore the 26th character has the index 25
```

Listing A.5 A Text Is a String

With the .format() method, we're getting ahead of ourselves because we won't explain how objects and methods work until later, but it's a great text formatting function that we want to use earlier.

Using the example of strings, we can demonstrate another special feature of the Python programming language known as *slicing*. This is a simple way of cutting out parts from a sequence of characters, as shown in Listing A.6. The slicing indexes always refer to the gaps between the characters of a string (or a list, an array, etc.), as shown in Figure A.2. Thus, [2:5] refers to everything between gap 2 and gap 5.

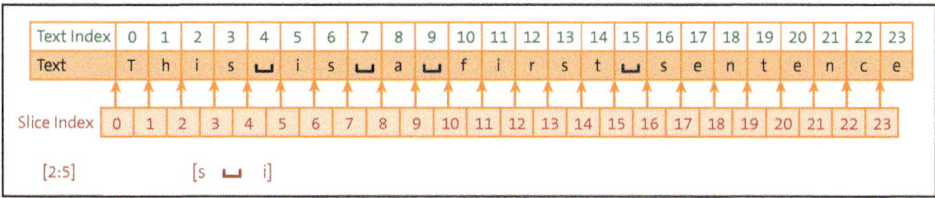

Figure A.2 Slicing Indexes Referencing the Gaps between the Letters

```
# With regard to strings, slicing means cutting out character strings
theSentence = "This is a first sentence"
theSentence[0:4]   # returns the characters from text index 0 to 3 or
                   # from gap 0 to gap 4
# Output: 'This'

# ... and a few more examples of slicing
theSentence[5:7]
# Output: 'is'

# you can also cut from the beginning to a certain position
theSentence[:7]
# Output: 'This is'

# or cut to the end of a sequence
theSentence[8:]
# Output: 'a first sentence'

# and this is what happens when a position is chosen that does not exist
theSentence [99]
---------------------------------------------------------------------------
IndexError                                Traceback (most recent call last)
<ipython-input-5-003d8dbd3e73> in <module>()
      1 # and this is what happens when a position is chosen that does not exist
----> 2 theSentence[99]

IndexError: string index out of range
```

Listing A.6 Slicing: Cutting Out Parts of Text

A.2 At the Top of the Hill: if-then-else and Loops

As already mentioned, a program code consists of a sequence of statements. This doesn't mean that the same sequence should always be executed, but rather that what is to be done is interactively dependent on input from a person or a machine, on calculations, or on measured values. To achieve this, tools for process control are needed, namely the *control structures*.

A.2.1 This Had Better Work: The "if" Statement

One of the most important control structures is the `if` statement, which first queries a condition that returns a Boolean value, namely the already known `True` or `False`, and then executes the corresponding additional statements, as shown in Listing A.7.

```python
# Example of the if statement
age = '5'

if age < 4:
    print("The value of the age variable is less than 4!")
elif age==4:
    print("The value of the age variable is 4!")
else:
    print("The value of the age variable is greater than 4!")
# Output: 'The value of the age variable is greater than 4!
```

Listing A.7 Example of an "if" Statement

In Listing A.7, we suddenly see indentations—no, these are not printing errors. In Python, commands are combined into blocks by indenting them. In our example, this means that all equally indented commands belong to one block. In many other programming languages, such commands are combined using brackets—very often using curly brackets {}. Summarizing by indentation generally makes the program code easier to read—at least according to Python fans.

A.2.2 Again and Again: "for" and "while" Loops

Loops allow a statement or a block of multiple statements to be executed several times in succession, either through a fixed number of passes (`for` loops) or until a certain condition is no longer fulfilled (`while` loops). This condition in the `while` loop is often referred to as a *termination condition*, although it's more of a run-through condition.

In Python, the `for` loop is used to iterate over a sequence of objects. `for x in [0,1,2,3]:` means that x assumes the values from 0 to 3 one after the other (see Listing A.8). These are four different values, so the subsequent statement or block of statements is executed exactly four times. The current value of x can also be used in the statements of the for loop.

```python
# a simple "for" loop
myValues = [0,1,2,3] # a sequence of objects
for x in myValues:
    print("Number: {} and square number: {}".format(x,x*x))
# Output:
# Number: 0 and square number: 0
# Number: 1 and square number: 1
# Number: 2 and square number: 4
# Number: 3 and square number: 9


# myValues can also be generated using the range() function
myValues = range(0,4) # as for slicing: including 0, excluding 5
for x in myValues:
    print("Number: {} and square number: {}".format(x,x*x))
# Output:
# Number: 0 and square number: 0
# Number: 1 and square number: 1
# Number: 2 and square number: 4
# Number: 3 and square number: 9


# range() is even more powerful, you can also enter an interval or
# specify an increment: range(start,stop,interval)
# print() automatically adds a line break at the end and
# using end='' you can prevent this
# It doesn't always have to be x - you can use any
# variable name
myValues = range(0, 10, 2)
for evenNumber in myValues:
    print(evenNumber, ", ", end='')
# Output:
# 0 , 2 , 4 , 6 , 8 ,

# or backwards, and this time without the detour via the myValues variable
for y in range(500,0,-100):
    print(y,", ", end='')
# Output:
# 500, 400, 300, 200, 100 ,
```

Listing A.8 "for" Loops with Number Sequences

However, the sequences for a `for` loop can consist of any different object types, for example, strings, Boolean variables, numbers—in other words, all object types that exist in Python (see Listing A.9).

```python
# Of course, this also works with other objects, here with a sequence
# of strings with a block of two statements
programLanguages = ['Ada','C','C#','Pascal','Modula','PYTHON']
row = 0
for lang in programLanguages:
    row = row + 1
    print(row,". ",lang)
# Output:
1 .  Ada
2 .  C
3 .  C#
4 .  Pascal
5 .  Modula
6 .  PYTHON

# This even works with a sequence of mixed object types.
# The type() function outputs the data type
mixedObjects = ["Text", 1, 4.2, "another text"]
for obj in mixedObjects:
    print("Value: {}, data type: {}".format(obj, type(obj)))
# Output:
Value: Text, data type: <class 'str'>
Value: 1, data type: <class 'int'>
Value: 4.2, data type: <class 'float'>
Value: another text, data type: <class 'str'>
```

Listing A.9 Various Object Sequences

But what if we want to have a loop that doesn't go through a list of objects? This is what the `while` loop is for. The condition of a `while` loop returns a Boolean value, and as you already know, this can be `True` or `False`. All statements within the `while` loop are executed until the condition is no longer fulfilled—that is, the value is `False`, which leads to the `while` loop being terminated (see Listing A.10). It's obvious why this condition is also referred to as the *termination condition*.

```python
# a simple example:
nrLoops = 10
i = 1
sum = 0

while i <= nrLoops:
```

```python
# This condition is fulfilled as long as i is less than or equal to 10,
# therefore returns the value True, and the statement block is
# performed with the two statements
    sum = sum + i
    i = i + 1

print("The sum is:", sum)
# Output:
The sum is: 55


# You can also place the termination condition at the end of the loop.
# The break command "jumps" out of a while loop
nrLoops = 15
i = 1
product = 1

while True:
    product = product * i
    i = i + 1
    if i > nrLoops:
        break
print("The product is:",product)
# Output:
The product is: 1307674368000
```

Listing A.10 "while" Loops with Termination Condition at the Beginning or End

A.2.3 List Comprehension

With list comprehension, Python has created a method for creating loops more elegantly and concisely. Basically, it's a compact method for creating lists from other objects that can be iterated over—iterables.

> **Iterable**
>
> In Python, an *iterable* is an object that can iterate over its elements one after the other, that is, run through them element by element. The most well-known iterables are lists, tuples, character strings (text), and dictionaries. In the following example, we see a list and how we can iterate over it using a for loop:
>
> ```python
> my_list = [ 10, 20, 30, 40]
> for element in my_list:
>     print(element)
> ```
>
> All objects that can be used in a for loop are iterables.

Let's make a list of squares with the first eight square numbers. In programming, this is usually done by using a loop (see Listing A.11).

```python
squares = []
for x in range(1,8):
    squares.append(x**2)
print(squares)
# Output:
[1, 4, 9, 16, 25, 36, 49, 64]
```

Listing A.11 Creating a List Using a Loop

We can do the same using list comprehension, as shown in Listing A.12. The basic structure of a list comprehension is as follows: [expression for element in iterable if condition].

```python
squares = []
squares = [x**2 for x in range(1, 8)]
```

Listing A.12 Creating a List Using List Comprehension

The range() function provides us with an iterable—more precisely, with all integers between 1 and 8.

If we only wanted the squared numbers of all odd numbers between 1 and 8, we can also do this easily (see Listing A.13).

```python
odd_squares = []
odd_squares = [x**2 for x in range(1,8) if x % 2 == 1]
print(odd_squares)
# Output:
[1, 9, 25, 49]
```

Listing A.13 List Comprehension with Condition

We've simply added the condition x % 2 == 1. For this purpose, the modulo operator % is used, which returns the remainder of a division. For odd numbers, the remainder when dividing by 2 is 1.

A.3 Downhill: Functions, Classes, Modules, and Some Greek

"Downhill" isn't meant negatively here, but describes the part of the Python roller-coaster ride that some passengers find hilarious (and good).

A.3.1 Functions Work

Functions are usually used when certain pieces of code occur repeatedly. These parts are then defined exactly once as a function and can be used (i.e., called) again and again. We've already used some representatives of such functions in our examples—namely, the print, range, format, and type functions. These functions belong to the *built-in functions*, as they are original components of Python. This means that someone has already put a lot of thought into the implementation, and we, as new programmers, can use these functions.

A function call always consists of a function name and parameters enclosed in parentheses, which we also refer to as *arguments*. What can be used for these arguments? First of all, it depends on how a function is defined, which can also mean that a function doesn't need an argument at all. But the arguments can consist of our known data types, variables, and other functions.

As is so often the case with programming languages, the functions are best explained using examples, as shown in Listing A.14.

```python
# Function definitions are introduced by the def keyword.
# Functions may or may not contain arguments.
# The statements a function is supposed to execute are indented
def greetTheWorld():
    print("Hello world!")

# The function is called as before
greetTheWorld()
# Output:
# Hello world!

# ... now we greet someone by name,
# that is, we pass a name argument
def greetings(name):
    print("Hello {}!".format(name))

greetings("Betty")
greetings("Robert")
# Output:
# Hello Betty!
# Hello Robert!

# We can also assign a default value for the name argument.
# This is also called the default value
def greetingsWithDefault(name="everyone"):
    print("Hi {}!".format(name))
```

```python
greetingsWithDefault("Jenny")
greetingsWithDefault()
# Output:
# Hi Jenny!
# Hi everyone!

# Now we write a function with numbers that returns a value
def convertCelsius2Fahrenheit(temp_in_Celsius):
    return (temp_in_Celsius *9 / 5) + 32

# Now we combine a for loop with the function call
for temp in [0.0, 36.0, 100.0, -273]:
    print("°C: {}  : °F: {}".format(temp, convertCelsius2Fahrenheit(temp)))
# Output:
# °C: 0.0    : °F: 32.0
# °C: 36.0   : °F: 96.8
# °C: 100.0  : °F: 212.0
# °C: -274   : °F: -461.2
```

Listing A.14 Functions Work

You already know how to enter comments in Python. For functions, it's usual to write a comment at the start of the definition explaining what the function does and how it can be used. This is intended as an instruction manual for programmers who use the function, which doesn't always have to be you.

Python provides its own syntax for this purpose. Enclose the comment in three pairs of double quotation marks to mark it as a short description of the function. This description can then be called interactively during programming or output as in Listing A.15.

```python
# Now a function which
# contains a short description (docstring)
def convertFahrenheit2Celsius(temp_in_Fahrenheit):
    """ returns the temperature in degrees Celsius"""
    return (temp_in_Fahrenheit - 32) / 1.8

# Output of a short description with the help function
help(convertFahrenheit2Celsius)

# or direct access to the docstring with __doc__
print("convertFahrenheit2Celsius:", convertFahrenheit2Celsius.__doc__)
# Output:
# Help on function convertFahrenheit2Celsius in module __main__:
```

```
convertFahrenheit2Celsius(temp_in_Fahrenheit)
    returns the temperature in degrees Celsius
```

```
convertFahrenheit2Celsius: returns the temperature in degrees Celsius
```

Listing A.15 Short Description ("help" Function or Docstring) of a Function

Sometimes, when defining a function, you may not even know how many parameters are required or what data type the parameter has (see the `print()` function). But there is also an elegant solution to this problem in Python: you can provide an argument and precede its name with an *asterisk* (*) in the definition. Python then assumes that any number of parameters can be entered when the function is called, and the parameter with the asterisk can be used as a list (see Listing A.16).

```python
# calculate the mean value of any number of values
def computeAverage(*arguments):
    """This function calculates the mean value of any number of numbers"""
    sum = 0
    count = len(arguments)
    for x in arguments:
        sum = sum + x
    return sum/count

print(computeAverage(3.0,5.0))
print(computeAverage(50.0,43.75,10.33,100.35))
# Output:
4.0
51.1075
```

Listing A.16 Any Number of Arguments

A.3.2 Some Greek: Lambda Functions

The value of the lambda functions, which are also referred to as *anonymous functions*, isn't immediately obvious when you're just starting out with programming. To put it very simply, lambda functions help to make programs easier to read, or rather, they are used to shorten the programming code. Because we use lambda functions in the first chapters, they have also found their way into this Appendix.

But let's take a look at their basic syntax in Listing A.17.

```
lambda arguments : expression
```

Listing A.17 Syntax of the Lambda Function

The number of arguments is arbitrary, but there can only be one expression.

Because you already know how to define a conventional function, we'll start with a simple addition function and then show you how to implement it with lambda functions in Listing A.18.

```python
def add_values(x, y, z):
    return x + y + z

print(5,4,7)
# Output:
16
```

Listing A.18 A Simple Addition Function

A lambda function would be used here to assign the `add_values` variable (see Listing A.19).

```python
add_values = lambda x, y, z: x+y+z
print(add_values(x,y,z))
# Output:
16
```

Listing A.19 A Simple Lambda Function

We don't want to say much more about this because while lambda functions can shorten the program code, they can also make it more difficult to understand.

A.3.3 Classes Are Great

Programming code for complex software contains many lines, and you need a way to structure the code better. We've already discussed one possibility of using the previous definition of functions. Another option for clearly structuring the programming code is to combine related data and functions into objects, as is suggested in object-oriented programming (OOP).

> **Object-Oriented Programming**
> OOP is a programming paradigm whose main idea is to combine data and associated functions into one object. Concepts such as classes, inheritance, polymorphism, and late binding are needed to implement OOP. We provide more detailed information on this in the course of the book and also refer you to the relevant "Further Reading" sections.

Classes are templates or descriptions of objects and naturally have a name and associated data, properties, and functions. A class is therefore something like a cookie cutter, and the instances of this class are the different cookies, as shown in Listing A.20.

```python
# We use the "class" statement to create a class
class Human:

    # Class attributes that all instances of this class share
    species = "Homo sapiens"

    # Each class needs an initialization, if by means of this
    # class definition an object or instance is created.
    # In Python, this is a special method called __init__, which is available for all
    # classes. If you do not write any, Python will generate them.
    def __init__(self, name):
        # The attributes are assigned values during initialization.
        # This is done with a special parameter: self
        self.name = name
        # The _age variable is "encapsulated", i.e. it is only
        # known within the class.
        # To assign or output this variable, separate methods are required
        self._age = 0

    # A first method of this class.
    # Every method needs "self" as its first argument!

    def greet(self, person):
        print("I, {greeter}, greet {welcomed}!".format(greeter=self.name, welcomed=person.name))

    # A class property that gives us the encapsulated age variable
    @property
    def age(self):
        return self._age

    # This class property assumes the value of the _age class variable
    @age.setter
    def age(self, ageInYears):
        self._age = ageInYears
```

Listing A.20 A Simple Class Definition for a Human

Now that we've defined our own class, we need to fill it with life. This means that we have to create objects using this template or description. These objects are referred to as *instances* of the class (see Listing A.21).

```python
person1 = Human("Janet")
person2 = Human("Brad")
```

```
person1.age = 40
person2.age = 35

print(person1.age)
person1.greet(person2)

# Output:
40
I, Janet, greet Brad!
```

Listing A.21 Instances of Our "Human" Class

In the course of the book, we repeatedly use classes and point out their different aspects.

A.3.4 Modules Are Modern

As already mentioned at the beginning of this appendix, Python is an open programming language. For this reason, it's freely available, and there's a large community constantly contributing to the expansion of Python. One expansion comes in the form of *modules* for a wide variety of topics. Modules each contain a collection of data types, functions, and classes. In the course of this book, we use a large number of modules, so we'll briefly describe how to use modules here.

To use modules, you must import them first. In Listing A.22, we import a math module called math. As soon as we've done this, all the associated data, functions, and classes are available to us.

```
import math
print(math.sqrt(16))
# Output:
4.0
```

Listing A.22 Import of Modules or Python Libraries

You can also import only selected functions of a module, as shown in Listing A.23.

```
from math import sin, cos, pi
# The sine function sin() expects an angle, but how?
# As a degree or radian?
# How do you find out? You already know that.
print(sin.__doc__)
print(sin(0), sin(pi/2), sin(pi), cos(pi/2))
```

```
# Output:
Return the sine of x (measured in radians).
0.0 1.0 1.2246467991473532e-16 6.123233995736766e-17
```

Listing A.23 Import and Use of Selected Functions

We would expect `sin(pi)` or `cos(pi/2)` to equal 0. Although we get very, very small numbers, why aren't they exactly 0? Well, `pi` can only be mapped up to a certain decimal place, so this is a rounding error.

Some module names can be very long, but you can choose a shortened alias and write m instead of math, for example. The `import` statement in Listing A.24 is used for this purpose.

```
import math as m
# the square root
m.sqrt(25)
# Output:
5.0
```

Listing A.24 Importing and Using Modules

You can already guess that there is an incredibly large number of different modules. This is another reason why the Python programming language is so powerful.

A.4 Right in the Middle: Special Data Types for Vectors, Matrixes, and Tensors

Because we have to work a lot with vectors and matrixes when dealing with neural networks, we'll dedicate a separate section to these two mathematical objects and their implementation as data types in Python. While the mathematical background is explained in Appendix B of this book, we'll restrict ourselves here to a simplified but sufficient description.

Simply put, vectors are a sequence of numbers:

$$v = \begin{pmatrix} 2,1 \\ 0 \\ 3,1 \end{pmatrix}$$

This vector v consists of a sequence of three numbers. The number of numbers is also referred to as the *dimension* of a vector. Vectors are also known as *tensors of the 1st level*, which also gives us an indication of the naming of the `tensorflow` Python module.

A matrix is an array of values. Another way of looking at it is to regard a matrix as a sequence of vectors of the same dimension:

A   Python in Brief

$$A = \begin{pmatrix} 2{,}1 & 0 & 6{,}21 & 2{,}1 \\ -3{,}5 & 3{,}45 & 9{,}2 & 1{,}55 \\ 22 & 0{,}45 & 3{,}14 & -32{,}1 \end{pmatrix}$$

In this case, matrix A has $m = 4$ columns and $n = 3$ rows, that is, a total of 12 elements.

> **What Are Tensors?**
> - A number is a tensor of the 0th level and is also referred to as a scalar.
> - Vectors as a sequence of numbers are also known as tensors of the 1st level.
> - A matrix as a sequence of vectors is a tensor of the 2nd level.
>
> You could go on like this, and everything that comes after that are hypermatrixes or 3rd-, 4th-, 5th-level tensors, and so on.

A.4.1   Data Types in Core Python

Core Python is everything that Python provides without using any modules—and that's a lot. To start with, there's the *list* data type, which allows for a sequence of numbers to be displayed, as shown in Listing A.25.

```python
# A pre-filled list as a representation of a vector
myValues = [2.1, 0 , 3.1, -5]

# You can also append something
myValues.append(11)
print("myValues: \t", myValues)

# and remove something
myValues.pop() # removes the last element of the list

# Access to individual elements of the list
# as is often the case in computer science, numbering starts at 0
print("myValues[1]: \t", myValues[1])

# You can change individual elements
myValues[1] = 33

# You can also highlight areas from a list.
# This sounds familiar -> slicing.
# The range goes from gap 1 to gap 3
print("myValues[1:3]: \t",myValues[1:3])   # -> [33, 3.1]
# Leave out the beginning
print("myValues[2:]: \t", myValues[2:])    # -> [3.1, -5]
# Take only every second entry
print("myValues[::2]: \t",myValues[::2])   # -> [33, -5]
```

A.4 Right in the Middle: Special Data Types for Vectors, Matrixes, and Tensors

```
# Turn the list around
print("myValues[::-1]: \t",myValues[::-1])   # -> [-5, 3.1, 33, 2.1]
# Output:
myValues:         [2.1, 0, 3.1, -5, 11]
myValues[1]:      0
myValues[1:3]:    [33, 3.1]
myValues[2:]:     [3.1, -5]
myValues[::2]:    [2.1, 3.1]
myValues[::-1]:   [-5, 3.1, 33, 2.1]
```

Listing A.25 Lists as a Data Type for Representing Vectors

A strange character combination has crept into the `print()` function, namely `\t`—an *escape sequence*—which is used for nonprintable elements. This character stands for a tab and enables us to output strings and variables in an attractive form.

There are other interesting data types in core Python as well, such as tuples, dictionaries, or sets, but we don't need them for the topic of this book. For details and references to further information, see Section A.6.

A.4.2 The "numpy" (Numerical Python) Module

This module not only provides us with the `array` data type, which is ideal for the representation of vectors, matrixes, and tensors in general, but also the special arithmetic operations and other useful functions that are defined and possible with these mathematical objects. This module can also be implemented very quickly and efficiently.

It's possible that the array module hasn't yet been installed. In Chapter 2, you can read how the installation of modules works.

Listing A.26 shows how a list gets created and converted into a NumPy array.

```
# The already known import of a module
import numpy as np

# We first create a list (core Python) of values = vector
myValues = [10, 120, 250, 50, 88, 99, 600]

# Conversion to numpy.array
myArray = np.array(myValues)
print(myArray)
# Output:
[ 10 120 250 50 88 99 600 ]
```

Listing A.26 The "numpy" Module

A Python in Brief

Let's assume that the values in the NumPy array shown in Listing A.27 are distances in kilometers, and we want to calculate the equivalent in miles (1 km = 0.621371 miles). This demonstrates the power of Python in general and NumPy in particular.

```
print(myArray * 0.621371)
# Output:
[ 6.21371 74.56452 155.34275 31.06855 54.680648 61.515729 372.8226 ]
myArrayInMiles = myArray * 0.621371
```

Listing A.27 Conversion of Entire Lists/Arrays

Listing A.28 shows the representation and some operations of matrixes (i.e., tensors of the 2nd level). Of course, the numpy module provides much more, but this information is sufficient to get you started.

```
type(myArrayInMiles)
# Output:
numpy.ndarray


# This is how a matrix is displayed in NumPy
A = np.array([ [ 2.1, 0, 6.21, 2.1], [-3.5, 3.45, 9.2, 1.55], [22, 0.45, 3.14, -32.1]])
print(A)
# Output:
[[  2.1    0.     6.21   2.1 ]
 [ -3.5    3.45   9.2    1.55]
 [ 22.    0.45   3.14 -32.1 ]]

# You can also create "special" arrays
B = np.ones((3,4))   # creates a 3(rows)x4(columns) matrix with 1's

# and some array operations
C = A - B # element-wise subtraction
print(C)
# Output:
[[  1.1   -1.     5.21   1.1 ]
 [ -4.5    2.45   8.2    0.55]
 [ 21.    -0.55   2.14 -33.1 ]]

# Transpose a matrix, i.e. mirror the matrix,
# so that a 3x4 matrix becomes a 4x3 matrix
print(A.T)
# Output:
[[  2.1   -3.5   22.  ]
 [  0.     3.45   0.45]
```

```
[ 6.21   9.2    3.14]
[ 2.1    1.55 -32.1 ]]

# Matrixes can be multiplied as follows
Crandom = np.random.random((4,3))*3 # Create a matrix with random numbers
C = np.matmul(A,Crandom)
print(C)
# Output:
[[ 18.02296147  13.15311799  19.28537819]
 [ 19.77867607  13.74555499  12.58145763]
 [-37.34231983 -30.03962606  37.20194038]]
```

Listing A.28 The "array" Data Type of the "numpy" Module for the Representation of Vectors and Matrixes

A.5 At the Finish: A Complete k-nearest neighbor Classifier

We've now studied many elements of Python programming and are rapidly approaching the end of our first roller-coaster ride—the *k-nearest neighbor classifier* (k-NN).

Before you can implement this procedure, we need to explain what this method does and how it works. *Classification* means the division of objects into different categories or classes. Of course, you need object descriptions for this, which are specified in ML as an object's characteristics. We use numbers as characteristics for better explanation. In Figure A.3, objects (e.g., squares and triangles) are described by two characteristics, namely numbers. The figures are plotted on the x-y axes. A point therefore represents two numbers, which in turn represent an object.

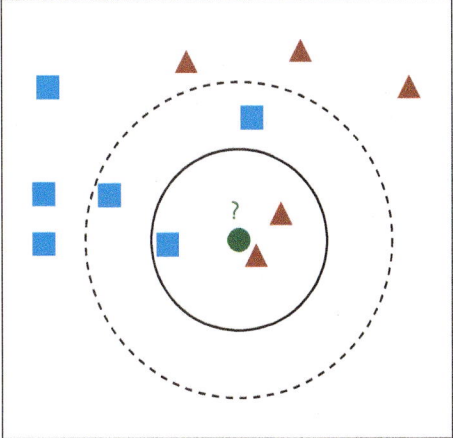

Figure A.3 Two Object Types (Blue Squares and Red Triangles) and One Object of Unknown Class (Green Circle) (Antti Ajanki, Example of k-nearest neighbor classification, https://commons.wikimedia.org/wiki/File:KnnClassification.svg)

In Figure A.3, representatives of two classes are shown: blue square and red triangle. In Listing A.30, we could write "blue square" and "red triangle", but we'll use "0" and "1" as class names instead.

An object whose class isn't yet known is shown as a green circle. The task of the classifier is to determine the class for this unknown object.

Supervised classification means that we provide the classifier with a set of objects of a known class. In the preceding example, these are the blue squares and red triangles. Based on this list of objects of a known class, we can train the classifier and teach it the patterns of the objects. This is why these processes are also referred to collectively as *pattern recognition*.

A *k-NN classifier* now looks at the *k* locally nearest known neighbors of the unknown object and assigns to the unknown object the class that occurs most frequently among the *k* locally nearest neighbors. In our illustration, we would assign the x "red triangle" class for *k* = 3 because the three nearest neighbors (within the solid circle) consist of two red triangles and only one blue square. If we were to choose *k* = 5 (within the dotted circle), our unknown object would have the "blue square" class. In this example, you can see a challenge of ML, namely the choice of an optimal *k* or, more generally, the optimal parameters, which is also referred to as the *parameterization problem*.

But let's now move on to the implementation of our classifier. In the previous paragraphs, we've provided a qualitative description of the process. However, a formal description of the algorithm is required for implementation. Pseudocode is often used for this purpose. *Pseudocode* is a formal list of the necessary steps that lead to a result, but without taking into account any formal specifications of a particular programming language. This can look like Listing A.29, for example.

```
# Pseudocode for a k-nearest neighbor classifier
Do the following for all unknown objects i:
   Calculate the distance from each known object to (unknown object) i.
   Sort the distances by size, starting with the smallest distance.
   Determine the frequency of the occurring classes of the first k distances.
   Assign the class with the highest frequency to the unknown object i.
```

Listing A.29 Pseudocode for a k-NN Classifier

Based on this pseudocode, we now create a class for the k-NN classifier, as shown in Listing A.30. We use existing modules for this purpose. We use module functions here without further explanation, as this would go too far at this point. To show the essentials, we don't implement any error queries or help.

As names of the class functions for training and classification, we use designations that are common in Python modules for ML (e.g., scikit-learn, Keras): *fit* for training and *predict* for classification.

A.5 At the Finish: A Complete k-nearest neighbor Classifier

```python
# the following modules provide us with ...
import numpy as np           # functions for sorting the distances
import numpy.linalg as nl    # functions for distance calculation
from collections import Counter  # Counting and determining the most frequent classes

# our kNN classifier class
class myKNN:
    """k-nearest neighbor classifier"""
    
    def __init__(self, k_neighbors=3):
        """
        Is called to initialize the classifier
        """
        self.k_neighbors = k_neighbors
    
    # Training phase of the classifier
    # The kNN classifier does not actually need any training, since
    # class membership is determined by the nearest neighbors.
    # We only transfer the Xb list of known objects to the class
    def fit(self, Xb, y=None):
        """
        Training of the classifier with list of known objects Xb
        """
        self.Xb = Xb
        self.y = y
        
        return self
    
    def predict(self, Xu, y=None):
        """
        Classification of the list of unknown objects Xu
        """
        # first we prepare the result vector
        classindices = []
        
        for i in Xu:
            # This statement determines the distances of the unknown object
            # to all known objects on our list
            distances = nl.norm(np.transpose(i - self.Xb), axis=0)
            
            # Sorting of the distances and list of the first k_neighbors indexes
            indicesSortedDistances = np.argsort(distances)[:self.k_neighbors]
```

```
        # Determine the frequencies of the classes and return the most
frequent one
        mostfrequentClass = Counter(self.y[indicesSortedDistances]).most_
common(1)[0][0]

        # Add the result to the result vector
        classindices.append(mostfrequentClass)
    return classindices
```
Listing A.30 Our First Custom Class of a k-NN Classifier

Now that we've defined our class, it's time to "cut out the cookies"; that is, we create some instances and carry out classifications.

We use the dataset known from Figure A.3 with 10 objects, divided into two classes: blue objects and red objects. The classes are usually numbered (blue = 1; red = 0), which is why the results are also shown in Listing A.31 as a number. Each object is therefore described by two characteristics.

```
# Initialization of two "cookies" with different parameters
classifier1 = myKNN(k_neighbors=3)
classifier2 = myKNN(k_neighbors=5)

# Here are the x-y coordinates of our red and blue objects
# and the corresponding class
Xb = np.array([[1, 8.8], [1, 11], [1.2, 15.9], [3.7, 11], [6.1, 8.8], [9.8,
14.5], [7, 17], [10, 8.1], [11, 10.5], [11.8, 17.5], [16.4, 15.8]])
y = np.array([1, 1, 1, 1, 1, 1, 0, 0, 0, 0, 0])

# then the training
classifier1.fit(Xb,y)
classifier2.fit(Xb,y)

# To test the classification, we use the green object
# with the characteristics (9,9) with
# unknown class and save it as the Xu variable.
# Here, too, we use the numpy module that we already imported in the definition
# of the class
Xu = np.array([[9, 9]])

# And now for the classification
print("classifier 1: ", classifier1.predict(Xu))
print("classifier 2: ", classifier2.predict(Xu))
```

```
# Output:
classifier 1:   [0]
classifier 2:   [1]
```

Listing A.31 The Application of Our "myKNN" Class

We may have been a little overzealous here, but we've successfully created our first custom class and at the same time implemented a simple ML method that succinctly introduces ML as a concept. This should whet your appetite for more in this book.

A.6 After the Ride Is Before the Ride

You now have a "compact" roller-coaster ride behind you. Needless to say, programming or learning to program requires multiple rides on increasingly wild roller coasters. This first ride should get you used to the more complex but also more exciting roller coasters featured in this book.

There are, of course, countless tutorials on Python available on the internet, some of which are included in the following section. If you prefer a printed book, you'll find two recommendations in the list as well.

A.7 Further Reading

- *Getting Started with Python* by Thomas Thies (2024, *www.rheinwerk-computing.com/5876*)
- Learn Python (*www.learnpython.org*)
- "Learn X in Y Minutes: Where X Is Python" (*https://learnxinyminutes.com/docs/python3*)
- Python (*www.python.org*)
- *Python 3* by Johannes Ernesti and Peter Kaiser (2022, *www.rheinwerk-computing.com/5566*)

Appendix B
Mathematics in Brief

This appendix isn't for cowards. You'll relive the complete trauma of your school days, and as soon as a mathematical symbol appears, you'll be dripping with sweat. This is amply provided for in this chapter.

This appendix contains a few interesting facts from mathematics that are relevant to neural networks and that you'll need to understand when programming algorithms.

B.1 Linear Algebra

Since Google made the TensorFlow framework available for artificial neural networks (ANNs), the term *tensor* has suddenly become quite well known. It's one of the most difficult terms in linear algebra. It's "only" about 180 years old, and Albert Einstein used it to develop the theory of relativity. Linear algebra initially deals with *scalars*, *vectors*, and *matrixes*, among other things.

Under a *scalar*, you can simply imagine a number, for example:

5

Of course, you can also implement a scalar using numpy in a small program like the one in Listing B.1.

```
import numpy as np
a = np.array(5)
print(a)
print('Scalar a has the dimension %d' %(a.ndim))
# Output
5
Scalar a has the dimension 0
```

Listing B.1 A Scalar in "numpy"

Think of a *vector* as a summary of numbers (for details, see Section B.1.1):

$$\begin{pmatrix} 0 \\ 1 \\ -1 \end{pmatrix}$$

That was the column-based notation. The same vector looks like this in row-based notation:

$(0, 1, -1)$

Listing B.2 shows what it looks like when we implement it in numpy.

```python
import numpy as np
v = np.array([0,1,-1])
print(v)
print('Vector v has the dimension %d' %(v.ndim))
# Output
[ 0  1 -1]
Vector v has the dimension 1
```

Listing B.2 A Vector in "numpy"

A *matrix* can be thought of as a similar object that needs one more dimension:

$$\begin{pmatrix} 0 & 2 & -2 \\ 0 & 0.5 & 2 \end{pmatrix}$$

We also have a small program for this, as shown in Listing B.3.

```python
import numpy as np
m = np.array([[0,2,-2],
              [0,0.5,2]])
print(m)
print('Matrix m has the dimension %d' %(m.ndim))
# Output
[[ 0.  2. -2. ]
 [ 0.  0.5  2. ]]
Matrix m has the dimension 2
```

Listing B.3 A Matrix in "numpy"

Linear algebra deals with the relationships between these types of objects. They can be interpreted geometrically, for example, as points or changes in three-dimensional space. They can also be used to solve systems of linear equations, to visualize (and calculate with) physical forces, and to help capture large amounts of data.

Linear algebra is an important basis for AI because it provides established algorithms for learning (e.g., the backpropagation algorithm) and computing operations can be optimally mapped to hardware.

Now, imagine this paper had a third dimension, then we could continue the series scalar, vector, matrix, and so on. You could even imagine a fourth or fifth dimension. Welcome to the world of tensors! At least that's how you can imagine a tensor.

B.1.1 Points, Vectors, and Matrixes

A *point* can be conceived as a concrete position in a *coordinate system*, which is described by its position information, that is, the *coordinates*. For example, the point $P = (1,2)$ has the coordinates $x = 1$ and $y = 2$ in two-dimensional space (see Figure B.1).

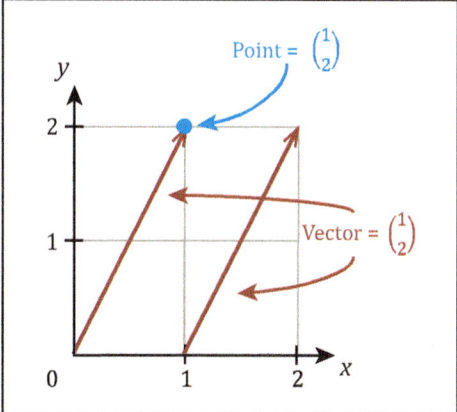

Figure B.1 Point and Vector

A *vector* can be described by an arrow that leads from a starting point to an endpoint. The vectors can also be represented by pairs of numbers in the coordinate system we use. The pairs of numbers don't describe the position, as is the case with the point, but the direction of the vector. All parallel arrows with the same length and the same orientation describe the same vector.

Here's an example of a vector:

$$\vec{v} = \begin{pmatrix} 1 \\ 2 \end{pmatrix}$$

Vectors can be added, subtracted, and multiplied in multiple ways, as we'll show in the next section.

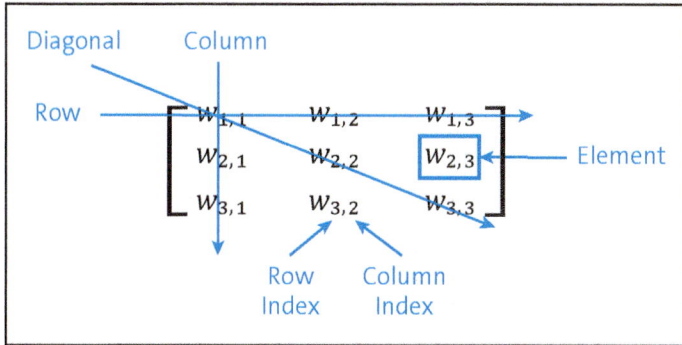

Figure B.2 Matrix

B Mathematics in Brief

A *matrix* (see Figure B.2) is a rectangular arrangement of elements, such as numbers:

$$A = \begin{pmatrix} 0.4 & 2 \\ 2 & 1 \end{pmatrix}$$

A matrix has rows and columns; for example, the matrix in Figure B.2 has three rows and three columns. To name an element and its position in the matrix exactly, *indexes* are used; in our example, these are the row and column indexes. An index is a counter that is incremented in whole steps and starts with 0 or 1, depending on the situation. The element $w_{3,2}$, for example, is located in the third row and in the second column.

You can use matrixes for calculations such as multiplications or additions. Matrixes are of particular importance to us, as the calculations in the ANN are carried out using matrix operations.

B.1.2 Vector Addition

Calculating with vectors is like lining up matchsticks, as shown in Figure B.3. The ignition head of the match plays the role of the arrow, and this shows the direction in which the match is going. The other end of the match is placed on the ignition head of the previous match.

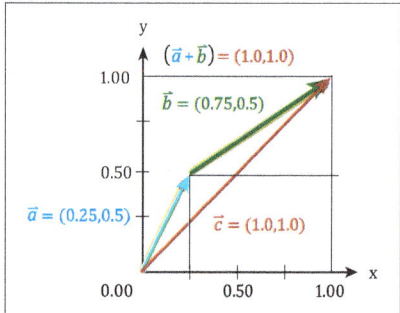

Figure B.3 Vector Addition

For example, if the vector is $\vec{a} = (0.25, 0.5)$ and $\vec{b} = (0.75, 0.5)$, then the sum of the vectors is calculated by adding the individual components:

$$\begin{aligned}(\vec{a} + \vec{b}) &= ((a_1 + b_1), (a_2 + b_2)) \\ &= ((0.25 + 0.75), (0.5 + 0.5)) \\ &= (1.0, 1.0)\end{aligned}$$

The sum of the two vectors $\vec{a}$ and $\vec{b}$ is again a vector $\vec{c} = (1.0, 1.0)$.

B.1.3 Multiplying Vectors by Scalars

If you want to stretch or shorten a vector $\vec{x} = (x, y)$, for example, you multiply it by a scalar k, that is, by a value. For a two-dimensional vector, this looks as follows:

$$\vec{x} \cdot k = \begin{pmatrix} x \\ y \end{pmatrix} \cdot k = \begin{pmatrix} x \cdot k \\ y \cdot k \end{pmatrix}$$

An arrow representing the vector becomes k times as long. In Figure B.4, for example, vector $\vec{c} = (1.0, 1.0)$ is multiplied by scalar $k = \frac{1}{2}$.

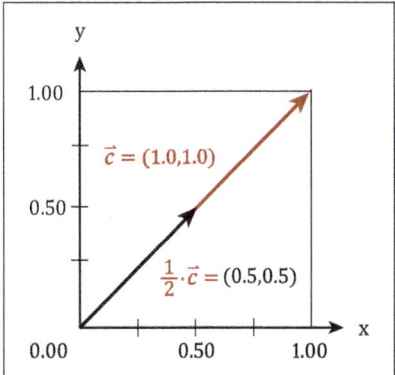

Figure B.4 Scaling a Vector

Each of the components is multiplied by scalar k. A concrete application example for *scalar multiplication* is the *normalization* of an input vector for training in the ANN. You can speed up learning by making the input vector's length 1. To do this, each component of the vector must be divided by the length of the vector. You can calculate the length of a vector $\vec{x} = (x, y)$ in the same way as you may have seen in school when studying right-angled triangles and the Pythagorean $a^2 + b^2 = c^2$ or $c = \sqrt{a^2 + b^2}$ theorem:

$$|\vec{x}| = \sqrt{x^2 + y^2}$$

Figure B.5 shows the length calculation for vector $\vec{c} = (1,1)$.

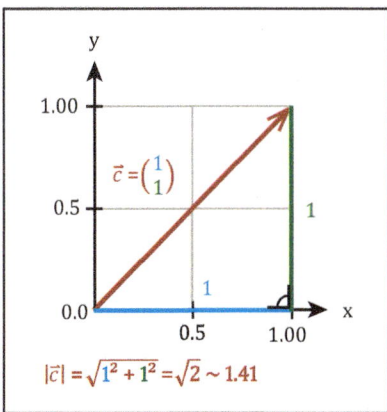

Figure B.5 Length Calculation for a Vector

If you then divide the vector by this value, it has the length 1. Because a division by $|\vec{x}|$ is the same as a multiplication by $\frac{1}{|\vec{x}|}$, the same applies here: Each component is treated individually:

$$\vec{e}_x = \begin{pmatrix} \frac{x}{|\vec{x}|} \\ \frac{y}{|\vec{x}|} \end{pmatrix}$$

Vector $\vec{e}_x$ is referred to as the *unit vector*. In our example for vector $\vec{c}$, the unit vector is

$$\vec{e}_c = \begin{pmatrix} \frac{1}{1.41} \\ \frac{1}{1.41} \end{pmatrix} \sim \begin{pmatrix} 0.71 \\ 0.71 \end{pmatrix}$$

B.1.4 Subtracting Vectors

In the vector subtraction of two vectors $\vec{c}$ und $\vec{b}$, the *inverse vector* is formed from the second vector $\vec{b}$ by multiplication with the scalar -1, that is, $(-1) \cdot \vec{b}$, and then the two vectors are added, that is, $\vec{c} + (-\vec{b}) = \vec{c} - \vec{b}$. The inverse vector still has the same position, but now points in the opposite direction.

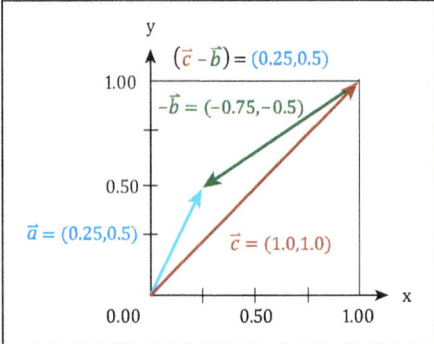

Figure B.6 Vector Difference

For example, if the vector is $\vec{c} = (1.0, 1.0)$ and $\vec{b} = (0.75, 0.5)$, then you determine the difference between the vectors by subtracting the individual components:

$$(\vec{c} - \vec{b}) = ((c_1 - b_1), (c_2 - b_2))$$
$$= ((1.0 - 0.75), (1.0 - 0.5))$$
$$= (0.25, 0.5)$$

B.1.5 Multiplying Vectors with Each Other

There are several variants for the multiplication of vectors with vectors: the result can again be a vector or a zero-dimensional value, that is, a scalar, as we discussed in the introduction to this section.

Here, we're only interested in the last case. This variant of multiplication is called the *scalar product*; sometimes, it's also referred to as the *inner product* $\vec{a} = (a_1, a_2)$. The scalar product turns two directed variables, the vectors, into an undirected variable, a scalar. The scalar product of two vectors $\vec{a} = (a_1, a_2)$ and $\vec{b} = (b_1, b_2)$ in two-dimensional space is defined as follows:

$$\vec{a} \cdot \vec{b} = \begin{pmatrix} a_1 \\ a_2 \end{pmatrix} \cdot \begin{pmatrix} b_1 \\ b_2 \end{pmatrix} = a_1 \cdot b_1 + a_2 \cdot b_2$$

The components are therefore multiplied in pairs!

Let's take a look at some properties of the scalar product, which are shown in Figure B.7:

- $\vec{a} \cdot \vec{b} > 0$ applies exactly when $\vec{a}$ and $\vec{b}$ form an acute angle.
- $\vec{a} \cdot \vec{b} < 0$ applies exactly when $\vec{a}$ and $\vec{b}$ form an obtuse angle.
- $\vec{a} \cdot \vec{b} = 0$ applies exactly when $\vec{a}$ and $\vec{b}$ are at right angles to each other.

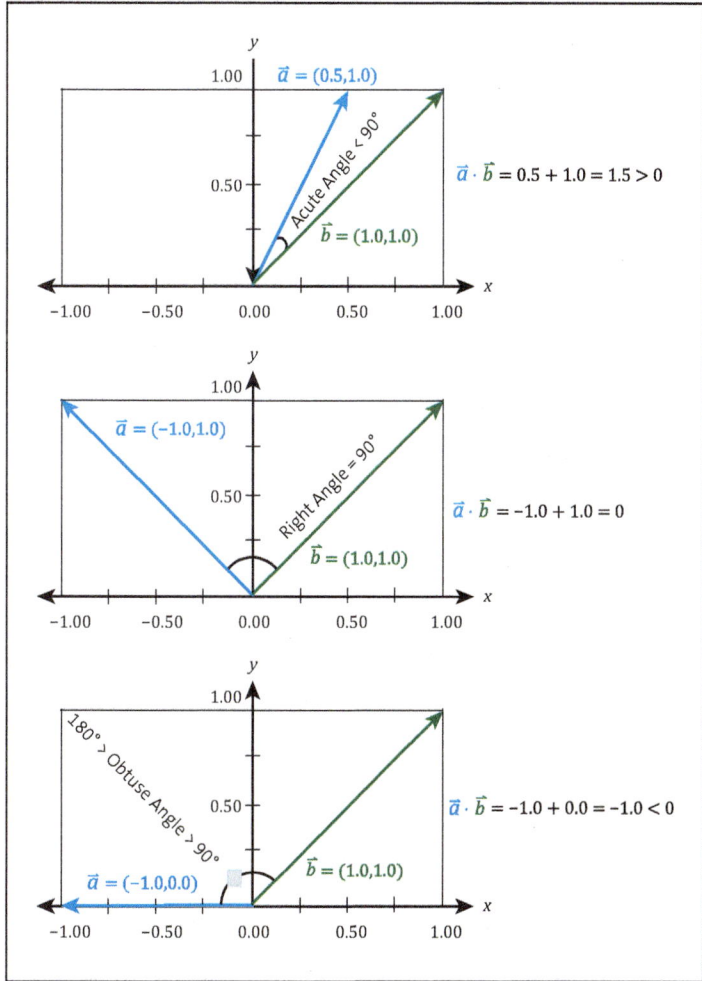

Figure B.7 Some Properties of the Scalar Product

B Mathematics in Brief

You'll see an application of the scalar product when determining the dividing line for two classes in Section B.1.9.

B.1.6 Multiplying Matrixes and Vectors with Each Other

A matrix W can be multiplied by a vector $\vec{x}$ if the number of columns in the matrix matches the number of components in the vector. The result of the multiplication is a vector that has as many components as there are rows in the matrix.

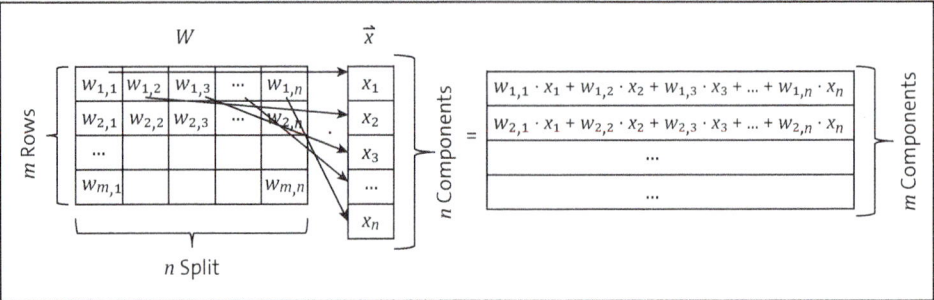

Figure B.8 Multiplication of Matrixes and Vectors

You multiply each row of the matrix by the entire vector, element by element. The results of the multiplication are added together, and the result is entered for each matrix row as a value in the newly created vector, as shown in Figure B.8.

We use matrix-vector multiplication, for example, to calculate the weighted sum for a neuron or multiple neurons. Chapter 5, Section 5.3, describes how you can use this tool.

B.1.7 Transposing a Matrix

The *transpose* of a matrix W, denoted by W^T, means that the matrix is tilted along the matrix diagonal, that is, along those entries for which the row and column index is the same. To achieve this, you need to swap the row index with the column index, as shown in Figure B.9. As a result, the first row of the transposed matrix on the right-hand side is created from the first column of the original matrix on the left-hand side, the second row from the second column, and so on.

Here's an example to illustrate this:

$$W = \begin{bmatrix} 1 & 2 & 3 \\ 4 & 5 & 6 \end{bmatrix}$$

$$W^T = \begin{bmatrix} 1 & 4 \\ 2 & 5 \\ 3 & 6 \end{bmatrix}$$

We use the transposing of a matrix from Chapter 6, Section 6.4, to determine the error fraction per neuron used in the backpropagation algorithm to adjust the weights.

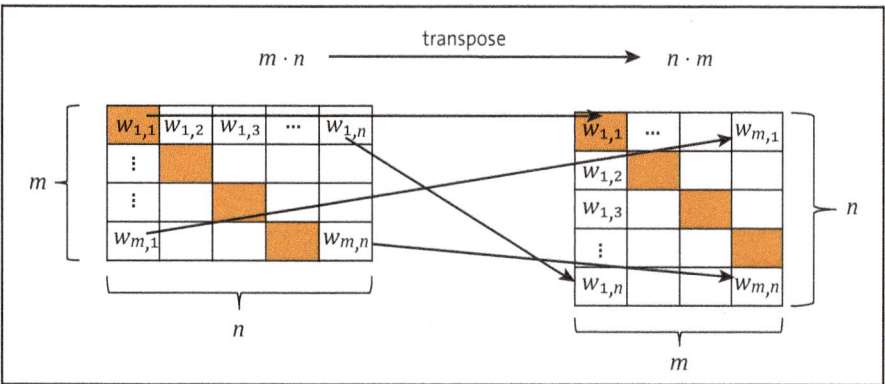

Figure B.9 Transposing a Matrix

B.1.8 Line Equation

Even if the mathematical concepts explained here look simple, it's worth taking a brief look at them. The concept of a *line* is very important for the perceptron, for example, because it enables the perceptron to learn to distinguish between two categories:

$f(x) = y = k \cdot x + d$,

for example,

$y = -2 \cdot x + 1$

Mathematicians refer to this representation as the *standard form* or *normal form*. A value y is calculated for each value x using the line equation. In our example, $f(x)$ has the value $y = 1$ at the position $x = 0$, and you can write this as a value pair: A = (0,1). The value $y = 0$ is assigned at position $x = 0.5$, which would be B = (0.5,0) as a value pair. The points for these pairs of values are located on the line.

Take a look at Figure B.10, where we've drawn the line and the two points.

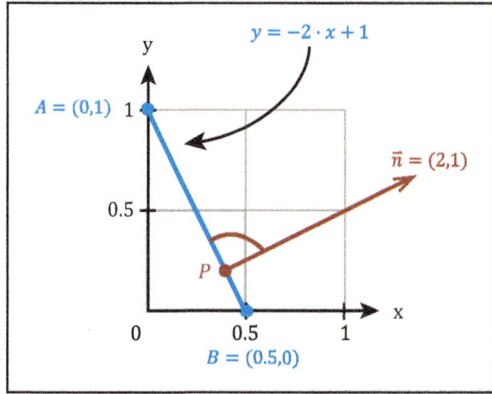

Figure B.10 A Line in Blue, Two Points, and the Normal Vector in Red

425

B.1.9 From Weight to Line

A line can also be described using a vector $\vec{n}$ and a point P; this is called the *normal vector form*. The vector is *normal* (mathematician's term for "at right angles") to the line, as shown in Figure B.10:

$$\vec{n} \cdot X = \vec{n} \cdot P$$

Task

Multiply the vectors using the scalar product, where the vector is $\vec{n} = \begin{pmatrix} n_1 \\ n_2 \end{pmatrix}$ and the points are $X = \begin{pmatrix} x_1 \\ x_2 \end{pmatrix}$ and $P = \begin{pmatrix} p_1 \\ p_2 \end{pmatrix}$.

Solution:

The solution involves computing the scalar (dot) product of the normal vector $\vec{n}$ with the position vector X and equating it to the dot product of $\vec{n}$ with the given point P, resulting in the following equation:

$$n_1 \cdot x_1 + n_2 \cdot x_2 = n_1 \cdot p_1 + n_2 \cdot p_2$$

Let's take another look at the equation from Chapter 3, Section 3.10, in which we derived a standard form for the weighted sum with threshold value comparison:

$$\sum_{i=0}^{n} w_i \cdot x_i \geq 0$$

We're interested in determining the separating line:

$$\sum_{i=0}^{n} w_i \cdot x_i = 0$$

We could of course convert this equation, which is also used in the perceptron to calculate the *net input*, back to the normal form by transforming it:

$$0 = w_0 \cdot 1 + w_1 \cdot x_1 + w_2 \cdot x_2 \qquad | - w_2 \cdot x_2$$

$$- w_2 \cdot x_2 = w_0 \cdot 1 + w_1 \cdot x_1 \qquad | : - w_2$$

$$x_2 = -\frac{(w_0 + w_1 \cdot x_1)}{w_2}$$

$$x_2 = -\frac{w_1}{w_2} \cdot x_1 - \frac{w_0}{w_2}$$

You may be wondering why we're transforming around here? The answer is that with this representation, we can wonderfully draw the *separating line* at least in the two-dimensional space between the categories.

Here's another small but important detail: if we look at the multiplied *normal vector form* shown earlier, you'll immediately experience a eureka moment:

$$n_1 \cdot x_1 + n_2 \cdot x_2 = n_1 \cdot p_1 + n_2 \cdot p_2$$

Still no eureka? Wait, let's write the equation a little differently and use the weights w_i instead of components n_i:

$$w_1 \cdot x_1 + w_2 \cdot x_2 - (w_1 \cdot p_1 + w_2 \cdot p_2) = 0$$

What about now? Exactly—this is the normal form from earlier, which means that the weights in the network are the components of the normal vector.

B.2 Calculus

Calculus deals with functions and, for example, with differentiating functions.

B.2.1 What Is a Function?

So, what is a *function*? Quite simply, you can say that a function is a *mapping* that assigns an element $y \in Y$ (Y is the *range*) to each element $x \in X$ (X is referred to as the *domain*), as shown in Table B.1, for example. The elements x and y in this example originate from the natural numbers with 0.

x	y
0	0
1	2
2	4
3	6
4	?
...	...

Table B.1 A Small Mapping Example

The question mark isn't in the table by mistake. Do you have any idea what the number is and how the mapping table continues? Can you possibly even provide a calculation rule for this?

These calculation rules represent what we consider as the function, whereby a calculated value y is assigned to a value x. In our example, multiplying x by 2 results in the value y, and in contrast to the table, this can be described in the following compact and space-saving way:

$$y = f(x) = 2 \cdot x$$

f is the function, and x is used as the input. We say that y is a function of x and is calculated using $2 \cdot x$.

B.2.2 Differential Calculus

Using *differential calculus*, we can determine the *rate of change* or *derivative* of a function, that is, how much one reference variable changes depending on another. For functions of the form $y = f(x) = \cdots$, the question is as follows: How does y change when x changes? In Figure B.11, you can see the function graph of the function $f(x) = y = x^2$ in gray color. It first shows how y depends on x. For example, y has the value 0.25 for $x = 0.5$.

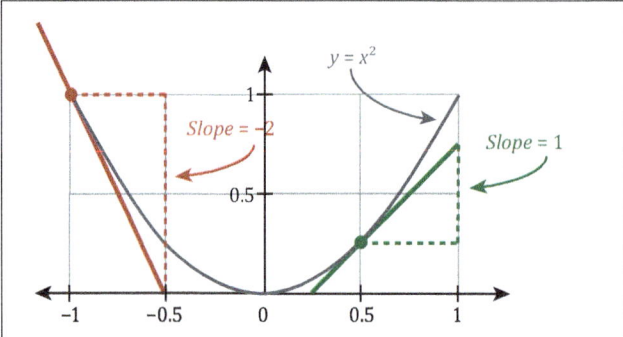

Figure B.11 Slope

Differential calculus is now about how x changes when y changes.

The derivative for our example $f(x) = x^2$ at one point can best be represented by a line that passes through a point on the function's graph and has the same slope as the graph at that point. Such a line is referred to as a *tangent*. Two tangents, for $x = -0.5$ and $x = 0.5$, are shown in red and green in the figure. You can then read the value of the derivative from the *slope of the tangent*. You can do this quite simply by going one unit to the right from the point (red and green) and then measuring how far you have to go up or down until you reach the line. In the case of the red line, you go one unit to the right and then two units down, so the *slope* and therefore the value of the derivative is -2.

B.2.3 Rules from Differential Calculus

If you're now wondering what differential calculus has to do with ANNs, we can tell you that it has a lot to do with them because the point is that networks should learn, and to learn means, for example, that errors get reduced during learning. Now comes the crucial question: How can an error be reduced?

Imagine a hill on which you've thrown a ball. If you stop holding the ball, it will roll down the hill and come to a stop in the valley, where there is no longer a slope. We can determine the direction in which the ball rolls, that is, into the flat, where there is no longer a slope, using differential calculus as it deals with the behavior of the slope.

Let's take a look at some specific *derivation rules*. Basically, we only need the following rules (a is a concrete number, e.g., 2, 3, or ...):

R1: If $f(x) = a$, the first derivative is $f'(x) = 0$

R2: If $f(x) = a \cdot x$, the first derivative is $f'(x) = a$

R3: If $f(x) = x^a$, the first derivative is $f'(x) = a \cdot x^{a-1}$

R4: If $f(x) = g(x) + h(x)$, the first derivative is

$f'(x) = g'(x) + h'(x)$

R5: If $f(x) = g(h(x))$, the first derivative is

$f'(x) = g'(h(x)) \cdot h'(x)$

R6: If $f(x) = \dfrac{h(x)}{g(x)}$, the first derivative is

$f'(x) = \left(\dfrac{h(x)}{g(x)}\right)' = \dfrac{h'(x) \cdot g(x) - h(x) \cdot g'(x)}{g(x)^2}$

Imagine the graph for $f(x) = a$ as a horizontal line—the function always has the same value, and the graph has no slope. In other words, the slope is always zero. This illustrates the first rule. The second rule and perhaps the other rules become clear when you draw graphs and look at the slopes. We take the liberty here of using the rules and notation f' for the derivation without further justification. If you want to do some research or resort to a textbook, the keywords *differential calculus* and *derivation rules* will help you.

B.2.4 Example of the Chain Rule

The fifth rule is referred to as the *chain rule* and isn't obvious straight away, which is why we've provided this example for you:

$h(x) = a \cdot x + b \cdot x$
$g(x) = x^2$

Thus, we can write $f(x)$ as follows:

$f(x) = g(h(x)) = (a \cdot x + b \cdot x)^2$

So far, so good, but now comes the derivation:

$f'(x) = \underbrace{\underbrace{g'(h(x))}_{R3} \cdot \underbrace{h'(x)}_{R4,\,R2}}_{R5} = 2 \cdot (a \cdot x + b \cdot x) \cdot (a + b)$

We have another example from Chapter 6 in which we used the following error function:

$E = \dfrac{1}{2} \cdot (y - \hat{y})^2$

To create the first derivation with regard to the weights, we must carry out the substitutions $\hat{y} = w \cdot x$ for the calculation. This results in the following:

$$E = \frac{1}{2} \cdot (y - w \cdot x)^2$$

The minimum error with regard to the weight is determined by applying the chain rule:

$$E'(w) = 2 \cdot \frac{1}{2} \cdot (y - w \cdot x) \cdot (-x) = (-1) \cdot x \cdot (y - \hat{y})$$

We've thus demonstrated the chain rule using two examples, which also plays a role in the calculation in the backpropagation algorithm, as we've just seen.

B.3 Derivative of the Sigmoid Function

In Chapter 6, we succinctly stated that the 1st derivative $f'(x)$ of the sigmoid function $f(x) = \frac{1}{1 + e^{-x}}$ reads as follows:

$$f'(x) = f(x)(1 - f(x))$$

Now that we're familiar with some of the rules of differential calculus, let's try to prove this statement, but first we need to reshape the sigmoid function a little:

$$f(x) = \frac{1}{(1 + e^{-x})} = \frac{e^x}{e^x} \cdot \frac{1}{(1 + e^{-x})} = \frac{e^x}{(1 + e^x)}$$

Now, we derive the sigmoid function according to all the rules of mathematics using rule R6, the *quotient rule*:

$$\left(\frac{h}{g}\right)' = \frac{h'g - hg'}{g^2}$$

The following values are given:

$$h(x) = e^x, \quad g(x) = e^x + 1$$

Here's what it looks like:

$$\frac{1}{(e^x + 1)^2}\left((e^x + 1)e^x - e^{2x}\right) = \frac{e^x}{(e^x + 1)^2}$$

Let's take another look at the first derivative of a sigmoid function, $f(x)$:

$$f'(x) = f(x)(1 - f(x))$$

Does our calculated derivation match the property? Let's see:

$$\frac{e^x}{(e^x + 1)^2} \text{ XXX } \frac{e^x}{(1 + e^x)}\left(1 - \frac{e^x}{(1 + e^x)}\right)$$

We transform the right-hand part of the previous formula:

$$\frac{e^x}{(1+e^x)}\left(1-\frac{e^x}{(1+e^x)}\right)$$
$$=\frac{e^x}{(1+e^x)}\left(\frac{1+e^x-e^x}{(1+e^x)}\right)$$
$$=\frac{e^x}{(1+e^x)}\frac{1}{(1+e^x)}=\frac{e^x}{(e^x+1)^2}$$

And, in fact, it's identical! Now the mathematician would say: *quod erat demonstrandum*, q.e.d. for short.

B.4 Key Statements

Although our main focus in this book is, of course, on the practical side of things, it's still interesting to know the properties of the networks we use. We've therefore compiled some general statements that are also mathematically proven (in the language of mathematicians: some *theorems*).

B.4.1 Perceptron Convergence Theorem

The *convergence theorem* by Frank Rosenblatt from 1959 (see Section B.7) states that for every linearly separable data record, the perceptron learning rule is guaranteed to find a solution in a finite number of steps.

B.4.2 Function Approximation

Loosely translated, Andrei Nikolayevich Kolmogorov proved in 1957 that a function with multiple variables can be represented by a linear combination of many functions with one variable.

George Cybenko adapted the theorem for our ANN context in 1989 (see Section B.7), so that a sufficiently large hidden layer with sigmoid activation functions and an output layer is basically sufficient to approximate any function.

Further interesting reading is provided by Kurt Hornik, Maxwell Stinchcombe, and Halber White (see Section B.7), who show that multilayer networks are universal approximators. This means that any function from a large function class can be approximated with arbitrary accuracy—provided that enough hidden neurons are available.

B.5 Notation

How do you write all this? The answer is provided in Table B.2, which we stick with in the book.

B Mathematics in Brief

Mathematical Objects	
x	Scalar value = a number
$\vec{x}^{(k)}, \vec{w}^{(k)}$	The k-th vector x or w in column notation (column vector)
$x_i^{(k)}$	The vector component at the i-th position of the k-th vector x
$\vec{x}^T, \vec{x}^t, \vec{x}'$	Input vector transposed, that is, in row notation (row vector)
$\vec{w}^T$	Weight vector transposed (row vector)
$X, W, \ldots$	Matrix
X^T	Transposed matrix: the mirroring of the values on the diagonal, that is, x_{ij} becomes x_{ji}
$f(x)$	Function: Calculation rule that calculates a value for a value x
Σ	Summation character
$\sum_{i=0}^{n} w_i \cdot x_i$	Weighted sum, counted from 0 to n, that is, $w_0 \cdot x_0 + w_1 \cdot x_1 + \cdots + w_n \cdot x_n$
η	Learning rate: the value that controls the extent to which changes occur during learning
θ	Threshold: the value that determines when changes take place
Significance for ANN	
X	Input matrix: training data
$\vec{x}$	Input vector: a data record from the training data
$\vec{w}$	Weight vector
$w_{j,i}$ or even w_{ji} or $w_{i,j}$ or even w_{ij}	Weight from node i to node j The index sequence occurring as ij or ji
$\vec{y}$	Desired output vector (training data)
$\hat{\vec{y}}$	Calculated output vector (of the neural network)
W	Weight matrix
$\text{sgn}(x), \text{step}(x), \text{sigma}(x)$	Step functions
$\sigma(x)$	Sigmoid function
A Few Greek Letters	
α	Alpha
β	Beta
γ	Gamma

Table B.2 Notation: Dry, but Necessary!

Mathematical Objects	
δ	Delta
ε	Epsilon
η	Eta

Table B.2 Notation: Dry, but Necessary! (Cont.)

B.6 Summary

In this appendix, we've dealt with some aspects of mathematics as gently as possible (we hope you agree) and looked at them in detail. The choice of topics was determined by the ANN theory. Analysis and algebra are the main areas that are used with select aspects. Differential calculus from analysis is a suitable means of determining the changes in weight when learning in networks. The calculations themselves were implemented using matrix calculations from algebra. Finally, we haven't missed the opportunity to at least mention Frank Rosenblatt's perceptron convergence theorem and Andrei N. Kolmogorov's function approximation.

B.7 Further Reading

- "Approximation by Superpositions of a Sigmoidal Function" by George Cybenko (1989, *http://cognitivemedium.com/magic_paper/assets/Cybenko.pdf*)
- "Multilayer Feedforward Networks are Universal Approximators" by Kurt Hornik, Maxwell Stinchcombe, and Halber White (1989, *www.cs.cmu.edu/~epxing/Class/10715/reading/Kornick_et_al.pdf*)
- "Two Theorems of Statistical Separability in the Perceptron" by Frank Rosenblatt (1959, *https://aitopics.org/download/classics:254C7499*)

Appendix C
TensorFlow 2 and Keras

TensorFlow(er Power)

The public provision of programming libraries by the major players in AI has made a significant contribution to its rapid development. In particular, Google's contribution with an entire program library ecosystem called TensorFlow has done a lot in this area, so that almost everyone can use AI and work with neural networks (especially after reading this book). The following sections present some important elements for understanding this ecosystem.

C.1 Introduction to TensorFlow 2

The meaning of *TensorFlow* has changed with version 2. While in version 1 it was still just the name for a program library, TensorFlow is now a collective term for an entire ecosystem of libraries that are responsible for training, storage, and deployment. You can find an overview of this in Figure C.1.

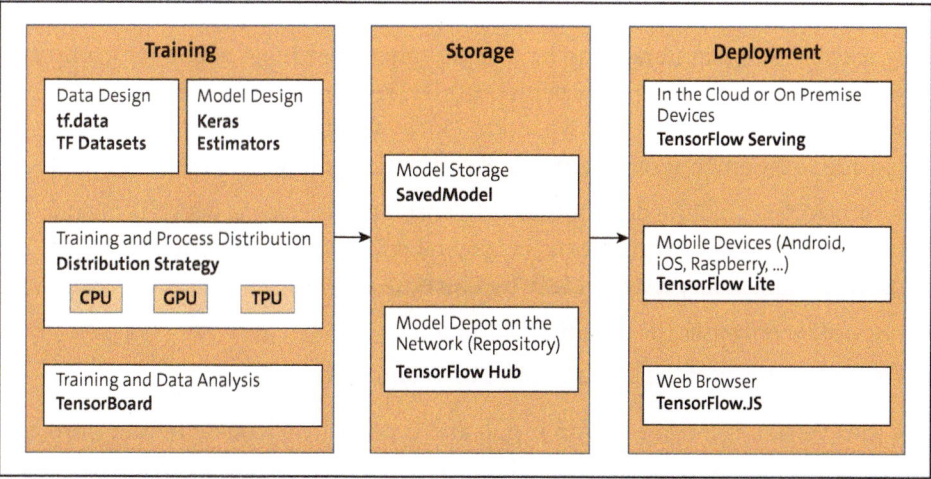

Figure C.1 The TensorFlow Ecosystem

> **TensorFlow Term Confusion**
>
> It's now somewhat confusing that *TensorFlow* can mean both the Python library and the entire ecosystem. We'll generally use TensorFlow to refer to the Python core library, and when we mean the ecosystem, we'll write *TensorFlow ecosystem*.

For the training phase, functions have been created that facilitate the handling of data (tf.data), significantly simplify the creation of the network architecture and parameters (integrated Keras library), and allow better control of computing resources (*distribution strategy*). In addition, an already tried and tested tool, *TensorBoard*, is used to analyze and control the training process and training accuracy. The storage of the constructed network structure, including trained weights, is now better supported, and you can use pretrained networks provided via the *TensorFlow Hub* for your own purposes.

If you want to use the trained models in a production system, the TensorFlow ecosystem also has a lot to offer here: in the cloud (*TensorFlow Serving*), for mobile devices (*TensorFlow Lite*), and for web-based applications (*TensorFlow JS*).

The core library of TensorFlow is specially designed for data-intensive and computation-intensive machine learning (ML) tasks. The basic idea behind TensorFlow is to visualize computing operations and data flows as graphs. This allows you to immediately see which tasks can be run in parallel, that is, can be performed simultaneously on different CPUs, GPUs, TPUs, or distributed computers in the network. Although TensorFlow is a Python library, the calculations for graph traversal are performed with very efficiently programmed C++ modules.

This scalable program library can be used to implement huge neural networks with millions of parameters for highly complex tasks. You suddenly have a tool that makes many ideas for neural networks feasible—ideas whose implementation previously failed due to the amount of data or lack of computing power.

The company behind the development of TensorFlow is Google, which is making the TensorFlow ecosystem available to the general public as an open-source system. The TensorFlow core library is available in two versions:

- tensorflow only uses the CPU of a computer.
- tensorflow-gpu uses the CPU and GPUs of a computer.

The CPU version is much easier to install and is perfectly adequate for our purposes. However, the following box contains information on how you can still obtain a GPU resource.

> **No GPU?**
>
> Many aspiring neural network builders construct their own GPU machine. There are excellent tutorials on this that you can find via Google search. GPU stands for *graphics processing unit* and is normally a graphics card with highly parallel computing units that enable fast screen display and changes, such as those required for computer games. GPUs are also ideal for ML tasks.
>
> However, it's much easier with hosting services from Amazon AWS (*https://aws.amazon.com*) or with the Google Cloud Service for Cloud Machine Learning (*https://cloud.google.com/ml*), which are available free of charge for experimentation up to a certain computing and data limit. Google provides a Jupyter Notebook environment with GPU or TPU support at *https://colab.research.google.com*, with minor restrictions on how long the GPU can be used. While GPUs were actually developed for graphic displays and then "misused" for ML, *tensor processing units* (TPUs) were developed by Google specifically for ML and work perfectly with the TensorFlow ecosystem.
>
> A dedicated GPU machine (also known as an on-premise solution) requires a relatively high initial investment, with almost no upper limit. However, you can then experiment with it as much as you like without having to think about the costs of using the GPU. Cloud solutions calculate GPU usage over time. The ideal strategy is one that carries out experimentation and program development on an on-premise solution. A cloud solution should be used for runs with large amounts of data and more computing resources.

C.2 Features of TensorFlow 2

The TensorFlow ecosystem offers a number of options that make it all the more interesting. We'll describe them briefly here.

C.2.1 Eager Execution

One of the biggest and most important changes in TensorFlow 2 compared to versions 1.10 to 1.18 (we'll abbreviate this as 1.10+) is the introduction of *eager execution*. Without eager execution, a calculation graph must always be created first, which is then executed in the second step within a session. In Listing C.1, eager execution is deactivated in the TensorFlow 2 library and thus corresponds to a TensorFlow code of versions 1.10+.

```python
import tensorflow as tf
from tensorflow.python.framework.ops import disable_eager_execution, enable_eager_execution
disable_eager_execution()
```

```
a = tf.constant(1)
b = tf.constant(2)
c = tf.Variable()
d = a + b
print(c)

# Output:
Tensor("add_1:0", shape=(), dtype=int32)
```

Listing C.1 TensorFlow without Eager Execution (Corresponds to TensorFlow Version 1.10+)

So, nothing is calculated here, only the calculation graph is created, and the form of the output sensor is specified in the `print` command. Let's run the same code again in Listing C.2, but now activate eager execution. At this point, it's important to mention that you should run this listing again in a separate Jupyter Notebook because the eager execution must be switched off or on at the start and can then no longer be changed for the running Jupyter kernel. We don't have to activate eager execution separately because this is automatically the case with TensorFlow 2, but we check it using the `tf.executing_eagerly()` function.

```
import tensorflow as tf

a = tf.constant(1)
b = tf.constant(2)
c = tf.Variable(5)
d = tf.multiply(a,c)+b

print("Is Eager Execution switched on? {}".format(tf.executing_eagerly()))
print(d)
# Output:
Is Eager Execution switched on? True
tf.Tensor(7, shape=(), dtype=int32)
```

Listing C.2 TensorFlow with Eager Execution (Switched On by Default in Version 2)

You can see immediately that not only does the output tensor get displayed here but also the result of the calculation—7.

C.2.2 Multi-GPU Management

TensorFlow naturally used GPU resources; however, when multiple GPUs were present in a machine, these resources weren't always used optimally. With the `MirroredStrategy()` module, the TensorFlow ecosystem now offers a way to better support synchronous distributed training on multiple GPUs.

```python
# Import the sequential class and the dense layer
from tensorflow.keras.models import Sequential
from tensorflow.keras.layers import

strategy = tf.distribute.MirroredStrategy()
print("Number of GPUs: {}".format(strategy.num_replicas_in_sync))

input_dimension = 4
# i.e. data with 4 features (e.g. iris dataset, chapter 3)
classes = 3 # so we expect three classes

with strategy.scope():
    model = Sequential([
    Input(shape=(input_dimension,)),
    Dense(32),
    Dense(64),
    Dense(classes)
    ])

model.compile(loss='categorical_crossentropy',
              optimizer='adam',
              metrics=['accuracy'])
# Output:
Number of GPUs: 1
```

Listing C.3 Using GPU Resource Management in TensorFlow 2

The output in Listing C.3 does, of course, vary from computer to computer depending on the number of GPU resources. It's important to note that the modeling must take place within `strategy.scope()`.

C.2.3 LiteRT

LiteRT was formerly known as TensorFlow Lite (TFLite), and you will likely find both terms in common use. You can use your custom network models on your own computer or in the cloud more or less unchanged. It gets exciting when you want to install and use your own trained network on a mobile device (Raspberry Pi, Arduino, Android or iOS smartphone, tablet, etc.). Note that no neural network is trained on a mobile device, as the computing power and memory are simply too small. Only classification or regression based on the trained network can take place on these devices.

```python
import tensorflow as tf

# Define a custom directory
directory = "test"  # Replace with any desired directory
```

```python
# Ensure the directory exists
if not os.path.exists(directory):
    os.makedirs(directory)

# Create a simple tf.keras model
x = np.array([-1, 0, 1, 2, 3, 4])   # Convert the list to a NumPy array
y = np.array([3, 1, -1, -3, -5, -7])  # Convert the list to a NumPy array

model = tf.keras.models.Sequential()
model.add(Dense(2, input_shape=[1]))
model.add(Dense(1))

# Define loss function and optimization method
model.compile(optimizer='sgd', loss='mean_squared_error')

# Start training
model.fit(x, y, epochs=50)

# Initialize the converter directly from the Python model
converter = tf.lite.TFLiteConverter.from_keras_model(model)

# Convert the model
tflite_model = converter.convert()

# Save the converted TFLite model to a file
with open(os.path.join(directory, "model.tflite"), "wb") as f:
    f.write(tflite_model)

print(f"TFLite model successfully saved in {directory}.")
# Output:
Train on 6 samples
Epoch 1/50
6/6 [==============================] - 0s 52ms/sample - loss: 45.3186
Epoch 2/50
6/6 [==============================] - 0s 2ms/sample - loss: 22.9742
Epoch 3/50
6/6 [==============================] - 0s 1ms/sample - loss: 15.4052
. . .
. . .
```

Listing C.4 Application of the TFLite Converter

In Listing C.4, a simple neural network is first created, then the loss function and optimization methods are defined, and the network is trained with simple data. The `tf.lite.TFLiteConverter.from_keras_model()` function is used to convert the trained model into a TFLite format.

So that we can also move the compressed, optimized network to a mobile device, we also need to be able to save it, as shown in Listing C.5.

```python
import pathlib

# Define a custom directory
directory = "test"  # Replace with any desired directory

# Save the TFLite model
tflite_models_dir = pathlib.Path(directory)
tflite_models_dir.mkdir(exist_ok=True, parents=True)
tflite_model_file = tflite_models_dir/"my_model.tflite"
tflite_model_file.write_bytes(tflite_model)
```

Listing C.5 Saving the Compressed TFLite Model

With this listing, the TFLite model gets saved as *my_model.tflite*, whereby this is done in a byte format. This means that the file isn't directly readable by a human when it's opened in a text editor, for example.

To use this model, you must of course install TFLite on the mobile device, whereby you must take certain differences into account depending on the device. However, refer to the *LiteRT Guide* (*https://ai.google.dev/edge/litert*).

C.3 Integrated Keras Library

We've already presented some libraries in Chapter 2, including the wonderful Keras library by Francois Chollet. This library was developed to better abstract the various neural network backends such as Theano, Microsoft Cognitive Toolkit, MXNet, and of course TensorFlow, and to simplify their use considerably. These backends actually do the necessary hard computing work, and Keras provides simplified and standardized access to the functionalities. Keras as a library still exists, of course, but Google has taken over not only Keras but also Francois Chollet—so it's a sustainable strategy to use TensorFlow 2 with the even more tightly and better integrated Keras.

TensorFlow 2 now offers three variants for implementing neural network models with Keras: *Sequential API*, *Functional API*, and *Model Subclassing*. We'll only take a closer look at the first two here because model subclassing is much harder to implement and is mostly used by researchers who want to have control over every little detail of the network model and learning. We don't want that yet, at least not in this book.

C.3.1 Sequential Model (Sequential API)

This method is the easiest way to build a network with Keras. For this book, the knowledge about this is completely sufficient because the reduced functionality of this method has no effect on our examples. With the sequential method, it's possible to implement the sharing of layers in the network, branching of layers, or multiple input or output branches, but it's very complicated.

But let's look at a simple example of how quickly a network architecture can be set up in Listing C.6.

```python
# Import the necessary packages
from tensorflow.keras.models import Model
from tensorflow.keras.models import Sequential
from tensorflow.keras.layers import Activation
from tensorflow.keras.layers import Dense

input_dimension = 4 # i.e. data with 4 features (e.g. iris dataset, chapter 3)
classes = 3 # so we expect three classes

model = Sequential()
model.add(Dense(32,input_shape = (input_dimension,)))
model.add(Dense(64))
model.add(Dense(classes))
model.add(Activation("softmax"))

model.summary()
# Output:
Model: "sequential"
```

Layer (type)	Output Shape	Param #
dense (Dense)	(None, 32)	160
dense_1 (Dense)	(None, 64)	2112
dense_2 (Dense)	(None, 3)	195
activation (Activation)	(None, 3)	0

```
Total params: 2,467
Trainable params: 2,467
Non-trainable params: 0
```

Listing C.6 Simple Neural Network according to the Sequential Model

This network could already be used to solve the classification of the Iris dataset from Chapter 3. All that is missing is the definition of a loss function and an optimization procedure as well as the data, as we've already shown in Listing C.4.

C.3.2 Functional Paradigm (Functional API)

The functional paradigm allows us to create more complex models with multiple inputs and multiple outputs (at different points in the network), to create branches in the network, and so on.

We now create the network from Listing C.6 in Listing C.7 according to the functional paradigm.

```python
Import of the necessary packages
from tensorflow.keras.models import Model
from tensorflow.keras.models import Sequential
from tensorflow.keras.layers import Activation
from tensorflow.keras.layers import Input
from tensorflow.keras.layers import Dense

input_dimension = 4
# i.e. data with 4 features (e.g. iris dataset, chapter 3)
classes = 3 # so we expect three classes

x_in = Input(shape=(input_dimension,))
x = Dense(32)(x_in)
x = Dense(64)(x)
x_inter = Dense(classes)(x)
x_out = Activation("softmax")(x_inter)


model = Model(inputs=x_in, outputs=x_out)

model.summary()
print("Output intermediate layer: ",x_inter)
print("Output last layer: ",x_out)
# Output:
Model: "model_6"
```

Layer (type)	Output Shape	Param #
input_7 (InputLayer)	[(None, 4)]	0
dense_24 (Dense)	(None, 32)	160

```
dense_25 (Dense)              (None, 64)              2112
_____
dense_26 (Dense)              (None, 3)               195
_____
activation_7 (Activation)     (None, 3)               0
=================================================================
Total params: 2,467
Trainable params: 2,467
Non-trainable params: 0

Output intermediate layer:  Tensor("dense_26/Identity:0", shape=(None, 3),
dtype=float32)
Output last layer:  Tensor("activation_7/Identity:0", shape=(None, 3), dtype=
float32)
```

Listing C.7 Neural Network Created with the Functional Paradigm

As you can see in Listing C.7, this allows us to access the output of intermediate layers (x_inter) in a very simple way, which could be reused in complex networks in later layers, something that is indeed done. This gives you a much greater flexibility when creating even more complex networks, which also make it easier to combine input data of different types (e.g., images, or audio and text data). Intermediate results can also be output.

C.4 Summary

With this information, you're now prepared to use TensorFlow 2, or at least you won't be surprised by it. Of course, there are plenty of other powerful libraries for neural networks, but we recommend that you familiarize yourself with one library first and build up expertise before trying out other variants.

C.5 Further Reading

- TensorFlow (*www.tensorflow.org*)

The Authors

Dr. Joachim Steinwendner is a scientific project leader specializing in data science, machine learning, recommendation systems, and deep learning. He has been involved in the development of neural networks, guiding them from a cutting-edge research topic to their current everyday relevance, and has worked across various industries.

Dr. Roland Schwaiger is a software developer, freelance trainer, and consultant. He has a PhD in mathematics and he has spent many years working as a researcher in the development of artificial neural networks, applying them in the field of image recognition. In his work, he places great importance on bridging the gap between theory and practice. Whether as an author, consultant, or lecturer, he is passionate about sharing his enthusiasm with others.

Index

A

Ability to abstract 321
Accuracy 346
Action 336
Action potential 21, 245, 248
Action selection 338
Activation function 31, 136, 137, 145
    linear 156
    sigmoid 158
Adaline 20, 93, 94, 115, 256, 322
    learning 115
    learning rules 148
Adaptive Linear Element → Adaline
Adaptive Linear Neuron → Adaline
Adaptive Moment Estimation (ADAM) 205
Adaptive resonance theory (ART) 326
Advertising campaign 324
Agent 336
AI ethics principles 281
AI laws 289
AI winter 257
Algorithm 417
    backpropagation 150
    smart 32
    SOM 327
Algorithmic fairness 281, 282
All-or-none law 29, 31, 248
AlphaGo 271
AlphaZero 271
Alternative function 329
Amazon SageMaker 62
Anaconda 20, 43
Anaconda Navigator 44, 47
Analysis tools, error curve 103
Anonymous function 403
Approximation 150
Area under ROC curve (AUROC) 347, 349
Array 102
    slicing 142
Artificial intelligence 19, 32
    connectionist 35
    strong 34
    subsymbolic 35
    symbolic 35
    weak 34
    workshop 34

Artificial neural network (ANN) 17, 32
    adaptive linear 115
    forgotten 115
    output 31
    stable solution 115
Assessment approaches 34
Association 320, 324–326
Association matrix 269
Associative abilities 261
Associative memory 258, 265
Assumption 321
Atari 319
Atom, electrically charged → Ion
Autonomy 33
Average pooling 185
Axon 245
Axon hillock 248

B

Ba, Jimmy Lei 205
Backend 59
Backpropagation 21, 147, 173
Backpropagation algorithm 148, 158
Backprop → Backpropagation algorithm
Backward path 165, 174
    diagram 163
Bacteria 39
Bag-of-words 308
Barto 344
Basic arithmetic operation
    for vectors and matrixes 420
    SOM 328
Batch gradient descent 216
Batch size 103, 216
Beekeeper 22
Bell curve, Gaussian 192
Bellman equation 337
Bengio, Yoshua 191
BERT transformer 233
Best matching unit (BMU) 327
Bias 26, 282
Bias neuron 85, 140
Bias trick 96
Big data 303, 304
Binary 254
Binary cross-entropy 361

Index

Binning 296
   logarithmic 298
Biological processes 242
Biological system 28
Biology 239
Bionics 28
Bird flight 28
Boolean algebra 128
Boolean variable 393
Boston Housing Prices dataset 294
Brain 28
Brain stem 243, 244
Bug 53

C

Calculation 95
   asynchronous 262
   MLP 134
   synchronous 262
Calculation error 95
Calculation formula 427
Calculation rule 68
   error backpropagation 157
Calculus 427
California Housing dataset 351
Cartesian coordinate system 76
Categorical cross entropy 216
Category 98, 324, 425
Causal prediction 323
C-cell 258
Cell
   complex 258
   photosensitive 256
   simple 258
Cell body 21, 29, 245
Central nervous system (CNS) 242
Central processing unit (CPU) 38
Cerebellum 243, 244
Cerebral cortex 241, 244, 269
Cerebrum 243
Chain rule 429
Characteristic 321
Charge 248
Charge difference 247
Cheat sheet 44, 57
Chi-square test 316
Chollet, François 210, 441
Circuit 254
Class 98, 324
Classification 23, 103, 107, 127, 320, 411
   accuracy 335
   behavior 345

Classification (Cont.)
   quality 345, 346, 348
   supervised 412
   task 347
Classification model
   comparison 349
   quality 347
Classifier 66, 236
   analyzing 108
   initializing 108
   learning 108
   linear 65
Cluster 324, 334
Clustering algorithm 324
Coding 248, 292
Coding block 181, 213
Cognitron 258
Column notation (vector) 82
Combination of properties 79
Combinations in the ANN 66
Combinatorial explosion 258
Combined prediction 323
Competitive learning 327
Complex cell 258
Comprehension space 201
Computational unit 24
Computer science 32
Computer scientist 66
Computer vision 179
Confusion matrix 345
Connection
   amplifying 30
   inhibitory 30
   weight 30
Connectionist AI 35
Connectome 243
Consciousness 33, 34
Control structure 396
Convergence 95
Convergence theorem 99, 256, 431
Convolution 181
Convolutional neural network (CNN) 18, 21, 35, 134, 179, 213, 244, 258, 270
Coordinate system 76, 419
   Cartesian 76
Copyright 289
Core 38
Core Python 408
Corpus callosum 243
Correction value 165
Correlation 321
Correlation matrix memories 258

448

Index

Cortex
   cerebral 269
   visual 258
Cortical area 244
Cost function 151, 204
Covariance matrix 313
Creativity 34
CRISP-DM 277
   business (process) understanding 278
   data preparation 279
   data understanding 279
   deployment 280
   evaluation 280
   modeling 280
Cross Industry Standard Process for
   Data Mining → CRISP-DM
Curse of dimensionality 311
Cybenko, George 431
Cybernetics 256

D

Dartmouth College 34
Data augmentation 310
Database 302
Data flattening 303
Data flow graph 58
Data mining 277
Data preparation 67, 279, 290
Data science 19
Data scientist 36, 66
Datasets 114
Data type, array 409
Debugger 56
Decision-making 35
   coin 349
Decision tree 284
Decoder 202
Decorator 379
DeepDream 373
Deep learning 17
DeepMind 319, 344
Deep neural network 21, 37, 134, 179, 271
Deep Q-network 339
Definition, operational 33
Definition area 427
Delta 96
   hidden 175
   output 175
Delta rule 125, 148, 257
Dendrite 21, 29, 245
Dendrite tree 30
Deployment 382

Derivation rule 429
Derivative 428
   first 150, 152
Descriptive 326
Development history of the ANNs 350
Diencephalon 243, 244
Difference 157
Difference in concentration 248
Differential calculus 428
Dimension 76, 140, 407
   fitting 141
Dimensionality 292, 311
Direction 419
   weight vector 330
Discount factor 337
Discriminative AI 179
Discriminator, linear 66
Distance, Euclidean 328
Distance (between vectors) 328
Docstring 50, 72
Dot product → Scalar product
Dropout 206
Dynamic regression 323

E

Eager execution 437
Early stopping 205
Ecological aspects 288
Edmonds, Dean 255
Eigenvalue decomposition 313
Eigenvector 314
Electrochemical equilibrium 248
Emotional intelligence 34
Encephalon 243
Encoder 200
Encoder-decoder attention 203
Energy consumption 288
Environment 45, 336
   interaction 336
Environmental impact 288
Episode 339
Epoch 103
Epsilon-greedy strategy 343
Epsilon limit 338
Equation 322
Error 94, 147
   backpropagation 157, 255
   calculation 115
   correction 116
   correction for SOM 327
   curve 111, 148, 167
   formula 157

Index

Error (Cont.)
  measure 147
  proportion 157, 162
  reducing 428
  reduction 155
  squared 20, 123, 148, 155, 167, 174
Escape sequence 357, 409
Estimator 106, 107
Eta 157
Ethical behavior 34
Ethics 34
Euclidean distance 328
Eureka moment 426
Evaluation field 346
Excitation 248
Excitation conduction 245
Excitatory 246
Excitatory input 254
Exclusive OR 20
Experience 36
Explainability 282, 284
Exploding gradient 191
Exploitation 338, 343
Exploration 338, 343
  incremental 336
Exploration-exploitation dilemma 338
Exponential linear unit function 193
Expression 269, 325
Extended weight vector 85

F

Fairness 283
Fall 149
False 18, 393
Feature 269, 321
  categorical 292
  deformation 258
  ordinal 292
Feature coding 292
Feature engineering 277, 279, 290
Feature extraction 302
Feature map 183
Feature normalization → Feature scaling
Feature scaling 299
Feature selection 316
  filtering methods 316
  wrapper methods 317
Feature transformation 312
  automatic 313
  knowledge-based 312
Feature vector 321
Feedback loop 261, 374

Feedback network 21, 261
Feed-forward calculation 147
Feed-forward network 27
Fifth generation 261
Filter concatenation 225
Fine tuning 365
First generation 260
Fishing net 326
fMRI 180
Forget gate 271
for loops 396
Formula 68
Forward delta 158
Forward path 163
Fourth generation 261
Frequency 325
Frequency modulation 249
Fukushima, Kunihiko 18, 258
Fukushima's Neocognitron → Neocognitron
Function 69, 427
  alternative 329
  anonymous 403
  Heaviside 31
  identical 143, 156
  linearly separable 156
  semilinear 31
Functional API 441, 443
Functionally dependent 323
Function approximation 431
Function notation 69

G

Gamification 336
General Data Protection
  Regulation (GDPR) 289
Generative adversarial network (GAN) 272
  discriminator 272
  generator 272
Generative AI 179, 384
Generative pretrained transformer (GPT) 38
GitHub 358
Glial cell 247
Glorot, Xavier 191, 192
Go 28
Golgi, Camillo 250
Golgi staining 250
Goodfellow, Ian J. 272
Google 19
Google Colab 20, 59, 61
Google DeepMind 271
Gooseneck function 137
Gradiens 149

Gradient 148, 150
   calculation 172
   exploding 191
   vanishing 191
Gradient descent 21, 147, 148, 257
   concept 149
   example 149
   implementation 152
   in pictures 149
   method 151
Gradient value 175
Graph 103
Graphics card 38
Graphics processing unit (GPU) 38, 437
   multi-GPU management 438
Green AI 288
Grid area 338
Group 324
   linear separation 65

H

Hadamard product 164
Hardware 38
He, Kaiming 192
Heart rate control 244
Heaviside, Oliver 73, 102
Heaviside function 31, 73
Hebb, Donald 93, 255
Hebbian learning rule 255
Hebbian synapse 256
Hemisphere 243
Hidden layer 38
Hidden neuron 27
Hierarchical Data Format 5 (HDF5) 221
High dimensionality data 292
Hinton, Geoffrey E. 206, 269, 271
Hochreiter, Sepp 271
Hoff, Marcian E. 115, 256
Hopfield, John 260
Hopfield network 21, 260, 262
   limits 262
   recognition 265
Hornik, Karl 431
Hubel, David H. 180, 258
Hugging Face 232
   model hub 235
Hunspell stemmer 308
Hyperparameter 185, 193
Hyperplane 103

I

Identical function 143, 156
if statement 396
Image classification 354
ImageNet 224
   dataset 224
ImageNet Large Scale Visual Recognition
   Challenge (ILSVRC) 224
Image recognition 17, 21
Imitation ability 33
Importing a module 74
Inception-v3 .... 224, 226, 227, 369
Independent influence 323
Independent variable 323
Index 24, 420
Inductive learning 36
Information, sensory 269
Inherent 324
Inhibitory 246
Inhibitory input 254
Initialization, random 106
Initialization value 110
Inner product → Scalar product
Input 23
   examples 103
   excitatory 254
   inhibitory 254
Input gate 271
Input layer 24, 326
Input neuron 24
Input-output relationship 319
Input value 23
Input vector 30
Institute of Electrical and Electronics
   Engineers (IEEE) 270
Integrated development
   environment (IDE) 43
Interactive Python (IPython) 49
Interconnectedness 244
Interface 29
Internal representation 324
Internal structure 36
Interpretability 282, 284
Inverse document frequency 309
Inverse vector 422
Ion 248
Iris 79

Index

Iris dataset 79
Iteration 103, 178, 329

J

JSON format 47, 49
Jupyter community 62
Jupyter Notebook 20, 39, 47
   cells 48
   cloud resources 62
   debugging 53
   documentation 50
   help 50
   starting 47

K

Kaggle 62, 366
Keras 19, 59, 441
Keras library 210
Kernel 49
Key figure 346
   context 347
Kigma, Diederik 205
k-nearest neighbor classifier (k-NN) ... 411, 412
Knowledge 35
Knowledge representation 33, 35
Kohonen, Teuvo 258, 269, 326
Kolmogorov, Andrei N. 431

L

Label 320
   missing 345
Label encoding 292
Lambda 102
Lambda function 403
Large language model (LLM) 17, 21, 179, 194
Latent sample 272
Lateral inhibition 185
Layer 23, 131
Leaky ReLU 193
Learning 82, 319
   algorithm 20, 95
   duration 338
   end 103
   from observations 36
   history 93
   inductive 36
   in the ANN 94
   method 19, 22
   online 103
   rate 116, 123

Learning (Cont.)
   semi-supervised 344
   strategies 319
   supervised 22, 36, 320
   temporal difference 337
   unsupervised 22, 36, 324, 325
Learning formula 330
   SOM 330
Learning rate (LR) 151, 154, 327, 328, 337
Learning rule 255
   delta rule 255
   error backpropagation 255
   Hebb 255
   Hebbian 255
   perceptron 255
Learning step 172
Least mean square (LMS) 123, 257
LeCun, Yann 181
Legal AI Act 289
Liability 290
Lighthill, James 258
Lighthill report 258
Line 31, 66, 98, 425
   linear regression 322
   rise 148
   ROC 349
Linear activation function 156
Linear algebra 417
Linear classifier 65
Linear discriminator 66
Linearly separable function 156
Linear separability 103, 257
Linear separation 65
Linear threshold unit (LTU) 75
Line equation 425
Line notation 82
List 102
LiteRT 439
Log file 279
Logical reasoning 33, 35
Logits layer 189
Long short-term memory (LSTM) 241, 271
   cell state 271
   gates 271
Loop 396
Loss function 151, 204

M

Machine learning (ML) 17, 19, 28, 33, 35
Machine learning community 347
Madaline 256
Magic function 49, 74

Map 326
Mapping 427
Market basket analysis 326
Marr, David 244
Masked self-attention 202
Math 406
Mathematical algorithm 28
Mathematics 417
   basic principles 18
Matplotlib 58, 73
Matrix 135, 137, 140, 418, 420
   diagonal 424
   output 141
   transpose 424
Matrix form 21
Matrix multiplication 135
Matrix operation 158
Matrix row 424
Maximizing profit 336
McCarthy, John 34, 256
McCulloch, Warren 28, 254
McCulloch-Pitts neuron 254, 261
Mean squared error (MSE) 157
Measure 147
Measuring method 147
Membrane potential 247
Membrane voltage 247
meshgrid 114
Messenger substance 245
Method of steepest descent 148
Microscope technique 250
Microsoft Azure 62
Mini-batch gradient descent 216
Minimum point 150
Minsky, Marvin 34, 255, 256
MNIST dataset 209
Model 36
   efficiency 350
   mathematical 241
Modeling 242
Model subclassing 441
Momentum 204
Moral 34
Mordvintsev, Alexander 373
Motor neuron 242
Movement of the weights 334
Multilayer network → Multilayer
   perceptron (MLP)
Multilayer perceptron (MLP) 131
   calculation 134
   predict 141
Multiplication, matrix with vector ... 424

Muscular system 242
Myelin sheath 247

N

Nabla 152
Namespace 74
Natural language processing (NLP) 33, 231
Neighborhood 327, 329
   calculation 330
   options 329
Neocognitron 18, 241, 244, 258
Nerve cell 28
Nervous system 242
Net input 123, 426
Network
   component 136
   feed-forward 27
   recurrent 27
Network architecture 137, 138
Network display 68
Network multilayer → Multilayer
   perceptron (MLP)
Network state, change 170
Network structure 144
Network with feedback 21
Neural network
   biological background 21
   models 274
   pretrained 382
Neurite 245
Neuroimaging 250
Neurology 242
Neuron 68, 242
   artificial 29
   biological 29
   hidden 27
   input 24
   postsynaptic 247
   predecessor 26
Neuron model 125
Neurons in the hidden layers 38
Neurotransmitter 245
Ng, Andrew 38
n-gram 308
NN conference 270
Noisy 262
Normal form 427
   vector 425
Normalized Difference Vegetation
   Index (NDVI) 312
NoSQL 303

Index

NumPy 58, 409
   array 79
numpy.random.RandomState 109
NVIDIA 38

O

Object attributes 110
Object-oriented programming 404
Object recognition 354
Object segmentation 354
Obstacles 338
One-hot encoding 211, 293
Open-source Python library 19
Operation, performance-optimized 140
Operational definition 33
Operator 256
Optical nerve 247
Optimal path 339
Osindero, Simon 271
Outlier 300
Output 23
   partial 345
   unknown 344
Output calculation 136
Output gate 271
Output layer 24
Output neuron 24
Output signal 31
Output value 23
Overall error 148
Overfitting 190, 205
   prevention 205
Overlap factor 124

P

Padding 186
pandas 58
Parameterization problem 412
Path, optimal 339
Pattern discovery 326
Pattern recognition 36, 412
Penalty term 363
Percentile binning 298
Perceptron 20, 65, 66, 75, 256, 257, 425
   components 75
   estimator 114
   learning 93, 94, 97
   learning algorithm 95, 99, 103
   learning rule 255, 431
   learning rule error 147
Peripheral nervous system (PNS) 242

Petal 79
Petalum 79
Photosensitive cell 256
Pitts, Walter 28, 254
Pixel attribution 287
Planning 65
   automatic 65
Point 76, 419
Pooling layer 185
Position 419
Positional encoding
   frequency-based 198
Position-invariant recognition 260
Potassium ion 248
Predecessor neuron 95
Prediction 320, 321, 323
   causal 323
   combined 323
   model 320
   time series 323
   visitor numbers 321
Prediction block 181, 213
Prediction model
   combined 323
Predictor 107
predict → Multilayer perceptron (MLP)
Pretrained neural networks 382
Primary area 244
Principal component analysis (PCA) 313
Privacy 289
Product, inner 83
Programming 417
Prompt 46, 195, 232, 384
Prompt engineering 195
Property, inherent 324
Pseudocode 412
Punishment 336
Purkinje cell 250
Pyramidal cell 244
Python 20, 39
   arguments 401
   basic principles 18, 389
   class 404
   function 401
   indentation 396
   module 406
   parameters 401

Q

Q-learning 271, 319, 337
   tabular 339
Quantile binning 297

Index

Queen bee 22
Quotient rule 430

R

Radius 330
   neighborhood 329
Ramón y Cajal, Santiago Felipe 249, 254
Random number generator (RNG) 109
Rate of change 428
Reality 322
Reasoning, logical 33
Recall error 263
Receiver operating characteristic (ROC) 347
Receiver operator (radar) 347
Receptive field 260
Receptor 246
Recognition, Hopfield 265
Rectified linear unit (ReLU) 137, 184
   dying 192
   function 184
   leaky 193
Rectifier function 137
Recurrent network 27
Recurrent neural network (RNN) 241, 261
Reduction, linear 329
Region 98
Regression 321, 322
   dynamic 323
   linear 321
Reinforcement algorithm 337
Reinforcement learning 22, 36, 319
Replay memory 343
Repository 358
Representation
   internal 324
Residual error 155
ResNet50 231
Resource requirements 288
Response 242
Resting potential 247
Reward 336, 337
   future 336
Right angle 426
Rise 149
RMSprop 352
Robot 22, 89
Rochester, Nathaniel 34, 256
Rodent hippocampus 254
Root mean square (RMS) 115
Rosenblatt, Frank 75, 256, 431
Rosenblatt's perceptron 125
Rules 256, 325
Rumelhart, David 157, 269
Running index 106

S

Samuel, Arthur 35
Saturation 192
Scalar 102, 417
Scalar multiplication 421
Scalar product 83, 164, 423
Scaling 258
Scaling-invariant recognition 260
Scatterplot 77, 322
S-cell 258
Schmidhuber, Jürgen 271
Scientific discipline 28
scikit-learn 58
scikit-learn.org 113
Second generation 260
Seed 110, 115
Segmentation analysis 324
Selection procedure 338
Self-attention 194, 201, 274
Self-organizing map (SOM) 241, 269, 319, 326, 350
   basic arithmetic operation 328
   class 332
   input layer 326
   iteration step 334
   learning formula 330
   Python code 331
Semilinear function 31
Semi-supervised clustering 345
Semi-supervised learning 319, 344
Semi-supervised regression 345
Sensitivity 347
Sensory information 242, 269
Sepal 79
Sepalum 79
Separability, linear 65, 103, 257
Separating line 98, 103, 426
   position 106
Sequential API 441, 442
Shannon, Claude 34, 256
Shift 258
Sigmoid function 31, 143, 156–158
   derivative 430
Signal 28
   chemical 28
   electrical 28
Signal amplitude 241
Signal effect
   excitatory 29

455

Index

Signal effect (Cont.)
   inhibitory 29
Signal frequency 241
Signaling molecule 245
Similarity 319, 324
Simple cell 258
Slicing (array) 142
Slope 149, 428
Softmax 347
Software agent 22, 37
Soma 21, 245
Spaces 76, 232
Specificity 347
Spiking neural network (SNN) 241
Spinal canal 243
Spinal cord 243, 244
SQL 302
Squared error 123, 148, 157, 167, 322
Srivastava, Nitish 206
Stable diffusion model 384
Stack trace 56
Staffing plan 65
Standard form 425
Standardization 421
State-action-value function 337
State transfer 170
Statistician 66
Stemming 307
Step function 73, 85, 125, 147
Step length 155
Stimulus 30
   auditory 28
   gustatory 28
   olfactory 28
   processing 242
   tactile 28
   visual 28
Stinchcombe, Maxwell 431
Stochastic gradient descent 216
Stochastic Neural Analog Reinforcement
   Calculator (SNARC) 256
Stop word 307
Storage, data coding 102
Strategy 37
Stride 187
String 394
Strong AI 34
Subscript 38
Subsymbolic AI 35
Suction cup 28
Sum
   vector 420
   weighted 24, 31, 84, 85

Supervised classification 412
Supervised learning 22, 36, 319
Sutton 344
Symbolic AI 35
Synapse 29, 245, 254
   Hebbian 256
Synaptic plasticity 255

T

Tabular Q-learning 339
Tactics 336
Tangent 428
Tangent slope 428
Taxes 256
Teh, Yee Whye 271
Temporal difference learning 337
Tensor 407, 417
   definition 408
   levels 408
TensorBoard 213, 436
TensorFlow 19–21, 39, 58, 417, 435
   features 437
TensorFlow 2 209
TensorFlow Hub 436
Tensor processing unit (TPU) 38, 437
Term frequency 309
Term frequency–inverse document
   frequency (TF-IDF) 309
Terminal 46
Termination condition 396, 398
Thalamus 244
Theta 85
Third generation 261
Threshold logic unit 254
Threshold potential 248
Threshold value 26, 31, 73, 96, 348
   evaluating 349
   setting 348
Time series prediction 323
Token 197
Token IDs 234
Tokenizer 234
Topographical organization 327
Topographic map 269
Topological representation 241
Topology 269
Topology-preserving 241
Total reward 337
Traceback 56
Training data 102
Training dataset 99
Training epoch 216

Transfer learning 223, 231, 366
Transformer library 232
Transformer model 274
Transformer network 35, 37
Transformer neural network 21, 37, 179
   output 203
   training 203
Transposing 82, 424
Troubleshooting 141
True 18, 393
Truth table 128, 130
Tuple 102
Two-dimensional 76

U

Understanding 33
Unit matrix 269
Unit vector 422
Unsupervised learning 22, 36, 319, 324
   learning algorithms 325

V

Value
   calculated 95
   desired 95
   estimated future 337
   future 337
   old 337
Value combination 131
Value range 427
Vanishing gradient 191
Variable 391
   Boolean 393
   dependent 321
   independent 321
Vaswani, Ashish 194
Vector 417, 419
   addition 420
   dimension 407
   length 421
   length calculation 421
   multiplication 165

Vector (Cont.)
   multiplication with matrixes 424
   multiplying by scalars 420
   multiplying with each other 422
   normal 426
   notation 76
   operation 158
   representation 76
   subtraction 422
   sum 420
Velcro fastener 28
Vesicle 246
Video game 37
Visual cortex 244, 258

W

Watkins, Chris 337
Weak AI 34
Weight 24, 30
   adjustment 103, 332
   change 96
   correction 152
   determination, Hopfield 263
Weighted sum 24, 31, 84, 85
Weight vector 30, 82, 96
   direction 330
   extended 85
   map 327
Werbos, Paul 258
while loops 396
White, Halber 431
Widrow, Bernard 115, 256
Widrow-Hoff rule 154
Wiesel, Torsten N. 180, 258
Williams, Ronald J. 269
Window size 323

X

XOR problem 20, 127, 129

Z

Zero padding 187

- Learn to program your own AI applications—even if you've never coded before!

- Get started with Python or use the KNIME platform for no-code development

- Work with neural networks, transfer learning, anomaly detection, reinforcement learning, and more

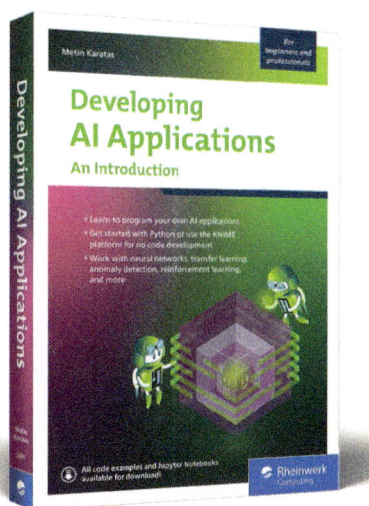

Metin Karatas

Developing AI Applications

An Introduction

It's time to get practical about AI. Move past playing around with chatbots and plugging your data into others' applications—learn how to create your own! Walk through key AI methods like decision trees, convolutional layers, cluster analysis, and more. Get your hands dirty with simple no-code exercises and then apply that knowledge to more complex (but still beginner-friendly!) examples. With information on installing KNIME and using tools like AutoKeras, ChatGPT, and DALL-E, this guide will let you do more with AI!

402 pages, pub. 06/2024
E-Book: $39.99 | **Print:** $44.95 | **Bundle:** $49.99

www.rheinwerk-computing.com/5899

- Work with pretrained LLM and NLP models on Hugging Face and LangChain
- Create vector databases and implement retrival-augmented generation
- Add an agentic system using frameworks such as crewAI and AutoGen

Bert Gollnick

Generative AI with Python

The Developer's Guide to Pretrained LLMs, Vector Databases, Retrieval Augmented Generation, and Agentic Systems

Your guide to generative AI with Python is here! Start with an introduction to generative AI, NLP models, LLMs, and LMMs—and then dive into pretrained models with Hugging Face. Work with LLMs using Python with the help of tools like OpenAI and LangChain. Get step-by-step instructions for working with vector databases and using retrieval-augmented generation. With information on agentic systems and AI application deployment, this guide gives you all you need to become an AI master!

392 pages, pub. 05/2025
E-Book: $54.99 | **Print:** $59.95 | **Bundle:** $69.99

www.rheinwerk-computing.com/6057

- Develop your own Python programs, step by step
- Work with variables, operators, loops, data types, functions, and modules
- Follow detailed exercises to build Python applications, GUIs, and more

Thomas Theis

Getting Started with Python

If you want to program with Python, you've come to the right place! Take your first steps with this Python crash course that teaches you to use core language elements, from variables to branches to loops. Follow expert guidance to work with data types, functions, and modules—and learn how to manage errors and exceptions along the way. Apply Python programming to develop databases, graphical user interfaces, widgets, and more. Practice your skills with example exercises, and start developing your own applications with Python today!

437 pages, pub. 07/2024
E-Book: $34.99 | **Print:** $39.95 | **Bundle:** $49.99

www.rheinwerk-computing.com/5876

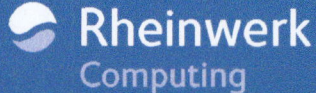